Adrian E. Scheidegger

Principles of Geodynamics

Third, Completely Revised Edition

With 126 Figures

Springer-Verlag
Berlin Heidelberg New York 1982

Dr. Adrian E. Scheidegger
Professor für Geophysik
Technische Universität Wien
Gusshausstraße 27–29
A-1040 Wien, Austria

ISBN-13:978-3-642-68459-3   e-ISBN-13:978-3-642-68457-9
DOI: 10.1007/978-3-642-68457-9

Library of Congress Cataloging in Publication Data. Scheidegger, Adrian E., 1925– .
Principles of geodynamics. Bibliography: p. Includes index. 1. Geodynamics. I. Title.
QE501.S3 1982 551 81-23329 AACR2.

© by Springer-Verlag Berlin Heidelberg 1958, 1963, and 1982.
Softcover reprint of the hardcover 3rd edition 1982

Typesetting: K + V Fotosatz, Beerfelden.

2132/3130-543210

To My Wife

Preface

Geodynamics is commonly thought to be one of the subjects which provide the basis for understanding the origin of the visible surface features of the Earth: the latter are usually assumed as having been built up by *geodynamic* forces originating inside the Earth ("endogenetic" processes) and then as having been degraded by *geomorphological* agents originating in the atmosphere and ocean ("exogenetic" agents). The modern view holds that the sequence of events is not as neat as it was once thought to be, and that, in effect, both geodynamic and geomorphological processes act simultaneously ("Principle of Antagonism"); however, the division of theoretical geology into the principles of geodynamics and those of theoretical geomorphology seems to be useful for didactic purposes. It has therefore been maintained in the present writer's works. This present treatise on geodynamics is the first part of the author's treatment of theoretical geology, the treatise on *Theoretical Geomorphology* (also published by the Springer Verlag) representing the second.

The present edition is third one of the book. Although the headings of the chapters and sections are much the same as in the previous editions, it will be found that most of the material is, in fact, new. Since the printing of the last edition, plate tectonics has come into its own, and it is now no longer possible to hold an "agnostic" standpoint in the face of the various theories of orogenesis: the evidence for some sort of "mobilism" in the Earth's evolution is now so overwhelming, that "fixist" theories can no longer hold more than a historical interest. Therefore, the entire philosophy of the treatment had to be completely changed.

The present book represents the current views of the writer on the subject of geodynamics. It is not a comprehensive literature survey, but rather a compilation of the writer's ideas on the matter. Thus, the first two chapters of the book give a brief summary of the physical facts about the Earth as far as they are known to date, the third puts together the principles of the theory of deformation of continuous matter which is the basic background of geodynamics, and the other chapters represent a review of the current views on various aspects of geodynamics, such as rotational phenomena, planetary problems, theories of orogenesis, tectonophysics and some special features.

Much of the material has been taken from the writer's own earlier efforts on the subject. He has drawn particularly heavily from those of his earlier articles which appeared in the Canadian Journal of Physics, in the Transactions of the American Geophysical Union, in Geofisica Pura e Applicata, in the Journal of the Alberta Society of Petroleum Geologists, in the Journal of Geology, in Canadian Oil and Gas Industries, in the Bulletin of the Geological Society of America, in Rivista Italiana di Geofisica e Scienze Affini, and in the Annali di Geofisica, Roma. Permission to do this has kindly been granted by the editors of the journals in question and this is here gratefully acknowledged. As always, Springer Verlag has been most efficient in producing the present book and has been willing to accede to the writer's numerous requests.

Adrian E. Scheidegger

Contents

1. Physiographic and Geological Data Regarding the Earth

1.1 Introduction

The present-day surface of the Earth is the outcome of the antagonistic action of two types of processes, caused in turn by endogenetic and exogenetic forces. The endogenetic forces are those that have their origin *inside* the solid Earth, the exogenetic forces those that have their origin outside the solid Earth[1]. Traditionally, it has been assumed that the endogenetic processes are the "primary" ones: Mountains are built up, high plateaus are created and continents are shifted by them. Then, the resulting structures would be broken down by the action of the exogenetic forces: by the weathering, denudation, and erosion caused by wind, water, and ice. In this fashion, one could speak of "cycles" in the Earth's development. However, the modern view is rather inclined toward the idea that both, endogenetic and exogenetic phenomena, are occurring concurrently, the present-day state of the Earth representing nothing but the instantaneous equilibrium reached by the action of the two types of antagonistic forces mentioned (*principle of antagonism*[1]).

In the study of the development of the Earth, separating the two types of processes mentioned above is therefore not necessarily a straightforward matter. Nevertheless, it is useful for didactic purposes to consider them separately. In this, one is helped by the fact that endogenetic processes lead to effects of a *systematic* (non-random) nature, whereas exogenetic processes lead to effects that exhibit a *random* character; thus, a separation of "endogenetic" and "exogenetic" phenomena is possible upon phenomenological grounds. The study of the endogenetic processes has been called "geodynamics", that of the exogenetic phenomena, "theoretical geomorphology". This book is concerned with geodynamics, the exogenetic processes having been treated by the writer in a separate volume[2].

---

1 Scheidegger, A. E.: Tectonophysics 55:T7 (1979)
2 Scheidegger, A. E.: Theoretical Geomorphology, 2nd edn. Berlin-Heidelberg-New York: Springer 1970

1.2 Geological Evolution of the Earth

1.2.1 The Basic Rock Types

A study of geodynamics of necessity has to start with a review of some basic
observational facts about the Earth that have been established by field investigations. The collection and classification of such facts, i. e., the taxonomic
part of the Earth sciences, is primarily the domain of geology. Through the
incessant efforts of generations of geologists, many facts have been learned
about the constitution of the rocks in many parts of the world. For a detailed
description of these facts, the reader is referred to any one of the many
excellent textbooks on physical geology that are in existence; in the present
context, our review must of necessity be held brief.

The appearance of rocks is the result of their geological past. Among the
great wealth of rock types, however, a broad classification can be made. The
two main rock types are *sedimentary* rocks and *igneous* rocks. Sedimentary
rocks are separated into more or less distinguishable parallel layers, wheras no
such structure is evident in igneous rocks.

Among *igneous* rocks, we encounter lava, which may be thought to have
been exuded from the deeper parts of the Earth during volcanic activity. Other
types of igneous rocks, such as the granites and granodiorites, were at one
time thought[3] to have a similar history as lavas, with the difference that the
cooling process had a much longer duration and took place at great depth.
Hence the name "batholiths" (from Greek $\beta\alpha\vartheta o\varsigma$, depth and $\lambda\iota\vartheta o\varsigma$, stone)
for masses of such granites found in the interior of mountain ranges.
However, the present-day[4] view inclines toward assuming that many
batholiths were formed in situ by a process called *metamorphose*. In the case
of batholiths, this process must have been very complete as it must have
involved melting of the present rocks in order to give them the igneous
appearance. In other metamorphic rocks, it has been less complete.

The rocks on the surface of the Earth are continuously subject to detrition
by the action of wind and water. Ground down by atmospheric influences, the
débris is carried in rivers to larger bodies of water where deposition takes
place. The accumulation of such débris, under further consolidation, gives
rise to the sedimentary rocks mentioned above. The process of accumulation
itself is called *sedimentation*. Sedimentary rocks, in accordance with their
mode of formation, are "stratified". Corresponding types of strata can often
be traced to various parts of the world.

One thus arrives at a *cycle* of evolution of rocks. Sedimentary rocks
become gradually metamorphosed, possibly even entirely molten, until they

---

3 Neumayr, M.: Erdgeschichte, 3rd edn. (ed. F. E. Süess). Leipzig: Bibliographisches Institut
 1920
4 Holmes, A.: Principles of Physical Geology. New York: The Ronald Press Co. 1945

have the appearance of igneous rocks. Then the process of detrition starts, the débris is deposited somewhere and, eventually, new sedimentary rocks are formed.

1.2.2 The Scales of Geological Phenomena

The geological phenomena encountered on the Earth range spatially and temporally over vast scales.

Beginning with the spatial scales, we note that Carey[5] has made a classification into pertinent orders of magnitudes, which reach from global to atomic. A modified tabular representation of Carey's classification is shown in Table 1. In this classification, the traditional division of geology into such subjects as "petrofabrics", "global tectonics", etc. is given a precise meaning. The resulting nomenclature is basic for geodynamics studies.

A nomenclature has also been developed for the temporal scale of geological phenomena. The fact that sedimentary rocks have been formed by deposition of debris yields a powerful means of dating them, at least relative to each other. During the process of deposition, it is inevitable that living and dead organisms become entrapped which are then preserved as fossils. It is thus possible not only to obtain an idea of the age of a stratum in which a fossil is found, but also to obtain a picture of the evolution of life. The nomenclature that evolved was originally entirely phenomenological. Type-strata were named after the localities where they occurred, such as the Jurassic after the Jura mountains in Switzerland or the Permian after the city of Perm in Russia. By an analysis of the development stage of the fossils contained in these strata, a standard sequence of layered rocks was obtained. The various geological "epochs" were then classified into "periods", the latter into "eras".

Table 1. Spatial scales of geological phenomena. (Modified after Carey[5])

Order	Characteristic extension	Traditional nomenclature
I	$10^4 - 10^1$ km	Continental and global structures. "Global tectonics"
II	$10^4 - 10^1$ m	Structures seen on regional maps. "Structural geology"
III	$10^1 - 10^{-2}$ m	Structures seen on field exposure. "Minor structures"
IV	$10^{-2} - 10^{-5}$ m	Structures seen under the microscope. "Petrofabrics"
V	$10^{-5} - 10^{-8}$ m	Atomic and lattice structures

5 Carey, S. W.: J. Geol. Soc. India 3:97 (1962)

Table 2. Traditional geological time scale. (Modified after Longwell[a], ages after Van Eysinga[b])

Era	Period	Epoch	Absolute age before present (m. y.)
Cenozoic	Quaternary	Recent Pleistocene	
			1.8
	Tertiary	Pliocene Miocene Oligocene Eocene Paleocene	
			65
Mesozoic	Cretaceous	Upper Lower	
			141
	Jurassic	Upper Middle Lower	
			195
	Triassic	Upper Middle Lower	
			230
Paleozoic	Permian	West Texas Ochoa Guadeloupe Leonard Wolfcamp	
			280
	Carboniferous	Pennsylvanian Mississippian	
			345
	Devonian		
			395
	Silurian		
			435
	Ordovician		
			500
	Cambrian	Upper Middle Lower	
			570
	Precambrian		

[a] Longwell, C. R.: Geotimes 2(9):13 (1958)
[b] Van Eysinga, F.: Geological Time Table. Amsterdam: Elsevier 1975

A drawback of this method of dating is that it is naturally confined to such times from which traces of life have been preserved to the present day. The traditional geological time scale, therefore, begins with that epoch from which the oldest fossils were found. Table 2 gives the traditional sequence of the geological eras, periods, and epochs. The absolute ages also given in this Table have been determined by radioactive dating methods (cf. Sect. 2.5.3).

1.2.3 Paleoclimatic Data

From a geological investigation of the various sedimentary strata it becomes evident that various parts of the Earth must have undergone large climatological changes[6]. Thus, it is well known that during the Pleistocene, large parts of Europe and North America were covered by sheets of ice, a condition which is commonly referred to as "Pleistocene ice age". The stratigraphic evidence for the recognition of this ice age has come from many sources[7]. Most striking are the results from investigations of deep sea cores[8,9]. It is now generally believed that the most recent (Pleistocene) ice age had three phases[10]: The first phase involved climatic processes that led to the last glacial maximum (30,000 – 18,000 years BP [Before Present]); the second phase was characteristic of the "equilibrium climate" at the last glacial maximum (about 18,000 years BP); the third phase was that of deglaciation starting about 14,000 years BP and ending at 5,000 years BP.

Ice ages have also occurred at other epochs. Best known is the Paleozoic ice age, in which South Africa, Brasil and other areas appear to have been covered by ice caps[11-14]. The reasons for the occurrence of ice ages have not yet been definitely established. Theories have been advanced, for instance, by Milankovitch[15], by Ewing and Donn[16,17], Emiliani and Geiss[18], Lungersgausen[19] and others. A review of the various possibilities has also been given by the writer[20].

6 Frakes, L. A.: Climates through Geologic Time. Amsterdam: Elsevier 1979
7 Emiliani, C.: Ann. N. Y. Acad. Sci. 95:521 (1961)
8 Broecker, W. S., Turekian, K. K., Heezen, B. C.: Am. J. Sci. 256:503 (1958)
9 Broecker, W. S., Ewing, M., Heezen, B. C.: Am. J. Sci. 258:429 (1960)
10 Hecht, A. et al.: Quat. Res. 12:6 (1979)
11 Bain, G.: Yale Sci. Mag. 27:No. 5 (1953)
12 Bain, G.: Rep. Int. Geol. Congr. Norden 21:pt 12, 84 (1960)
13 Martin, R.: Can. Oil Gas Ind. 15:No. 10 (1961)
14 Stehli, F. G.: Am. J. Sci. 255:607 (1957)
15 Milankovitch, M.: Kanon der Erdbestrahlung und seine Anwendung auf das Eiszeitenproblem. Belgrade: Kgl. Serb. Akad. 1941
16 Ewing, M., Donn, W. L.: Science 127:No. 3307, 1159 (1958)
17 Ewing, M., Donn, W. L.: Science 129:463 (1959)
18 Emiliani, C., Geiss, J.: Geol. Rundsch. 46:576 (1959)
19 Lungersgausen, G. F.: Dokl. Akad. Nauk SSSR. 108:707 (1956)
20 Scheidegger, A. E.: Theoretical Geomorphology. 2nd. edn. Berlin-Heidelberg-New York: Springer 1970

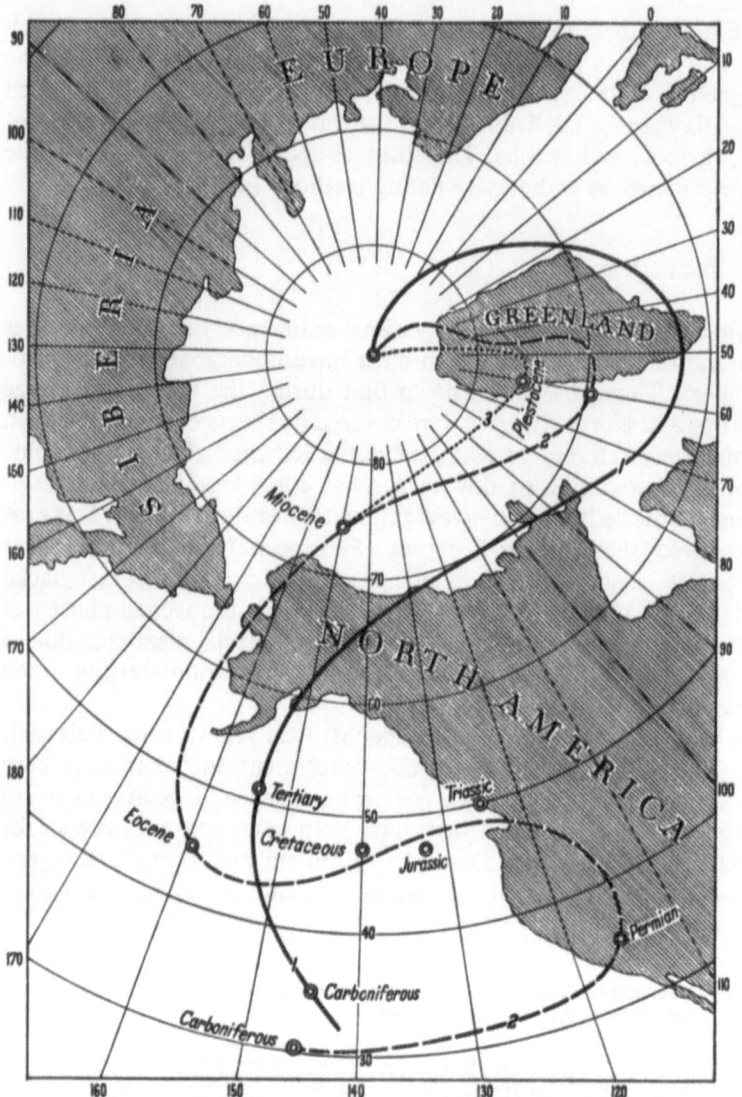

Fig. 1. Path of the North Pole as indicated by paleoclimatology. *1* after Kreichgauer[21], *2* after Köppen and Wegener[22], *3* after Köppen[23]. (After Köppen[23])

If one combines the observations on paleoglaciations, one is led to assuming as a plausible explanation that the geographic position of the North Pole underwent changes during geologic history.

21 Kreichgauer, D.: Die Äquatorfrage in der Geologie, 1. Aufl. Steyl 1902
22 Köppen, W., Wegener, A.: Die Klimate der geologischen Vorzeit. Berlin: Gebr. Bornträger 1924
23 Köppen, W.: Meteor. Z. 57:106 (1940)

The first to investigate the climatological evidence comprehensively in this fashion was Kreichgauer[21]. Later Köppen and Wegener[22] and Köppen[23] made thorough investigations of paleoclimatic data. This yielded three attempts at a reconstruction of the polar paths which are shown in Fig. 1. The trace of the pole runs in all three attempts from somewhere near Hawaii in the Carboniferous to its present position.

It is therefore seen that it is possible to postulate a reasonably coherent path of the pole to explain various geological and climatological observations. However, it should be noted that the observed effects could also be explained by a drift of the continents with regard to a fixed pole. This type of interpretation permits of an additional degree of freedom, inasmuch as the continents may also be assumed to have shifted with regard to each other and not only with regard to the pole. In effect, the uniformity of climates in the now widely separated southern continents may well be better explained by "continental drift" than by "polar wandering".

1.3 Geography of Continents and Oceans

1.3.1 Global Arrangement

An inspection of a globe shows that the latter is mostly covered by the blue color of water. It is a fact that the land areas cover less than one-third of the surface of the Earth, the rest is covered by the sea. Many land areas form chunks of considerable size which are called *continents*. It appears that most of the continents are antipodic to oceans and, with some good will, one may say that four old continental areas have their position, roughly speaking, at the corners of a tetrahedron (see Fig. 2). In addition, the continents are all

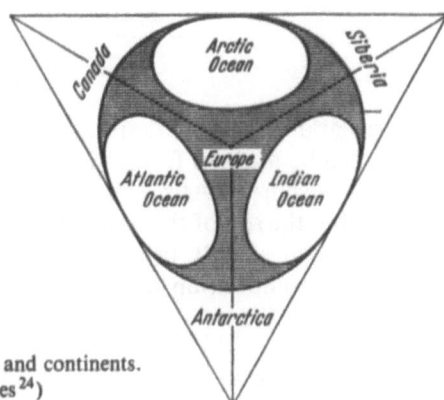

Fig. 2. The tetrahedral distribution of oceans and continents. Pacific ocean on concealed face. (After Holmes[24])

24 Holmes, A.: Principles of Physical Geology. New York: Ronald Press 1945

roughly triangular in shape, touching each other in the North and pointing southward.

The question naturally arises whether the above features are the expression of some sort of intrinsic regularity or not. Particularly, one might think that there is a fundamental significance in the fact that continents are antipodic to oceans. However, it has been shown by Evison and Whittle [25] that, if one were to take a series of areas making up one-third of the surface of a sphere, and if one where then to place these areas on the sphere in a random fashion, the most probable position of these areas would, in fact, be non-antipodic to each other (i.e., antipodic to the non-covered areas).

Fig. 3. Baker's [26] composition of the continents. (After Du Toit [27])

A remarkable observation is that the continental structures on the Earth's surface can be made to fit together rather well like a jigsaw puzzle. The fit of the Western shore of Africa with the Eastern shore of South America is quite obvious, but the rest of the continents can also be made to fit with more or less ease. This had already been observed as early as in 1911 by Baker [28] who showed the composition of the continents reproduced in Fig. 3. However,

25 Evison, F. F., Whittle, R.: Geol. Mag. 98:377 (1961)
26 Baker, H. B.: See Du Toit [27]
27 Du Toit, A. L.: Our Wandering Continents. Edinburgh: Oliver & Boyd 1937
28 Baker, H. B.: Cited in Du Toit [29]

more recently[29-31] it has been the practice to fit the Earth's continents together into *two* groups, called *Laurasia* and *Gondwanaland*, rather than into one as done by Baker. Laurasia is the complex of Europe, Asia, and North America, which is even at the present time not very widely dispersed; Gondwanaland is the combination of all the southern continents fitted together. The geometrical validity of this fit was tested by Carey[32] and was subsequently established beyond any doubt by Bullard et al.[33]. Finally, it was improved by Le Pichon et al.[34] by showing that the 2,000 – 3,000 m isobath has to be taken for the purpose of making the fit, rather than the present-day coast or "continental slope-line".

From the artificial arrangement of the continents into two big blocks, it is an easy step to postulating that the continents actually *were* formed originally as such blocks and that they subsequently "broke up" and "drifted" into their present position[35]: We shall discuss the dynamical possibilities for this having occurred later, and at the present time only mention the physiographic evidence bearing thereupon as exhibited by the fit of the continents. In addition to this physiographic indication, many geological data have been collected[35,36] with the intention to find features common to the various continents which might indicate whether or not and, if so, when the continents moved apart from the two original blocks. It is possible to state the following points in favor of the hypothesis of continental drift:

a) The orogenetic activity in the southern continents is localized in a belt that can be followed continuously through Gondwanaland as the *"Samfrau"* geosyncline.
b) Glaciation in the Carboniferous and Permian period seems to radiate from a point corresponding to the position of the South Pole as postulated by Köppen (cf. Sect. 1.2.3) for that epoch, but appears to cover parts of the southern continents in such a fashion as to suggest that the latter were close together at that time.
c) Paleobiological evidence seems to indicate that the southern continents had, even in comparatively recent times, some land connection between each other. Otherwise the simultaneous occurrence of e.g., marsupalia in South America and Australia would appear as difficult to explain. Other

29 Du Toit, A. L.: Our Wandering Continents. Edinburgh: Oliver & Boyd 1937
30 King, L.: Geology 8:111 (1980)
31 Rickard, M. J., Belbin, L.: Tectonophysics 63:1 (1980)
32 Carey, S. W.: In: Continental Drift, A Symposium (ed. Carey), p. 177. Dept. Geol. Univ. Tasmania 59:177 (1958)
33 Bullard, E. C., Everett, J. E., Smith, A. G.: Philos. Trans. R. Soc. London Sec. A 258:41 (1965)
34 Le Pichon, X., Sibuet, J.-C., Francheteau, J.: Tectonophysics 38:169 (1977)
35 Wegener, A.: The Origin of Continents and Oceans. Translated from 3rd German edn. by J. G. A. Skerl. London: Methuen 1924
36 Du Toit, A. L.: Our Wandering Continents. Edinburgh: Oliver & Boyd 1937

examples of this kind are the distribution of the Scorpionidae[37] and of the Fusulinacean Foraminiferidae[38].

The above arguments are based entirely on *geological* evidence. During the last decades, a vast amount of *paleomagnetic* evidence was added to the former, generally supporting the ideas of continental drift. The body of this type of data will be discussed in Sect. 2.7.2 of this book.

1.3.2 The Hypsometric Curve

An informative way to represent the distribution of continents and oceans is obtained by calculating the percentage of the Earth's surface above or below a certain height level. By differentiation, this leads to a statistical distribution curve of heights which indicates what percentage of the Earth's surface lies at a certain level. This distribution curve has been termed *hypsometric curve* of the Earth. It is shown in Fig. 4 (after Kossinna[39]).

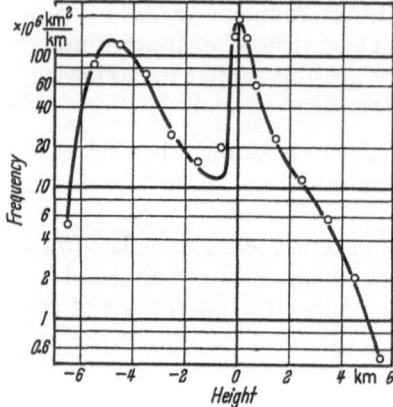

Fig. 4. The hypsometric curve. (After Kossinna[39])

With the hypsometric curve one can do some statistical analyzing. From the data of Fig. 4 it is at once obvious that the distribution of heights has two maxima, corresponding to the mean levels of the continents and that of the ocean floors. However, a more exact investigation of the data by Joksch[40] showed that the hypsometric curve is not the sum of two, but of *three* elementary distributions. The constituent distributions are logarithmico-normal defined by the equation

37 Du Toit, A. L.: Our Wandering Continents. Edinburgh: Oliver & Boyd 1937
38 Ross, C. A.: Geology 7:41 (1979)
39 Kossinna, E.: Die Erdoberfläche. In: Handbuch der Geophysik, Bd. 2, S. 875. Berlin-Heidelberg-New York: Springer 1933
40 Joksch, H. C.: Z. Geophys. 21:109 (1955)

$$w(x) = \frac{1}{s(x-a)} \frac{1}{\sqrt{2\pi}} \exp\left\{-\frac{[\log \text{nat}\,(x-a) - m]^2}{2s^2}\right\}. \qquad (1.3.2-1)$$

Here, a is called the vanishing point of the distribution. Instead of using the parameters m and s above, it is often convenient to use the "median" M defined by

$$\int_a^M w(x)\,dx = 0.5 \qquad (1.3.2-2)$$

and a "normal deviation" S indicating that interval of x around the median which contains 90% of all the area underneath the distribution curve.

In the above terms, Joksch showed that the hypsometric curve can be represented as follows (all lengths in km):

a) by a logarithmico-normal distribution of weight 61/100 with $a = -7.5$, $M = -4.5$ and $S = 3.7$
b) by a logarithmico-normal distribution of weight 23/100 with $a = -0.5$, $M = 0.2, S = 1.2$
c) by a logarithmico-normal distribution of weight 16/100 with $a = -7.5$, $M = 0.5$ and $S = 3.5$.

The tripartite composition of the hypsometric curve suggest that the three levels might have been created each by an individual process of the same nature.

1.3.3 A Comparison of Continents and Oceans

From the few brief remarks made thus far, it is evident that there is a fundamental geological difference between continents and oceans[41]. The uniform great depth of the ocean basins alone serves as an indication that continental and oceanic areas are not alike. Naturally, continental "structure" may continue for some distance offshore as a "shelf", but the transition from continental to oceanic structure is rather abruptly marked by the sudden sloping of the shelf so as to reach the depths of the oceanic abysses over a short distance.

A true comparison between continents and oceans can be made only after the various geological and geophysical aspects of the two areas have been reviewed. This will be done below.

---

41 Bullard, E. C.: Proc. R. Soc. London Ser. A 222:403 (1954)

1.4 Physiography of Continental Areas

1.4.1 General Features

Turning first to continental areas, we note that the most conspicuous irregularities thereon are undoubtedly mountain ranges. It will of course be necessary to consider as continental mountain ranges also such occurrences as offshore island chains in the sea; the latter are nothing but the peaks of submerged mountains.

If one examines the mountains somewhat more closely, a few remarkable facts become apparent. Thus, we observe that mountains are not scattered at random over the Earth's surface, but, firstly, that they occur in ranges, secondly, that the ranges themselves form chains, and thirdly, that the chains of ranges seem to form worldwide systems. The series of mountains which belong to such one worldwide system are referred to as belonging to one *orogenetic system*.

We have already mentioned in Sect. 1.1 that endogenetic processes have often been considered as *cyclic*, the idea being that mountains have to be *first* built up before they can *then* be eroded[42]. the final remnant of a once mountainous area would become a shield consisting of rocky knolls alternating with puddles of water, and then the process could start over again. However, the recognition of the principle of antagonism (cf. Sect. 1.1) casts some doubt upon this view, inasmuch as erosion begins operating as soon as a mountain is created. The neat separation of orogenesis into building-up and denudation stages can therefore hardly be maintained any longer. Rather, it seems that orogenetic belts grow at their edges, being active at different times in different parts of the world, until a new system starts to develop somewhere else. Umbgrove[43,44] was probably one of the chief proponents of such periodicity and synchronism of orogenetic activity. However, a more extreme view has been taken by Stille[45,46], according to whom short, worldwide and synchronous orogenetic phases would alternate with long periods of acquiescencee. This, however, seems to be open to criticism[47] as it is difficult to establish an exact correlation in time for orogenetic movements in widely separated regions of the Earth. Thus, it appears that orogenetic activity occurs during active periods which are separated by comparable periods of lesser activity. In this instance, it is therefore still possible to identify individual orogenetic periods, which are often loosely referred to as "cycles".

---

42 Wegmann, E.: Rev. Géogr. Phys. Géol. Dyn. Sér. II 1:3 (1957)
43 Umbgrove, J. H. F.: The Pulse of the Earth. The Hague: M. Nijhoff Publ. Co. 1947
44 Umbgrove, J. H. F.: Am. J. Sci. 248:521 (1950)
45 Stille, H.: Grundfragen der vergleichenden Tektonik. Berlin: Gebr. Bornträger 1924
46 Stille, H.: Einführung in den Bau Amerikas. Berlin: Gebr. Bornträger 1940
47 Rutten, L. M. R.: Bull. Gol. Soc. Am. 60:1755 (1949)

It appears that, since the Paleozoic epoch, there have been at least two orogenetic "cycles", the one referred to as the "Appalachian-Caledonian-Hercynian" cycle which took place at the end of the Paleozoic, the second called "Alpine-Himalayan-Circumpacific" cycle which started at the end of the Mesozoic and has still not yet come to an end. Sometimes, the Paleozoic cycle is counted as two but it seems preferable to count is as one for the present purpose.

It is not known how many orogenetic cycles occurred before the beginning of the traditional geologic time scale. With the absence of fossils it becomes increasingly difficult to trace the various orogenetic movements. It is certain, however, that numerous cycles *did* occur during the last 2 billion years, the order of magnitude of their number is commonly given as eight. Before 2 billion years ago, indications are that orogenesis occurred in a different mode than that described above (cf. Sect. 2.5.3). A discussion of the means for arriving at the above number of cycles will be given in Sect. 2.5. In all, it is reasonable to assume that about ten orogenetic cycles occurred altogether since the beginning of present-day type orogenesis 2 billion years ago.

The development of a mountain system in an orogenetic cycle has been connected with the sequence of the occurrence of various "phases of the orogeny"[48-55]. At the beginning of a cycle, one generally assumes that a *geosyncline*[56,57] is formed, usually thought to occur at a continental margin[58]. By this is meant that a trough would develop in the Earth's crust where the orogenetic belt would finally appear. The trough would then be filled in with sediments which would eventually be lifted up by some mechanism to form the mountains. The occurrence of a geosyncline is an entirely hypothetical assumption, but it is made plausible by the observation that, in mountain belts, the thickness of the sediments must have been tremendous.

1.4.2 Mountain Ranges

It is striking feature[59] of many ranges of mountains and islands of our planet that they have the shape of curved arcs which at first sight appear more

48 Bemmelen, R. W. van: Mountain Building. The Hague: M. Nijhoff 1954
49 Bubnoff, S. N.: Byull. Mosk. Ova. Ispyt. Prir. Otd. Geol. 63, 33:No. 1, 3 (1958)
50 Gidon, P.: Bull. Soc. Geol. Fr. Ser. VI, 7:125 (1957)
51 Khain, V. E.: Proc. 21st Int. Geol. Congr. Norden Pt. 18:215 (1960)
52 Kraus, E. C.: Geol. Rundsch. 50:292 (1960)
53 Kraus, E. C.: Proc. 21st Int. Geol. Congr. Norden Pt. 18:236 (1960)
54 Pavoni, N.: Vierteljahresschr. Naturforsch. Ges. Zürich 105:181 (1960)
55 Schmidt, E. R.: Ann. Inst. Geol. Hung. 49:931 (1959)
56 Kraus, E.: Z. Dtsch. Geol. Ges. 106:431 (1954)
57 Bleissner, M. F., Teichert, C.: Am. J. Sci. 245:465, 482 (1947)
58 Brockamp, B.: Geologie 4:363 (1955)
59 Cf. Scheidegger, A. E., Wilson, J. T.: Proc. Geol. Assoc. Can. 3:167 (1950)

or less circular[60-62]. The most outstanding examples of this kind are the arcs of the Philippines, Riu Kiu, Japan, Kuriles, and Aleutians. Similar features are equally found in other places of the world. The curved structure of Persia and the Himalayas is obvious, and so is the curved structure of the mountains of the Pacific Coast of British Columbia, the United States, and México.

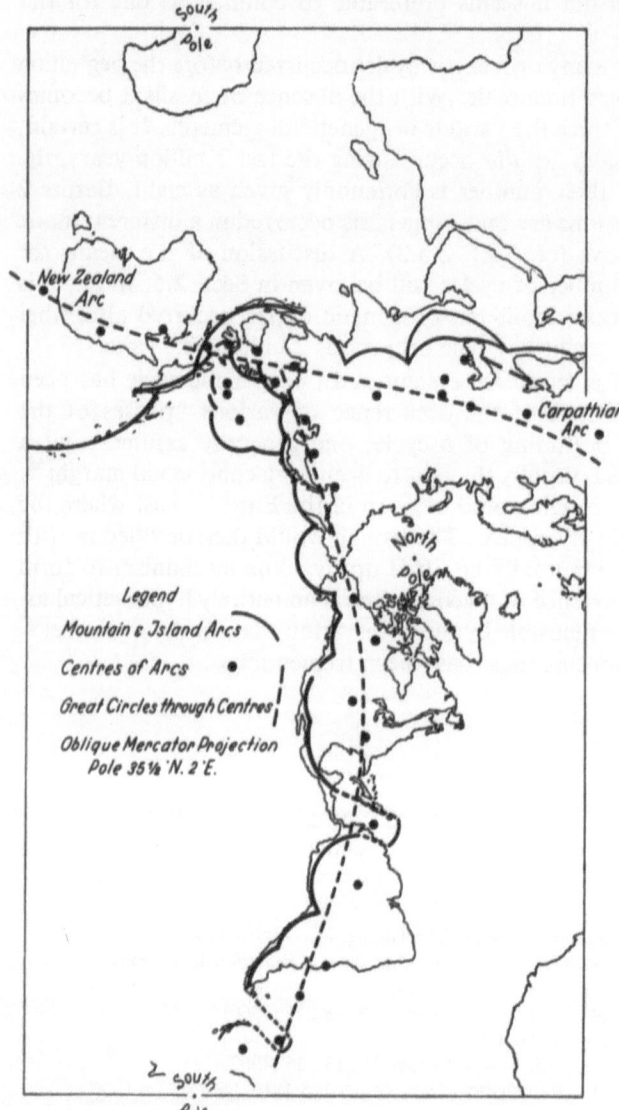

Fig. 5. The mountain and island arcs of the recent orogenetic system. (After Wilson[63])

60 Cf. Lake, P.: Geogr. J. 78:149 (1931)
61 Umbgrove, J. H. F.: The Pulse of the Earth, 2nd. edn. The Hague: Mart. Nijhof 1947
62 Eardley, A. J.: Proc. 8th Pac. Sci. Congr. 2A:677 (1956)

A mathematical analysis[59] shows that the present-day arcs are indeed nearly circular; their centers can be defined and are arranged along two great circles of the world. Thus the Mesozoic-Cenozoic orogenetic systems form two large belts, one around the Pacific and the other through the Alpine and Himalayan systems to Oceania (see Fig. 5). These belts can be traced very easily. They nearly follow two great circles intersecting each other at right angles. It may be suspected that the Paleozoic mountains formed a similar system, but it is not so easy to trace it owing to subsequent erosion and recent orogenetic activity. Nevertheless, Wilson[63] has traced part of it in North America.

It thus appears that the main orogenetic activity of the Earth has been concentrated at any one time in narrow belts. The orogenetic belts form worldwide systems. The recent ones very nearly follow two great circles and the contention is that *every* orogenetic system follows this pattern; it should be noted, however, that it is very difficult to substantiate this for the older systems.

The observation that mountain ranges are aligned in worldwide systems is one of the facts that led to the development of the so-called plate-tectonics or new global tectonics: It is assumed that the lithosphere is divided up into a number of plates (Fig. 6). Some of the plate margins are characterized by mountain ranges; others, however, are formed by features to be described later. We shall develop the argument for plate tectonics as the discussion of physiographic, geological, and geophysical features progresses in this book.

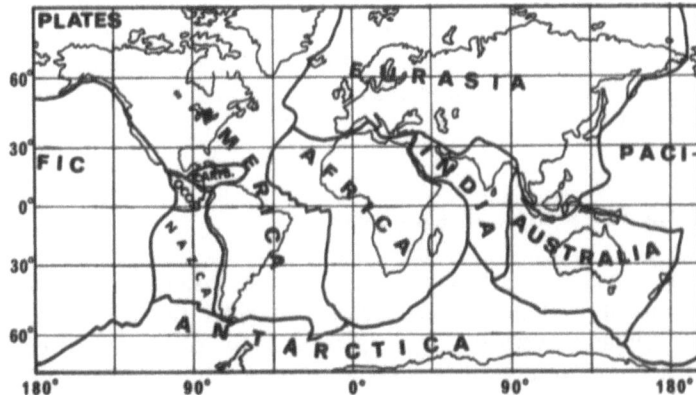

Fig. 6. The plates of the world. (After Scheidegger[63a])

63 Wilson, J. Tuzo: In: The Earth as a Planet (ed. Kuiper), p. 138. Chicago, Ill.: Univ. Chicago Press 1954
63a Scheidegger, A. E.: Rock Mech. Suppl 6:55 (1978)

A further interesting feature of mountain ranges is that some appear to be associated with a significant crustal shortening taking place in their neighborhood. The crustal shortening of mountain ranges is borne out in their ubiquitous "nappe" structure: Large portions of the sedimentary sequence are folded over or thrust from "behind", often over younger strata. Within the thrust sheets, the sequence may be in the original order or overturned.

The amount of crustal shortening can be estimated simply by a direct measurement of the strata in the great mountain systems. If one assumes that in a normal cross-section of a mountain range the length of the section of a stratum (which is a curved line) is equal to the length of that section before it was folded, i. e., when it was flat on the ground, one can determine how much shortening must have taken place. This, of course, assumes that the strata have undergone no deformation of area but only one of shape. Estimates of shortening obtained in this manner are quoted to be 50 to 80 km in the Appalachians, 40 to 100 km in the Rocky Mountains of Canada[64] and 17 km in the Coast Range of California[65]. For the Alps, Heim[66] quotes 240 to 320 km. Compared with the assumed (unfolded) cross-section of a mountain range, these values represent a shortening of up to 4 : 1.

For Tibet, even larger values has been quoted (from paleomagnetic studies; cf. Sect. 2.7.2). Thus, Molnar and Chen[67] postulated a crustal shortening of 2,500 – 3,500 km since the late Cretaceous-early Tertiary in that area.

The *form* of the individual peaks also seem to be conditioned by the presence of a compressive tectonic stress field. Thus, Gerber and Scheidegger[68] have collected the geomorphological evidence for the existence of a geophysical stress field in the Alps.

In summary, it can be stated that continental mountain ranges appear to be compression-type features. If they are taken as indicative of plate margins, it is where the plates move toward each other ("collide") that mountains are formed.

1.4.3 The Margins of Continents

The margins of continents, where oceanic and continental areas meet, are of utmost importance for geodynamic considerations.

Already a very cursory inspection of the physiography of the Earth shows that there are essentially two types of continental margins; these have been recognized very long ago and have been termed "Atlantic" and "Pacific" type of continental margins.

64 North, F. K., Henderson, G. G. L.: Alberta Soc. Pet. Geol. Guideb. 4:15 (1954)
65 Pirson, L. V., Schuchert, C.: Textbook of Geology. New York: Wiley 1920
66 Heim, A.: Geologie der Schweiz. Leipzig: Tauchnitz 1921
67 Molnar, P., Chen, W.-P.: Nature (London) 273:218 (1978)
68 Gerber, E., Scheidegger, A. E.: Riv. Ital. Geofis. Sci. Aff. 2(1):47 (1975)

In the Atlantic-type margin, the continental platform does not show much relief, its structures simply continue for some distance out to sea, until a sharp drop, termed "continental slope", occurs. The latter then becomes more gradual until it merges into an abyssal plane.

The Pacific-type margin, on the other hand, shows much relief. A large mountain range is adjacent to the sea. The continental shelf is much reduced and may quickly drop off into a very deep trench, where some of the greatest ocean depths on Earth are observed. Just before the descent to the trench, much volcanic activity may be observed. Further onward, the trench surface rises to the normal depths of an abyssal plane.

The interpretation of the two types of continental margins has been that the Pacific type is active, the Atlantic type, passive. At the Pacific-type margin, two lithospheric plates move towards each other (collide). At the Atlantic type, the continents are simply riding along, and no plate margins exist.

At the active, Pacific-type margins, there occurs evidently much orogenetic activity. It has been thought that the formation of the trench is the initial phase of the development of a geosyncline. As the trench becomes filled in, the tremendous thicknesses of sediments found in orogenic belts may become established. Finally, as the orogenesis occurs, it would appear that an addition is thereby created to the previously existing continent. In this fashion, one arrives at the idea of "continental growth" (outbuilding of the craton)[69-72].

1.5 Physiography of Oceanic Areas

1.5.1 General Remarks

Turning now to a description of the oceanic areas of the world, we noted that approximately two-thirds of the Earth's surface are covered by the sea. However, all those regions that are either continental shelves or that are situated between continents and marginal island arcs (including their foredeeps) are usually considered in conjunction with continental orogenetic processes. Subtracting these areas from the total area covered by the sea, one still ends up with over half of the Earth's surface as "true" oceanic area. It therefore becomes evident that the study of oceanic features is of extreme importance in the discussion of geodynamics.

69 Weeks, L. G.: Bull. Am. Assoc. Petrol. Geol. 43:350 (1959)
70 Maxwell, J. C.: In: Geology of Continental Margins (eds. Burk, C., Drake, D.). Berlin-Heidelberg-New York: Springer 1974
71 Smithson, S. B.: Contrib. Geol. 17(1):65 (1979)
72 Cook, F. A., Brown, L. D., Oliver, J. E.: Sci. Am. 243(4):124 (1980)

In the course of modern oceanographic investigations, it turned out that the deep sea is not just a bottomless abyss, but that there is a rather varied topography. The major structural elements that have been discovered are (1) abyssal plains, (2) mid-ocean ridges, (3) smaller features such as guyots, mid-ocean islands, and (4) archipelagic aprons. Finally, (5) extensive fracture zones have been discovered.

We shall now discuss the various oceanic features with emphasis on those characteristics that might have a bearing upon the understanding of geodynamic processes.

1.5.2 Abyssal Plains

The abyssal plains are usually thought to be the least disturbed areas of the oceans. In their original state, they show a hilly relief of roughly 300 m (particularly in the Pacific), but many parts are smooth. It is found that almost all the smooth parts are concentrated around islands or sea mounts, and have the form of "archipelagic aprons"[73,74]. These aprons seem to consist of volcanic rocks several kilometers thick. This material buries the original hilly relief of the sear floor so as to form a smooth plain. The total volume of material contained in the archipelagic aprons in the Pacific has been estimated by Menard[73] as 4×10^6 km^3.

The average depth of submersion of abyssal plains is approximately 5,000 to 6,000 m. At one time, it had been thought[75] that the Atlantic abyssal plains were different from the Pacific abyssal plains. This, however, does not seem to be the case. Since the original relief of abyssal plains appears to have been hilly, there is some doubt as to whether they really can be considered as "undisturbed".

The abyssal plains are covered with sediments. Exhaustive studies have been made of the nature and composition of these sediments[76-84]. It turns out that pelagic sedimentation rates are very slow, of the order of 2 mm per millenium[85]. The actual sediment thicknesses are about 0.5 km and it is there-

73 Menard, H. W.: Bull. Am. Assoc. Petrol. Geol. 40:2195 (1956)
74 Emery, K. O.: J. Geol. 68:464 (1960)
75 Kuenen, P. H.: Marine Geology. New York: Wiley 1950
76 Umbgrove, J. H. F.: The Pulse of the Earth. The Hague: Nijhoff 1947
77 Nafe, J. E., Drake, C. L.: Geophysics 22:523 (1957)
78 Sutton, G. H., Berckhemer, H., Nafe, J. E.: Geophysics 22:779 (1957)
79 Broecker, W. S., Turekian, K. K., Heezen, B. C.: Am. J. Sci. 256:503 (1958)
80 Shepard, F. P.: Science 130:141 (1959)
81 Hamilton, E. L.: J. Sediment. Petrol. 30:370 (1960)
82 Heezen, B. C. et al.: C. R. Acad. Sci. 251:410 (1960)
83 Arrhenius, G.: Pelagic Sediments. In: The Sea, Ideas and Observations (ed. Goldberg, E. D.). New York: Interscience 1961
84 Ericson, D. B. et al.: Bull. Geol. Soc. Am. 72:193 (1961)
85 Dietrich, G., Kalle, K.: Allgemeine Meereskunde. Berlin: Gebr. Bornträger 1957

fore seen that ocean floors must be very young. Indeed, the oldest rocks that have been found on ocean floors are Cretaceous in age.

1.5.3 Mid-Ocean Ridges

The most important features of the topography of the ocean bottom are mid-ocean ridges. These are huge "mountain" ranges rising above the abyssal plains, sometimes (in the form of islands) reaching to the surface of the sea. It has been conjectured as early as 1956 by Ewing and Heezen[86] that the system of mid-ocean ridges, with some exceptions, is a continuous one, reaching all around the globe. This conjecture has subsequently been proven to be correct[87-90]; the presently established picture of the worldwide ridge system is shown in Fig. 7. It is interesting to note that the ridges follow almost everywhere closely the median lines of the ocean basins. Inasmuch as the ocean ridges, like mountain ranges, form a worldwide system, they may also be considered as margins of lithospheric plates (cf. Fig. 6).

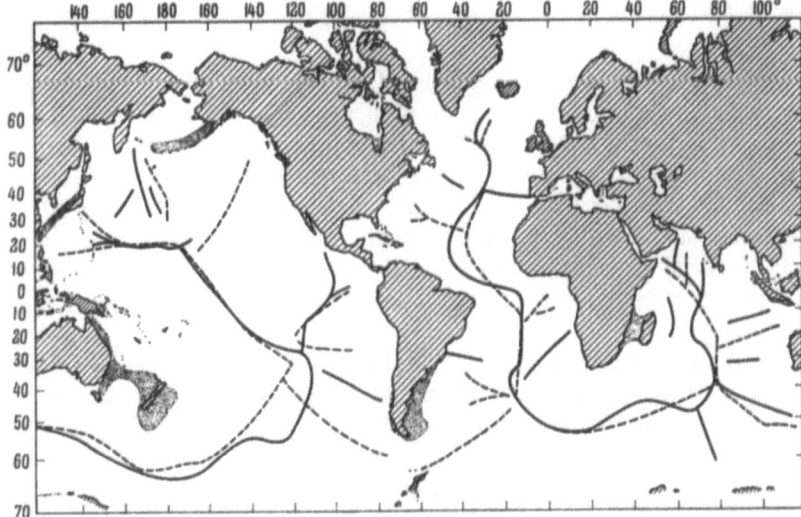

Fig. 7. Worldwide distribution of mid-ocean ridges (*solid lines*); the *dotted lines* indicate the position of the medians of the ocean basins. (After Menard[88])

86 Ewing, M., Heezen, B. C.: Am. Geophys. Monogr. 1:75 (1956)
87 Menard, H. W.: Bull. Geol. Soc. Am. 69:1179 (1958)
88 Menard, H. W.: Experientia 15:205 (1959)
89 Hess, H. H.: Preprints. Int. Oceanogr. Congr. Washington 33 (1959)
90 Ewing, M., Heezen, B. C.: Science 131:1677 (1960)

Best investigated is the Mid-Atlantic Ridge complex[91-93]. It is up to 2,000 km wide (1,500 is an average) and can be visualized as a broad swell of varied topography. There is also evidence of *fossil* ridges in the Pacific[5].

An interesting feature of mid-ocean ridges is that there appears to be a definite rift[94], approximately 100 km wide, at their crest. Depths in this rift exceed the maximum depths at the adjacent sides of the ridge out to 160 km or more. A typical profile across the Mid-Atlantic Ridge is shown in Fig. 8. The "rift"-like appearance of the crest of mid-ocean ridges is an indication that, at the margins indicated by them, the hypothetical lithospheric plates move *away* from each other. One thus arrives at the notion of a "plate-tectonic cycle": Plates are "created" at the rifts, drift apart from there and are subducted at the opposing margins.

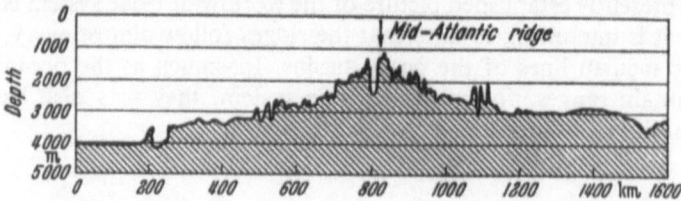

Fig. 8. Typical profile of the Mid-Atlantic Ridge. (After Elmendorf and Heezen[93])

For the directional distribution of the segments of the mid-ocean ridges several types of regularities have been proposed. Thus, Lutz and Foland[95] noted that there is a statistical preference for the ridges to run NS (a closer investigation yields that they may run preferentially meridionally to a center at 65°N, 40°W rather than meridionally to the North Pole). On the other hand, Pavoni[96] claimed that the statistical distribution of ridge-segment directions fits a conical distribution to two poles, situated on the equator at 170°W and 10°E. Such regularities, if they are real, would engender global kinematic conditions.

In summary, it may be said, therefore, that mid-ocean ridges are tensional features forming a system which reaches all around the globe. If one supposes the existence of lithospheric plates (cf. Sect. 1.4.2), the mid-ocean ridges, like the continental mountains, would be located at the plate margins. However, the mid-oceanic plate margins are *tensional* features and are thus the counterpart of the continental collision margins. This view is fully supported by paleomagnetic studies (cf. Sect. 2.7.2); the integrated evidence for the postulation of "plate tectonism" will be presented in Sect. 6.1.2.

91 Mammerickx, J., Herron, E., Dorman, L.: Bull. Geol. Soc. Am. I 91:263 (1980)
92 Tolstoy, I.: Bull Geol. Soc. Am. 62:441 (1951)
93 Elmendorf, C. H., Heezen, B. C.: Bell Syst. Tech. J. 36:1047 (1957)
94 Hill, M. N.: Deep Sea Res. 6:193 (1960)
95 Lutz, T. M., Foland, K. A.: Geology 6:179 (1978)
96 Pavoni, N.: Umsch. Wiss. Tech. 1971(9):318 (1971)

It may further be noted that rift-type features are, on occasion, also seen on land. Best known are the African Rift valleys and the Rhine graben. It has been postulated that such features are in fact related to mid-ocean ridge systems. Perhaps the lithospheric plates are moving apart at these locations just like at the ridges and some day an ocean will be formed there. It has been noted that the preferential orientation of continental rifts is meridional[97]; inasmuch as there has been a similar observation regarding oceanic ridges, an analogous origin of the two types of phenomena may be indicated.

Thus, at the mid-oceanic ridges and possibly at the continental rifts, oceans are "opening up" at the present time. Possibly, oceans can also close again. This idea of this type of cyclicity has been termed the "Wilson cycle"[98]. Inasmuch as the evidence for possibly closing oceans has been destroyed in the collision zones, the existence of such a cycle is highly hypothetical[99].

1.5.4 Smaller Features in Basins

Finally, we turn our attention to the remaining oceanographic features mentioned in Sect. 1.5.1. These features are mostly types of smaller irregularities that may occur within the confines of the oceans.

One of these is the *sea mount*. Sea mounts of various shapes and sizes are found in many basins. If they reach to the surface of the sea, they give rise to the development of coral atolls[100]. Some of the sea mounts are flat-topped; they are then called *guyots*. The flat top seems to indicate that these sea mounts once reached to the surface of the water and that wave action levelled their tops. They represent, thus, drowned ancient islands. They are particularly numerous in some parts of the Pacific[101-103]; some exist in the Atlantic as well[104].

The above observations point to a relative subsidence of many features in the ocean. Coral atolls seem to point in the same direction; for, on a subsiding sea mount, under favorable climatic conditions, coral growth could just keep up with subsidence. From a hole drilled on Bikini, Kuenen[105] estimated a rate of subsidence (before glacial times) of a little less than 5 mm/century. It would thus appear that sea mounts and atolls are of volcanic origin. After their formation, they start sinking owing to the extra weight caused by their presence.

97 Ranalli, G., Tanczyk, E. I.: J. Geol. 83:526 (1975)
98 Wilson, J. T.: Proc. Am. Philos. Soc. 112:309 (1968)
99 Dewey, J., Spall, H.: Geology 3:422 (1975)
100 Wiens, J. H.: Ann. Assoc. Am. Geogr. 49:No. 1, 31 (1959)
101 Dietz, R. S.: New Sci. 5:14 (1959)
102 Hamilton, E. L.: Sunken Islands of the Mid-Pacific Mountains. Geol. Soc. Am. Mem. No. 64 (1956)
103 Menzel, H.: Z. Geophys. 37:595 (1971)
104 Rabinowitz, P. D., Purdy, G. M.: Earth Plan. Sci. Lett. 33:21 (1976)
105 Kuenen, P. H.: Marine Geology. New York: Wiley 1950, p. 467

As noted, sea mounts are particularly numerous in the Pacific. A series of them is concentrated within a number of "fracture zones" off the shores of North America. The Pacific fracture zones were first discovered by Menard[106-108]; later, such fracture zones were also found elsewhere[109-112]. A drawing of Menard's fracture zones is given in Fig. 9. Within these fracture zones, the topography, apart from featuring an accumulation of sea mounts, is characterized by deep, narrow troughs, asymmetrical ridges and escarpments. The fracture zones also affect the bottom water circulation in a peculiar fashion[114].

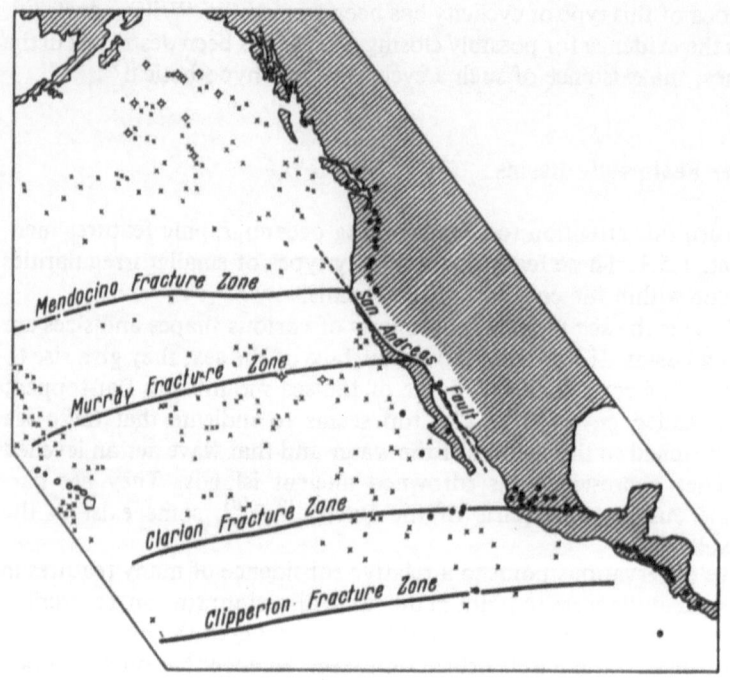

Fig. 9. Fracture zones off the Pacific Coast of North America. (After Menard[113])

106 Menard, H. W.: Bull. Geol. Soc. Am. 66:1149 (1955)
107 Menard, H. W., Fisher, R. L.: J. Geol. 66:239 (1958)
108 Menard, H. W.: Bull. Geol. Soc. Am. 70:1491 (1959)
109 Panov, D. G.: Bull. (Izv.) Akad. Nauk SSSR Ser. Geol. 1958:No. 9, 84 (1958)
110 Hersey, J. B., Rutstein, M. S.: Bull. Geol. Soc. Am. 69:1297 (1958)
111 Wright, J. B.: Tectonophysics 34:T43 (1976)
112 Francheteau, J., et al.: Can. J. Earth Sci. 13(9):1223 (1976)
113 Menard, H. W.: Bull. Geol. Soc. Am. 66:1149 (1955)
114 Vangriesheim, A.: Oceanolog. Acta 3(2):199 (1980)

1.6 Physiography of Tectonic Features

1.6.1 General Remarks

The rocks deposited by sedimentation or emplaced by igneous processes cannot generally be expected to have persisted in their original position throughout the geologic ages. Disregarding the effects of weathering and erosion (which are the subject matter of geomorphology), we note that the rocks are continuously affected by endogenic forces. The resulting effects on the Earth are called "tectonic features".

The most common tectonic features are *faults:* breaks occur in the continuity of the rocks due to the action of endogenic stresses. Small faults are called *joints*; these are "cracks" that are visible at every rock face or outcrop. The orientation structure of *valleys in plan* may also be the result of tectonic processes: While the deepening of valleys is certainly due to exogenic (erosive) forces, these may act preferentially in ways predesigned by tectonic forces.

If the tectonic stresses do not reach the breaking strength of rocks, one may be faced with continuous, rather than discontinuous, deformations. On a large scale, such continuous deformations lead to *folds*. On a microscale, individual constituents of the rock (such as pebbles or crystal grains) become deformed. These effects are treated in the study of *petrofabrics*.

We shall deal with the various tectonic effects in their turn below.

1.6.2 Faults

Starting with *faults*, one can say that these are fracture surfaces along which the rocks have undergone a relative displacement. They occur in parallel or subparallel systems which have usually a wide lateral distribution.

Physically, a fault is determined by first stating its position on the Earth (by giving latitude and longitude) and secondly by giving its "character" by fixing (a) the *fault plane* and (b) the *direction of relative motion* of the two sides of the fault thereon[115]. It is thus seen that, in order in determine the character of a fault, one needs *three* parameters, two determining the position of the fault plane, and one determining the direction of the motion vector. Sometimes more parameters are given, but then constraints must apply between them.

Most representations of faults aim at fixing (a) the fault surface, and (b) the direction of motion. The fault surface can be determined, e.g., by giving the direction of strike (i.e., the line of intersection of the fault surface with a horizontal plane) together with the dip (i.e., the angle between the fault plane and a horizontal plane) as shown in Fig. 10. Alternatively, the dip direction

---

115 Scheidegger, A. E.: Bull. Seismol. Soc. Am. 47:89 (1957)

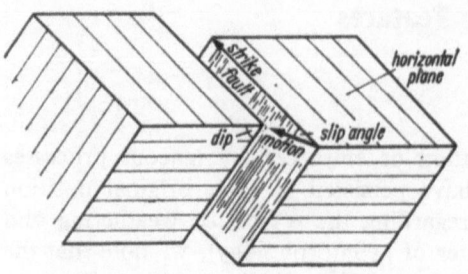

Fig. 10. A fault

and dip can be given. The direction of motion can be determined, e.g., by giving the angle between the motion vector and the strike direction of the fault. This angle is called "slip angle". Alternatively, introducing more than three parameters for the determination of a fault, the direction of the motion vector can be given by stating its azimuth (referring to the northward direction) and its inclination to the horizontal. Another way to give the direction of the motion vector is by stating strike and dip, or dip direction and dip, of a plane which is orthogonal to the motion vector. The latter plane is commonly called "auxiliary plane" of the fault.

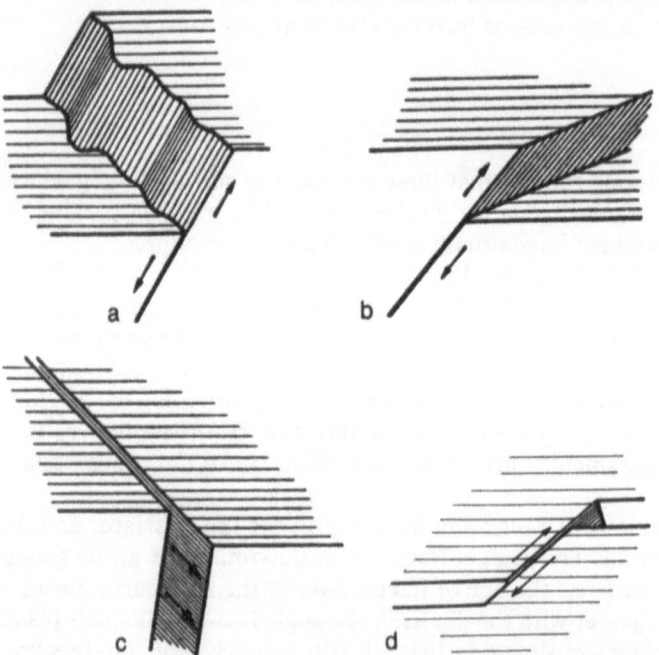

Fig. 11a – d. The four idealized types of faults. **a** Normal fault, **b** reversed fault, **c** transcurrent fault, **d** transform fault

The terminology of faults has been coined by geologists[116,117]. It is customary to call a fault transcurrent if the slip angle (that is the angle between strike and motion vector) is smaller than 45°. Transcurrent faults may be further classified as "dextral" or "sinistral". Imagine oneself standing on one side of the fault, and looking at the other side. If the other side appears to have moved to the right, the fault is dextral, if it appears to have moved to the left, the fault is sinistral. If the slip angle is greater than 45°, one also differentiates between two cases: normal faults, if the (over-) hanging wall moves downward, and reversed faults or thrusts if the hanging wall moves up. A special type of transcurrent fault is the transform fault which has the property that it is always associated with the opening up of an offset trench[118]. The offsets on mid-ocean ridges are effected by such transform faults; the trench corresponds to the rift atop the ridges. Figure 11 will serve to illustrate the geologists' conception of faulting.

Deep-seated faults are sometimes not directly visible at the surface of the Earth, but their presence may be recognizable in geomorphic features. Such features have been called "lineaments" by Hobbs[119] who defined the latter as "significant lines of landscape which reveal the hidden architecture of the rock basement". Later, other definitions of the term "lineament" (or "linear") have also been given, but it has been strongly suggested to use Hobbs' nomenclature[120] which will also be followed in the present treatise. Faults (and lineaments) often occur in systems and can easily be mapped from air photographs of remote sensing data[121]. In this instance, of course, only the strikes of the features can be determined. Inasmuch as the picture often is one of bewildering complexity, a statistical procedure has to be employed for an analysis. For this purpose, one commonly constructs "rosette diagrams" for a region. Such diagrams usually represent the number (or the relative number) of directions found within given intervals as a circular histogram. The maxima on such a circular histogram are the "preferred" directions of the lineaments (fault strikes). A possible digitization and computerization of this procedure will be discussed later in connection with the analysis of joints.

Lineament systems can also be analyzed statistically with regard to the distribution of the *lengths* of the individual "elements". Ranalli[122] has shown that the distribution is of lognormal type which implies that the faulting process can be regarded as a random process obeying the law of proportionate effect.

116 Holmes, A.: Principles of Physical Geology. London: T. Nelson 1944

117 Lensen, G. J.: New Zeal. J. Geol. Geophys. 1:307 (1958)

118 Wilson, J. T.: Nature (London) 207:343 (1965)

119 Hobbs, W. H.: Earth Features and their Meaning. New York: Macmillan (see p. 227 therein) 1912

120 O'Leary, W., Friedman, J. D., Pohn, H. A.: Bull. Geol. Soc. Am. 87:1463 (1976)

121 See e. g. Hodgson, A. (ed.): Proc. First Internat. Conf. on the New Basement Tectonics, Salt Lake City, 1974. Utah Geol. Assoc. Publ. No. 5 (1976)

122 Ranalli, G.: Can. J. Phys. 13(5):704 (1976)

Particularly puzzling phenomena are large thrust faults: Major sheets containing lithological units may be thrust over others, usually with stratigraphic separations measured in kilometers. The contact is knife-sharp and often no more impressive than a mere bedding plane. The phenomenology of this type of feature has been investigated by Gretener[123] who showed that there are four types of characteristic aspects: First, the already mentioned sharpeness of contact; second, the existence of basal tongues (shown schematically in Fig. 12); third, the occasional occurrence of small boudins (for the physiographic description of boudinage see Sect. 1.7.2), and fourth, the existence of side thrusts. As will be shown in Sect. 7.2.2, no theory of these features exists as yet. There are indications that the type of thrusting discussed above occurs at the present time. Thus, Schäfer[124] has investigated recent highway cuts in the Appalachians and has shown that the bores for the dynamite show progressive near-horizontal offsetting with time. The thrusting mechanism, therefore, must be held to proceed constantly.

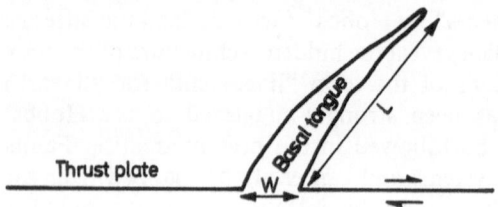

Fig. 12. Schematic presentation of a basal tongue structure. Usually the dimensions are within the following ranges: $5 < w < 100$ cm; $1 < L < 30$ m. (After Gretener[123])

1.6.3 Joints

Small faults, i.e., cracks in rocks, have been called *joints*. Such joints are found in every outcrop, where they form, in general, a complex picture. However, in spite of the fact that a rock wall may be dissected by many joints, it is often possible to recognize three sets as forming the surfaces of the fundamental rock "cell": Evidently, three fractures are needed to define a block of rock. Such a block is called the "fundamental joint parallelepiped"; very often it is visible as such in an outcrop. However, even if the picture is very complex, three main sets of joints are usually recognizable. Of the three main joint directions, the geologist will usually associate one (called "spurious") with some lithological factor, such as bedding planes, schistosity, etc. The remaining two directions must then be considered as tectonically induced.

The standard evaluation of joint orientation measurements is by means of plotting the pole of each joint surface investigated in an outcrop upon the Lambert (equal-area) projection of a unit sphere (Fig. 13)[125].

123 Gretener, P. E.: Bull. Can. Petrol. Geol. 25(1):110 (1977)
124 Schäfer, K.: Nature (London) 280:223 (1979)
125 Müller, L.: Der Felsbau, 624 pp. Stuttgart: F. Enke 1963

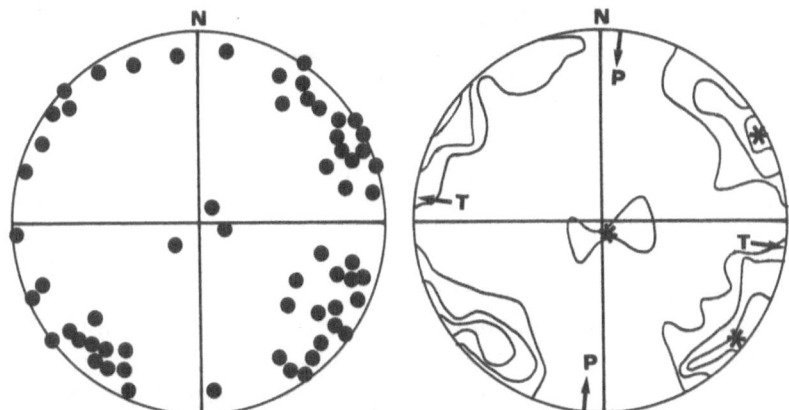

Fig. 13. Poles (*left*) and density lines (*right*) of joints at an outcrop near Maria Langegg, Bohemian Massif, Austria. The density maxima are indicated by *asterisks*, *P* and *T* indicate the maximum and minimum compression directions, respectively, constructed as the bisectrices of the "preferred" joint orientations (cf. Sect. 2.4.5)

It is customary to state the frequency of joints of a certain orientation in "per cent"; this is given by the percentage of the total number of joint directions found within 1% of the surface of the lower half of the unit sphere. Then, equidensity lines are drawn in the field of density numbers and the maxima are determined. These maxima represent the poles of the preferred joint orientations (Fig. 13). It is possible to computerize the above procedure. The computer can be programed to list the density values in a table and to plot the density lines automatically.

The standard procedure outlined above is evidently a non-parametric statistical procedure: No assumptions are made with regard to the distribution that is to be expected with regard to the orientation of the joints. Hence it is simply the maxima above an arbitrarily fixed cutoff which are chosen as representatives of the "preferred" joint orientations. No "averaging" or "best fit" of the joint orientations is attempted at this stage. The cutoff is simply raised until a series of maxima become evident.

The use of a non-parametric statistic precludes one from taking average or best-fitting orientations of joints, as well as from making a statement of confidence limits for the result, since the latter always refer to some theoretical distribution. Hence, it is most desirable to find a parametric representation for the theoretical distributions to which one would expect joint orientations to be subject. For this purpose, a computational method was developed by Kohlbeck and Scheidegger[126]. In that method, two statistical probability distributions (so-called "Dimroth-Watson distribution") of the type $\exp(k \cos^2 \vartheta)$ about a mean direction are fitted to the data; the two best-fitting mean

126 Kohlbeck, F., Scheidegger, A. E.: Rock Mech. 9:9 (1977)

directions are determined by computer using a function-minimization procedure. The computational procedure is nothing but a development of the older method of drawing density diagrams for the joint directions in some suitable projection of a unit sphere and picking two density maxima.

The above procedure for the determination of "preferred" joint directions has been carried out in many parts of the world. On the whole, one finds the following phenomenological features[127]:

1. At a single outcrop, one finds ordinarily three joint systems which are usually very definite: One system is near-horizontal (dips to 40°) and corresponds to some lithological factor; the other two systems are near-vertical and almost orthogonal to each other (angle of intersection 80° – 90°).

2. The non-lithological joints in fresh outcrops appear to cut clear across joint systems of obviously older age. Attempts have also been made to refer the surface joints to the "tectonic" coordinate system A, B, and C connected with the tectonic deformation (see Sect. 1.6.5). However, no consistency is usually obtained in this manner, indicating that the joints have not undergone the deformation but are of entirely recent origin. These joints are therefore interpreted as tectonic joints.

3. Several outcrops near to each other (within a few kilometers) usually show preferential joint orientations that are consistent which each other.

4. Outcrops within a region (10 – 20 km radius) commonly show, if treated together, definite preferential tectonic joint orientations. When the outcrops are considered singly, however, one often finds that about one-fifth of them show "anomalous" tectonic joint orientations which are rotated up to about 30° with regard to the "regionally" preferential orientations. The anomalous outcrops are not randomized, but show a consistency among each other.

5. Thus, the rotated joints may "come through" as secondary, weak maxima in a regional joint diagram. The latter, then, has the following features: One strong lithological maximum, two strong maxima corresponding to two near-orthogonal orientations, and two further, but weak, near-orthogonal maxima corresponding to the tectonic joints in the anomalous outcrops.

6. The well-developed regional joint system can commonly be explained in terms of global plate tectonics. This will be further discussed in connection with a discussion of the tectonic stress field (Sect. 2.4.5).

1.6.4 Valleys in Plan

An inspection of a hydrographic map of an area shows generally at first glance that the valley trends are not random. According to the "Principle of Antagonism" (Sect. 1.1), the non-random part must be expected to be due to

127 Scheidegger, A. E.: Riv. Ital. Geofis. Sci. Aff. 5:1 (1979)

the action of tectonic agents. The assumption that valley trends are predetermined by geotectonic phenomena stands in contrast to the assumption of valleys being solely caused by exogenic agents (i. e., water, ice, and wind). In fact, there is a certain controversy about this matter to this day, but the evidence in favor of *some* geotectonic control of the valley orientations is building up steadily. Thus, it has been shown that the main characteristic of exogenic agents is their randomness (cf. Sect. 1.1)[128]. Valleys caused by erosion alone should, therefore, be randomly oriented. Evidence from statistical analyses of valley directions shows that the latter are not random (see below). Furthermore, the often large vertical displacement rates (see Sect. 1.7.6) in mountain areas suggest that the surface features are of very recent origin. This makes it difficult to believe that modern river nets should have been determined by some ancient drainage pattern which is retained to this day.

The contention that valley orientations have been determined by tectonic features is by no means new. Thus, Frebold[129] found a correlation between joint and valley orientations in the Brocken area of Germany. A similar relationship was noted between valley (and fjord) directions and fracture patterns in Norway[130,131]. Gerber and Scheidegger[132] have qualitatively correlated trend patterns of Alpine valleys with the tectonic stress system, and Potter[133] has postulated a geotectonic origin for the direction of most big rivers of the world[133].

Naturally, it cannot be assumed that valleys are simply "fissures" in the ground; rather, the preferential orientation of river links is influenced by tectonic agents[134,135]. Superimposed upon the non-random tectonic part there is a random exogenic part, and, in order to make a proper analysis, statistical methods must be used to separate the non-random from the random effects. Thus, in order to make a quantitative evaluation of valley trends, the latter have to be "rectified" (i. e., straightened) by considering them as edges in a graph[136]. This may be a somewhat "brutal" procedure, but it is at least independent of the bias of the researcher. Otherwise, the fitting of straight (and therefore measurable) segments to the "wiggly line"[137] representing the river course on a map, would be extremely arbitrary. However, in the described fashion, the distribution of valley orientations can be represented numerically in a unique manner and, in consequence, can be analyzed statistically.

128 Gerber, E., Scheidegger, A. E.: Z. Geomorphol. Suppl. 18:38 (1973)
129 Frebold, G.: Geogr. Ges. Hannover 1932/33:89 (1933)
130 Gregory, J. W.: The Nature and Origin of Fjords. London: Murray 1913
131 Randall, B. A. O.: Geogr. Ann. 43:336 (1961)
132 Gerber, E. K., Scheidegger, A. E.: Verh. Geol. Bundesanst. (Austria) Wien 1977(2):165 (1977)
133 Potter, P. E.: J. Geol. 86:13 (1978)
134 Toynton, R.: Bull. Geol. Soc. Norfolk 30:39 (1978)
135 Holland, W. N.: Aust. Geogr. 13:338 (1977)
136 Scheidegger, A. E.: Rock Mech. Suppl. 9:109 (1980)
137 Ghosh, A. K., Scheidegger, A. E.: J. Hydrol. 13:101 (1971)

The determination of the best-fitting orientation to the histogram is best performed by the same procedure as with joints. For this purpose, a valley is regarded, so to speak, as a "vertical joint". Then, a series of superposed distributions of the type $\exp(k\cos^2\vartheta)$ are fitted to the data by the usual function-minimization program[138].

Let us illustrate the above procedure on an example: Austria[139]. Figure 14 shows (above) the hydrographic map of the country and (below) the "rectified" graph.

It is, then, possible to make a statistical analysis of the rectified directions. The polar diagram of the rectified river-trend poles is shown in Fig. 15a, which may be compared with the polar diagram of the joint poles for Austria (Fig. 15b). The correspondence is not immediately obvious, but the parametric computer evaluation yields

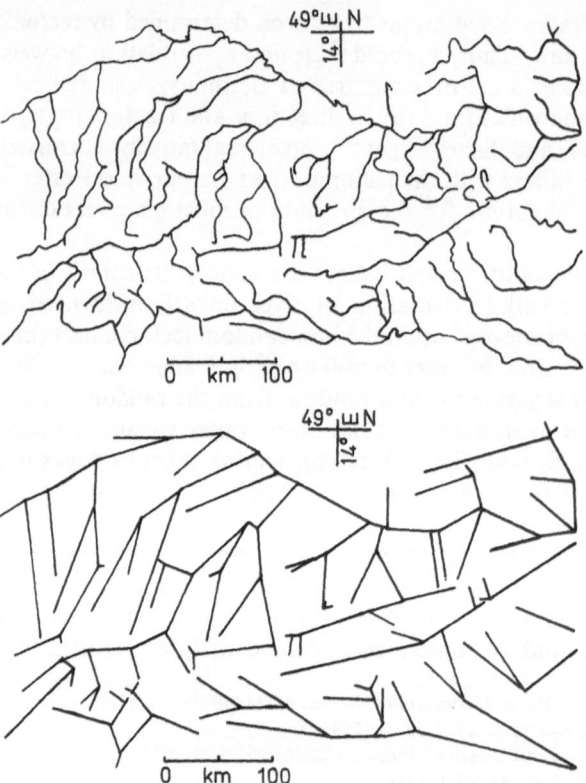

Fig. 14. Hydrographic map of Austria (*above*) and its rectification (*below*). (After Scheidegger[140]; reproduced by permission of the Austrian Geographical Society)

138 Kohlbeck, F., Scheidegger, A. E.: Rock Mech. 9:9 (1977)
139 Scheidegger, A. E.: Mitt. Oesterr. Geogr. Ges. 121:187 (1979)
140 Scheidegger, A. E.: Mitt. Oesterr. Geogr. Ges. 121:187 (1979)

valley (poles)
| | Maximum 1 | N 172° ± 3° E |
| | Maximum 2 | N 92° ± 3° E |

joints (poles)
| | Maximum 1 | N 5° ± 25° E |
| | Maximum 2 | N 92° ± 22° E |

There is evidently an excellent correspondence, at least in Austria, between joints and valley trends. This has been confirmed even on a local level in Tyrol[141]. Similar conditions have been found elsewhere, e.g., Switzerland[142,143], the Himalaya[144], and the Canadian Shield[145].

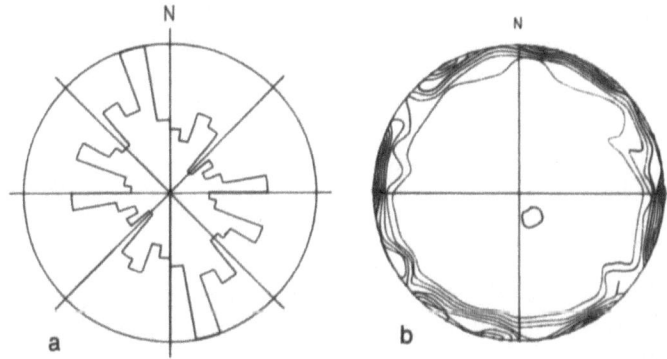

Fig. 15. a Polar histogram of river trend normals in Austria, **b** joint-pole density diagram for Austria. (After Scheidegger[146]; reproduced by permission of the Austrian Geographical Society)

1.6.5 Folds

In the formation of mountain ranges it is evident that not faulting, but a process properly called *folding* is of prime importance. It appears that the strata in such mountain ranges have been contorted to a fabulous extent[147]. It almost looks as if some supernatural giant took originally level strata, pulled

141 Drexler, O.: Einfluß von Petrographie und Tektonik auf die Gestaltung des Talnetzes im oberen Rißbachgebiet (Karwendelgebirge, Tirol). Muench. Geogr. Abh. 23: Univ. München (1979)
142 Scheidegger, A. E.: Geogr. Helv. 34:9 (1979)
143 Scherler, K. E.: Zur Morphogenese der Täler im südlichen Tößbergland. Dipl. Thesis, Geogr. Inst., ETH-Zentrum, Zürich, 1976
144 Scheidegger, A. E.: Arch. Meteorol. Geophys. Bioklimatol. Ser. A28:89 (1979)
145 Scheidegger, A. E.: Z. Geomorphol. 24(1):19 (1980)
146 Scheidegger, A. E.: Mitt. Oesterr. Geogr. Ges. 121:187 (1979)
147 Cf. e. g. Holmes, A.: Principles of Physical Geology. New York: The Ronald Press Co. 1945

them up and folded them over at his will, until they may become recumbent[148]. Naturally, the action of water and wind will erode some of the folded materials and the physiographic appearance of a mountain range is therefore one of high peaks and deep valleys. Nevertheless, the continuity of the folded strata can be traced from peak to peak and the original (i. e., undisturbed by erosion) position of the layers can be reconstructed. The horizontal distances over which the strata are folded over may be up to several score kilometers. The Western Alps give a classical example. The nomenclature of folds has again been created by geologists[147]. The possible types of simple folds are classified as anticlines, synclines, and monoclines. The three types are illustrated schematically in Fig. 16, which is self-explanatory.

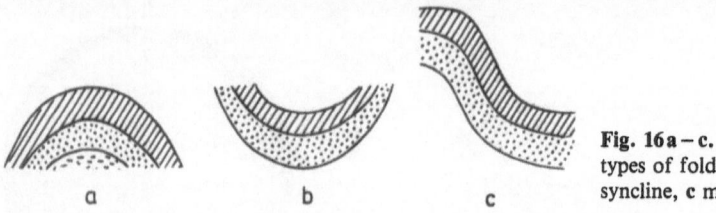

a   b   c

Fig. 16a – c. The three ideal types of folds: **a** anticline, **b** syncline, **c** monocline

The two sides of a fold are called its *limbs*. The bisecting plane of the two limbs is termed the *axial plane* of the fold. Finally, the *axis* of the fold is the intersection of the axial plane with the uppermost stratum. The angle of the axis with the horizontal plane is called its *pitch* (see Fig. 17).

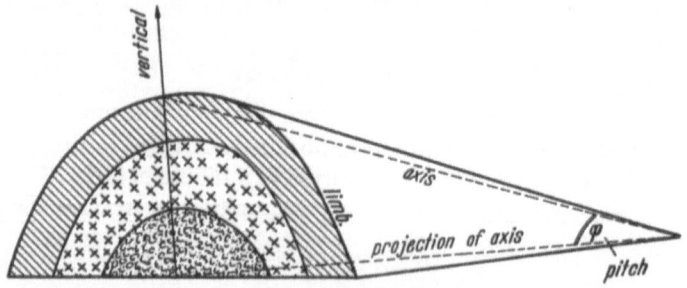

Fig. 17. Schematic drawing of an anticlinal fold

The quantitative description of a fold is not a simple matter, inasmuch as a whole series of parameters are required to effect it. Hansen[149], in fact, introduced 12 parameters. The most important of these are those describing the

148 Sitter, L. U. de: Bull. Soc. Belge Geol. 66:No. 3, 352 (1957)
149 Hansen, E.: Strain Facies. Berlin-Heidelberg-New York: Springer 1971

direction of the fold axis, commonly called the *b* axis; the latter is the line along which *no* deformation occurred during the folding process.

The simple types of folds discussed above form the elements of more complex features. A review of the quantitative morphology of folds has been given by Hansen[149]. The superposition of various fold systems can produce a picture of considerable complexity which may be difficult to unravel in field investigations. Many studies of this type have been published in the literature (cf. Brongoleyev[150] for Central Asia; Stauffer and Mukherjee[151] for the Canadian Shield; Laubscher[152] for the Jura mountains). The procedure is to observe the directions of *b* axes and to determine the most frequent orientations.

A further class of folds that are supposed to develop in shear zones are shear folds. These are closely spaced zigzag patterns between shear fractures[153]. A similar feature are drag folds (e. g., Ramberg[154]) which are folds "with quasi-monoclinic symmetry usually, but not exclusively, developed in relatively thin competent layers in the flanks of larger folds." In this, the drag folds are generally found in soft layers which are sandwiched between more competent ones.

In a folded area, it is possible to introduce a "tectonic" coordinate system. The *b* axis of this system is the fold axis (in a packet of folded layers, this is the only direction that has not undergone a curvature). The other two directions are normal to the *b* axis such that the *c* axis is normal to the layering and the *a* axis parallel to it. The units are chosen in such a manner that they encompass the shortening (if any) that took place during the deformation process. Thus in the pre-fold stage the *c* axis was vertical, the *a* and *b* axes were horizontal, and the units were the same on all the axes.

Linears are usually represented in the "tectonic coordinate system" by giving direction ratios (in tectonic units), planes by giving the (ratios of the) reciprocals of the segments which they cut off the three axes, in conformity with the Miller indices in crystallography. These quantities are usually denoted by *h, k,* and *l* as referring to the *a, b* and *c* axes, respectively.

It is convenient to refer all those quantities to the tectonic co-ordinate system which has been affected by the folding process.

1.6.6 Petrofabrics

Finally, tectonic events can also affect the "fabric" of rocks: By this is meant that from small to smallest components of the rock may become deformed by tectonic processes. Thus pebbles may appear to be squeezed in a conglom-

150 Brongoleyev, V. V.: Problema skladkobrazovaniya v zemnoy kore. Moscow: Nedra 1967
151 Stauffer, M. R., Mukherjee, A.: Can. J. Earth Sci. 8:217 (1971)
152 Laubscher, H. P.: Tectonophysics 37:337 (1977)
153 Esz, V. V.: Geotektonika (USSR) 1969(3):52 (1969)
154 Ramberg, H.: Geol. Mag. 100(2):97 (1963)

erate, crystal grains may be squashed in an igneous rock, and special features such as stylolites may form.

Stylolites are particularly interesting features, because they are pressure-solution phenomena which occur mainly in limestone and chert: they grow from a given surface in the direction of the largest pressure (see Fig. 18). Of great interest are those stylolites which are horizontal, as they are an indication of the fact that the maximum compression is or was horizontal.

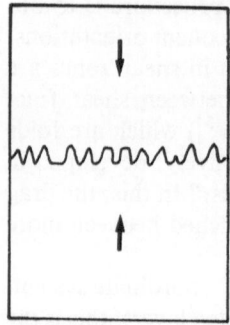 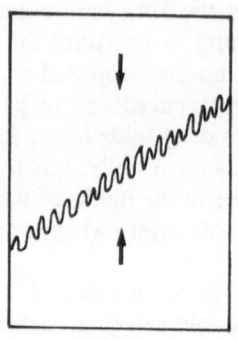

Fig. 18. Stylolite pillars growing orthogonally (*left*) and obliquely (*right*) to a given stylolitic seam in the direction of the largest compression. (After Schäfer[154a])

Petrofabric elements are generally linear or planar. Hence, in the investigation of a rock sample, the problem arises of a statistical analysis of the data. In general, like in the analysis of joints, one or more distributions (so-called Dimroth-Watson distributions) of the type $\exp(k\cos^2\vartheta)$ are fitted to the data and the best-fitting position of the center of the distribution is determined. For a single distribution, the problem is analytically solvable by an eigenvalue method[155]. A review of other, more approximated graphical methods has been given by Darot and Bouchez[156]. Similarly, methods for analyzing and representing the shape of pebbles (usually approximated by an ellipsoid) have been suggested by Hossack[157] and by Burns[158].

The theory of petrofabric formation leads to intricate problems of the genesis of rocks and structural geology, inasmuch as all phases of the rock deformation since its emplacement may be recorded. We shall deal in this book only with those aspects of this vast subject which are of concern to geodynamics, as the need arises to do so.

154a Schäfer, K. H.: Fridericiana (Karlsruhe) 23:30 (1978)
155 Scheidegger, A. E.: U.S. Geol. Surv. Prof. Pap. 525:C164 (1964)
156 Darot, M., Bouchez, J. L.: J. Geol. 84:239 (1976)
157 Hossak, J. R.: Tectonophysics 5(4):315 (1968)
158 Burns, K. L.: Tectonophysics 7(3):177 (1969)

1.7 Physiography of Some Local Features

1.7.1 Introduction

In this, the final section of the general description of the physiographic and geological background of geodynamic phenomena, we shall discuss some features of a somewhat local significance.

First of all, we shall describe some characteristic pinch-and-swell and "boudinage" structures which appear to be the result of local instability conditions. Then, we shall discuss diapirical structures which may be regarded as internal unstable flow phenomena as well.

More intensive "diapirical" structures, but with a significant internal energy source, are represented by volcanic effects which show characteristic cratering phenomena. These will be treated next.

Then, we shall consider features which are superficially similar to volcanic craters: these are impact craters caused by fallen meteorites. Finally, we shall consider displacements on the Earth's crust that have been determined by geodetic means.

1.7.2 Boudinage

A particularly intriguing feature is the occurrence of boudins. The term "boudinage structure" was introduced by Lohest[159]. It is French meaning "sausage structure" and refers to a fractured sheet of rock wedged in between non-fractured rocks, or, in geological terms, to a fractured competent layer situated between two incompetent layers. Each fragment of the fractured rock often looks like a sausage, and the whole array resembles in many cases a chain of sausages. Hence the name of these structures.

Thus, boudins are oblong bodies; they have been described, e. g., by Ramberg[160] and De Sitter[161]. Their thickness varies[160] from about 10 mm to approximately 20 m. The shortest boudins are little longer than thick; the largest are many times longer than thick.

The shape of boudins may be quite varied. Some of them are rectangular bodies with sharp corners, others are barrel-shaped or lenticular, and finally some of them have an appearance as if they would have been twisted around their central axis.

Boudinage structures are quite common in gneissic and regional metamorphic areas where the rocks are well exposed. They are also abundant in some low-grade schist regions. Usually, of course, only a two-dimensional

159 Lohest, M.: Ann. Soc. Geol. Belg. 36 B:275 (1909)
160 Ramberg, H.: J. Geol. 63:512 (1955)
161 Sitter, L. U. de: Geol. Mijnb. 20:277 (1958)

cross-section of the structures is exposed in rock outcrops so that the three-dimensional arrangement of the boundins is not always easy to infer.

A physiographic appearance similar to that of boudinage structures is exhibited by pinch-and-swell structures. In the latter the wedged-in "compentent" layer is not completely fractured but seems to pinch out and swell up regularly in a pattern which is also sausage-like. Such structures are very common in conformable pegmatites and in quartz veins. From the close physiographic similarity of such pinch-and-swell structures with boudinage one might expect that the physical explanation of the two phenomena should also be similar.

Related to boudinage are ptygmatic features[162,163]. These are small crumplings found in granitic, pegmatite, and aplite veins. Kuenen[164] specified the definition by adding that for a fold to be ptygmatic, "there must be an enveloping host which shows no crumplings. If this definition is accepted, then veins, dikes, and sedimentary beds can all be said to exhibit ptygmatic structures." According to Kuenen[164] "the meandering pegmatite and aplite veins in gneiss and migmatites are the type features."

1.7.3 Diapirs

An interesting phenomenon is the existence of diapirs (domes). Generally, a dome is a fold which is "anticlinal" in every direction, viewed from its crest. This can also be stated by saying that the "axis" of the fold degenerates into a point.

One usually calls domes only such features as are caused by intrusions from below[165-168]. The intrusive material may be salt, gypsum or some other material capable of plastic flow. The strata above the intrusive material are pushed upward and represent the "dome".

Of particular interest is the case when the intrusive material reaches the surface of the Earth. One speaks in such cases of "piercement domes". The physiographic appearance is then that of a circular structure (see Fig. 19). A number of such structures have been discovered in the Canadian Arctic and were described by Heywood[169]. A particularly beautiful example is the Isachsen Dome on Ellef Ringnes Island (see Fig. 19). In this particular dome, the core rocks are gypsum, anhydrite, limestone, and basalt. The intrusion of

162 Milch, L.: Neues Jahrb. Mineral. 2:29 (1900)
163 Sederholm, J. J.: Bull. Comm. Geol. Finl. 23:1 (1907)
164 Kuenen, P. H.: Tectonophysics 6:143 (1968)
165 Trusheim, F.: Z. Dtsch. Geol. Ges. 109:111 (1957)
166 Feely, H. W., Kulp, J. L.: Bull. Am. Assoc. Petrol. Geol. 41:1802 (1957)
167 Heim, A.: Eclogae. Geol. Helv. 51:1 (1958)
168 Weidie, A. E., Martinez, J. D.: Bull. Am. Assoc. Petrol. Geol. 54(4):655 (1970)
169 Heywood, W. W.: Trans. Can. Inst. Min. Metall. 58:27 (1955)

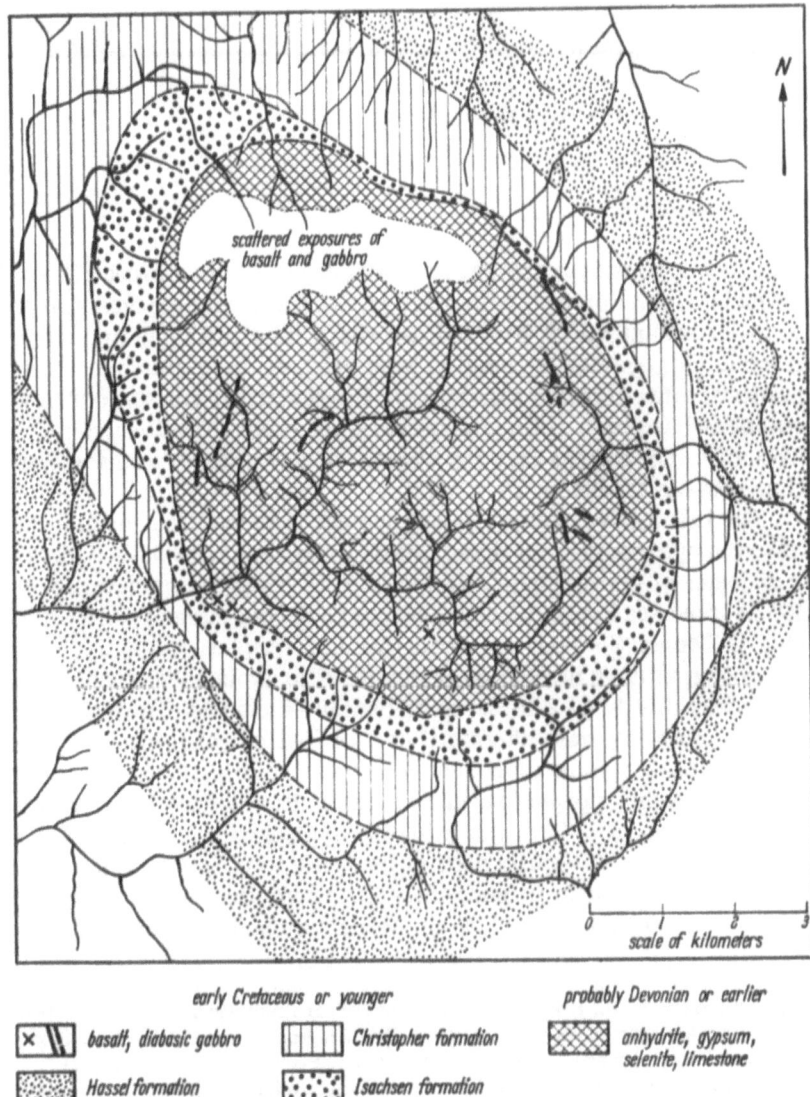

scattered exposures of
basalt and gabbro

N

0   scale of kilometers

early Cretaceous or younger

| | basalt, diabasic gabbro | | | | | Christopher formation

Hassel formation   Isachsen formation

probably Devonian or earlier

anhydrite, gypsum,
selenite, limestone

Fig. 19. Geological sketch of the Isachsen Dome in the Canadian Arctic. (After Heywood[169])

this material has upwarped the overlying sediments and caused faults radial
and tangential to the outer contact of the dome. The sedimentary rocks are
even sharply upturned and overturned in many places. The effects of the
diapirism are not apparent farther away than 3 to 5 km from the structure.

Particularly interesting are piercement structures that have been observed
in the deep ocean: A large diapir field has been discovered on the continental

slope west of Angola, Africa[170], others in the Gulf of Mexico[171], as well as elsewhere in the deep ocean[172]. Usually, these piercement structures have been explained as salt intrusions, but the explanation as igneous intrusions seems to be more likely in view of thermal data.

1.7.4 Volcanoes

A different type of circular structure that occurs on the surface of the Earth is found in connection with volcanism. It is well known that volcanism produces cone-like mountains of lava and ash which have one or more craters at the summit. Gas and lava is found to pour out from these craters in various proportions, often in cataclysmic spasms. Physiographic descriptions of volcanic eruptions have been given, for instance, by Rue[173], by Rittmann[174], and by Cotton[175]. The magma erupting when a volcano is active seems to come from chambers not far below[176].

The best-known volcano is undoubtedly Mount Vesuvius as it has been observed and investigated since the time of antiquity. Its basis is circular with a diameter of approximately 16 km. It rises gently to an elevation of about 595 m above sea level and hence abruptly to the two summits of the mountain. One is Monte Somma, the other Mount Vesuvius proper. The cone of Vesuvius rises at an inclination of about 30° to about 1,300 m above sea level. It consists mostly of a loose accumulation of cinders. Throughout history, periods of activity have alternated with periods of quiescence.

The above picture of Mount Vesuvius is fairly typical for an average volcano. Specific morphological studies of volcanoes have been made in many parts of the world[177-182]. A drawing of a typical volcano is shown in Fig. 20.

Volcanoes exist in many parts of the world (see Fig. 21). There is a general accumulation of them in plate margins. There are, however, plate margins without any volcanoes (such as the Himalayas and there are intraplate areas

170 von Herzen, R. P., Hoskins, H., van Andel, T. H.: Bull. Geol. Soc. Am. 83:1901 (1972)
171 Ewing, M., Ericson, D. B., Heezen, B. C.: Bull. Am. Assoc. Petrol. Geol. 42:995 (1956)
172 Lancelot, Y., Embley, R. W.: Bull. Am. Assoc. Petrol. Geol. 61:1991 (1977)
173 Rue, E. A. de la: L'homme et les volcans. Paris: Gallimard 1958
174 Rittmann, A.: Vulkane und ihre Tätigkeit, 2. Aufl. Stuttgart: Ferdinand Enke 1960
175 Cotton, C. A.: Volcanoes as Landscape Forms, 2nd edn. Christchurch: Whitcombe & Tombs Ltd. 1952
176 Gorshkov, G. S.: Bull. Volcanol. (2) 19:103 (1948)
177 Jaggar, T. A.: Origin and Development of Craters. Geol. Soc. Am. Mem. No. 21 (1947)
178 Grover, J. C.: Geogr. J. 123:298 (1957)
179 Machado, F.: Atlantida 2:225, 305 (1958)
180 Castello Branco, A. de, et al.: Le volcanisme de l'ile de Faial et l'éruption du volcan de Capelinhos. Memoria No. 4, Serviços Geológicos de Portugal, Lisboa 1959
181 Vlodavec, V. I.: Die Vulkane der Sowjetunion. Gotha: VEB Geogr.-Kart. Anst. 1954
182 Sato, H.: Distribution of Volocanes in Japan. Proc. Inst. Geog. Un. Reg. Conf. 1957, p. 184 (1959)

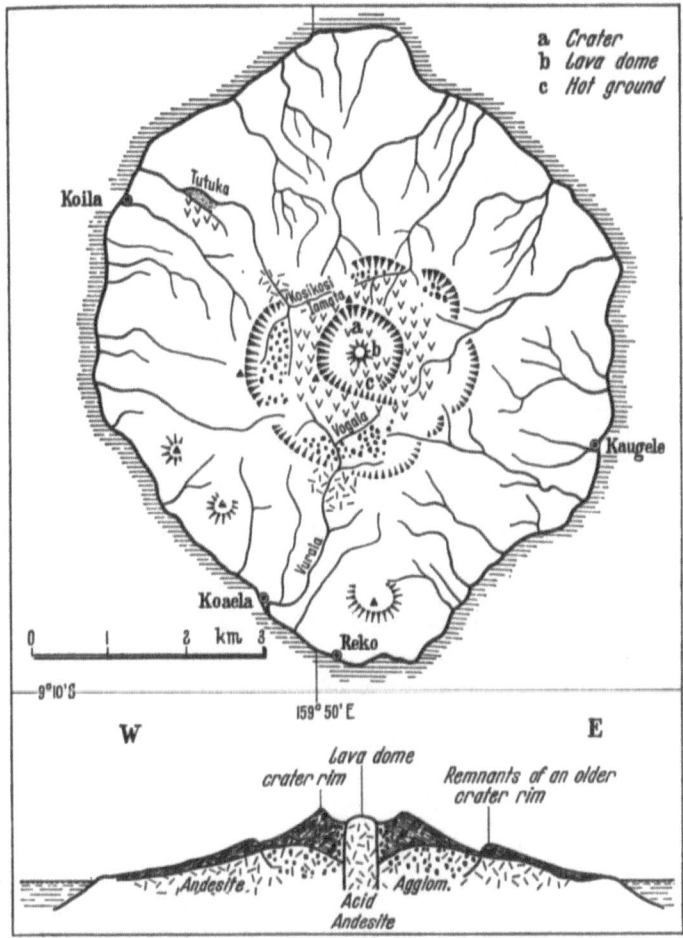

Fig. 20. Savo Volcano, in the Solomon Islands. (After Grover[178])

with prominent volcanoes (such as the Congo). Often, volcanoes are arranged in linear chains. The contention has been that these represent the record of "hot spots" which "burn holes" into the crust, thus recording plate motions[183,184].

1.7.5 Astroblemes

Another type of characteristic local features are crater-like holes in the ground, obviously caused by the impact of meteorites. Striking examples of

---

183 McDougall, I., Duncan, R. A.: Tectonophysics 63:275 (1980)
184 Bonatti, E., Harrison, C., Fisher, D. E., Honnorez, J., Schilling, J., Stipp, J., Zentilli, M.:
    J. Geophys. Res. 82(17):2457 (1977)

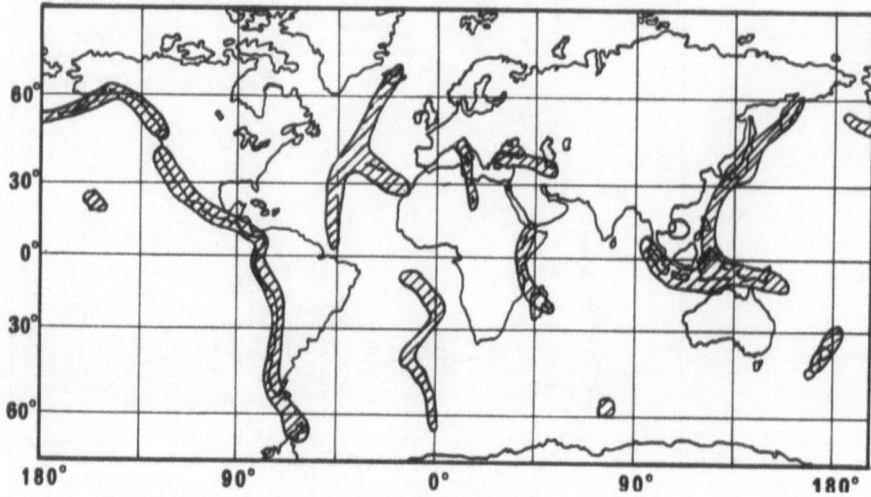

Fig. 21. World distribution of major volcanoes: areas with volcanic activity are shaded

such craters have been found in Arizona[185] (Barringer Crater), in Northern Quebec[186] (New Quebec Crater), and elsewhere[187-189]. A typical profile of such a crater is shown in Fig. 22.

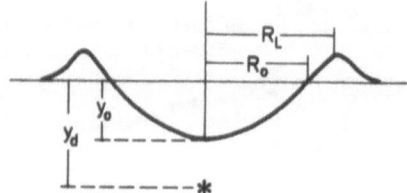

Fig. 22. Outline of crater shape. The notation: R_L is the lip radius, R_0 is the apparent radius, y_a is the apparent depth, and y_d is the depth of burial. (After White[189a])

Erosion and weathering are evidently affecting the physiography of craters very rapidly. Thus, it may be assumed that the Earth has been stricken by meteorites in the past but that the impact-origin of the features visible at

185 Barringer, D. M.: Proc. Acad. Natl. Sci. Philadelphia 66:556 (1914); 76:275 (1924). – Nininger, H. H.: Arizona's Meteorite Crater; Publ. Amer. Meteorite Museum, Sedona, Ariz. 1956
186 Millman, P. M.: Publ. Dom. Obs. Ottawa 18, No. 4:59 (1956)
187 Dietz, R. S.: J. Geol. 67:496 (1959)
188 Nininger, H. H.: Science 130:1251 (1959)
189 Roddy, D. J., Pepin, R. O., Merrill, R. B. (ed.): Impact and Explosion Cratering: Planetary and Terrestrial Implications. Proc. Symp. on Planetary Cratering Mechanics, Flagstaff 1976. Oxford: Pergamon Press 1977
189a White, J. W.: J. Geophys. Res. 78(35):8623 (1973)

present is not immediately obvious. In this instance, one speaks of "fossil craters", circular structures which have been filled in and covered by detritus and other foreign material. Further changes may have occurred during glaciation, and all that is seen at the present time is perhaps a change in the appearance of the vegetation due to the difference in soil above the originally shattered and unshattered area. Thus, the search for possible fossil craters becomes often a very difficult one. Nevertheless, it has been established beyond doubt for many circular features that they *are* of meteoritic origin.

This result has prompted one to search for other features on the Earth's surface that may have been caused by meteorite impact. One of the characteristics of meteorite craters is that they are almost perfectly circular. The hypothesis has therefore been advanced that the Gulf of St. Lawrence[190] or the Sudbury Basin[191] were created by the impact of huge meteorites. Such hypothetical meteorite structures have been called "astroblemes". An astrobleme event[192] has also been held responsible for the remarkably ubiquitous global faunal extinction that occurred at the end of the Cretaceous, the more extreme views ranging as far as the assumption of the impact of a comet[193].

The quest for the definitive identification of large-scale circular structures as true astroblemes is very difficult. A possible diagnostic feature is the existence of "shatter cones", conical products of shock metamorphism presumably caused by the impact of debris[194,195]. It should be stated, though, that a tectonic origin has also been claimed for such features[196]. Studies of gravitational anomalies around circular structures have also been helpful[197,198] in connection with their identification as impact phenomena. In this fashion, in addition to the features already mentioned earlier, the Ries[199] in Germany, the Bushveld Complex in South Africa[200], the Elgygtgyn structure in Siberia[201], the Île Rouleau structure in Quebec[202], and the Wilkes Land Anomaly in Antarctica[203] have been regarded as possible astroblemes. A number of additional possible astroblemes have also been collected by Saul[204].

190 Willmore, P. L., Scheidegger, A. E.: Trans. R. Soc. Can. 50:Ser. III, Can. Comm. Oceanogr. 21 (1956)
191 Dietz, R. S.: J. Geol. 72:412 (1964)
192 Smit, J., Hertogen, J.: Nature (London) 285:198 (1980)
193 Hsü, K. J.: Nature (London) 285:201 (1980)
194 Roy, D. W.: Tectonophysics 60:T37 (1979)
195 Mason, G. D.: Nature (London) 225:393 (1975)
196 Fleet, M. E.: Bull. Geol. Soc. Am. Pt. I, 90:1177 (1979)
197 Sweeney, J. F.: J. Geophys. Res. 83:2809 (1978)
198 Fudali, R. F.: J. Geol. 87:55 (1979)
199 Engelhardt, W. von: Geochim. Cosmochim. Acta 31:1677 (1967); also: Grau, W.: Geogr. Rundsch. 30(4):144 (1978)
200 Rhodes, R. C.: Geology 3:550 (1975)
201 Dietz, R. S., McHone, J. F.: Geology 4:391 (1976)
202 Caty, J. L., Chown, E. H., Roy, D. W.: Can. J. Earth Sci. 13(6):824 (1976)
203 Weihaupt, J. G.: J. Geophys. Res. 81(32):5651 (1976)
204 Saul, J. M.: Nature (London) 271:345 (1978)

A significant feature of large-scale astroblemes is that their profile, in contrast to that of smaller craters, shows a characteristic "hump" in the middle[205]. A typical structure is shown in Fig. 23.

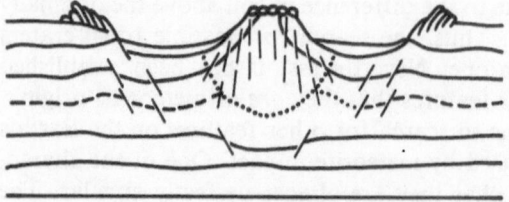

Fig. 23. Typical structure of an astrobleme. (After Sawatzky[207])

In view of the above results, it may be assumed that the Earth is hit by meteorites not less frequently than the Moon whose surface is studded by impact craters. The difference in appearance between the Earth and the Moon is imply due to the presence of an atmosphere (and therewith of exogenic agents) on the former and a lack thereof on the latter. Indeed a plot of all the more-or-less certainly identified fossil meteorite impact sites for the area of North America (after Sawatzky[205,206]), yields an impressive picture, as shown in Fig. 24.

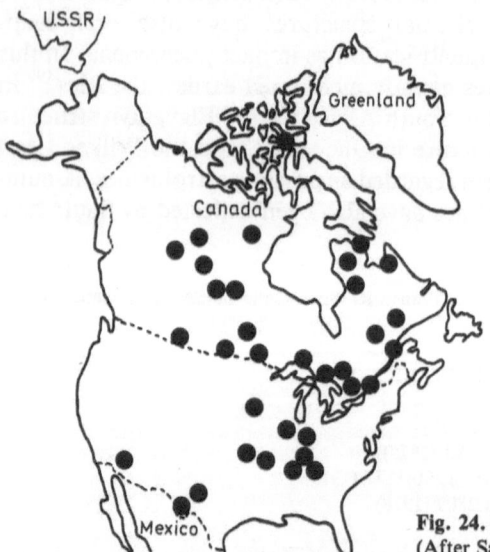

Fig. 24. Meteorite impact sites in North America. (After Sawatzky[208])

205 Sawatzky, H. B.: Bull. Am. Assoc. Petrol. Geol. 59(4):694 (1975)
206 Sawatzky, H. B.: Geophysics 41(6):1261 (1976)
207 Sawatzky, H. B.: Bull. Am. Assoc. Petrol. Geol. 59(4):694 (1975)
208 Sawatzky, H. B.: Geophysics 41(6):1261 (1976)

1.7.6 Recent and Contemporary Displacements

1.7.6.1 General Remarks

It is evident that the surface of the Earth is not in a static condition. The mere fact that mountains are constantly being attacked and eroded by wind, water, and ice, shows that geodynamic displacements must take place for there to be a stationary state. The rates of Earth deformation range temporally over large intervals. Lensen[209] has set up a nomenclature for this purpose. However, it is particularly the *recent* (last 10,000 years or so) and *contemporary* displacements that are of interest. One usually distinguishes between vertical and horizontal displacements since the methods of measurement and interpretation are fundamentally different in the two cases.

1.7.6.2 Vertical Displacements

1. Methods. Let us first consider vertical displacements. The methods for determining such displacements are geological, geomorphological, and geodetic.

The *geological* evidence is based upon an analysis of the strata in question. From the existence of fine-grained sediments such as silt, clay, and mud, which commonly occur at the bottom of the sea, it is inferred that such an area was depressed below sea level at one time. Similarly, lagoon and reef facies may be recognized, or the position of ancient shore lines may be ascertained. Thus, it may be possible to date the time of transition of a particular spot from sea bottom to dry land. De Geer[210] was a pioneer in such studies. Naturally, the global fluctuation of the sea level (*eustatic* changes) must be taken into account.

The *geomorphological* evidence may be based on the analysis of ancient river beds[211] or the depth at which the ground-water level occurred[212].

The *mareographic* evidence is based on contemporary sea-level records (tide gages). The problem of the elimination of noise in the records of tide-level gages is thereby not to be neglected[213].

Finally, the method of choice is *geodetic* re-leveling of a profile after a number of decades. This yields relative displacement rates of the points on the profile. Recently, substantial advances have been made with regard to the instruments as well as the interpretation of such measurements[214,215].

209 Lensen, G. J.: Bull. R. Soc. N. Z. 9:97 (1971)
210 De Geer, D. J.: C. R. Int. Geol. Congr. Stockholm 2:849 (1910)
211 Mike, K.: Tectonophysics 29:359 (1975)
212 Bylinskaya, L. N., Gorelov, S. K., Setunskaya, L. E., Filkin, V. A.: Tectonophysics 29:389 (1975)
213 Vanicek, P.: Can. J. Earth Sci. 17:265 (1978)
214 Whitcomb, J. H.: J. Geophys. Res. 81(26):4937 (1976)
215 Vanicek, P., Castle, R. O., Balazs, E. I.: Rev. Geophys. Space Phys. 18(2):505 (1980)

2. Results. The literature giving results of contemporary vertical displacement rates is extremely large. Some results have been summarized by Gopwani and Scheidegger[216]; others were presented at a symposium on the subject matter held in Zurich in 1975[217]. The large-scale global picture, such as it is known today[218] is shown in Fig. 25.

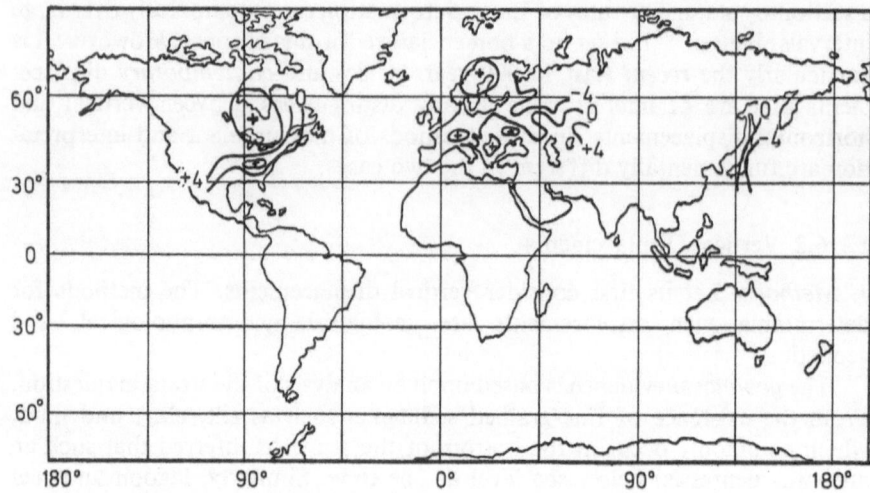

Fig. 25. The tendencies of recent vertical crustal movements determined by means of geodetic methods. Contours in mm/year. (After Vyskocil[218])

In the global picture, two pronounced uplift areas are immediately evident: This is the region around the Baltic and that around Hudson Bay. Paleoclimatological investigations show that these regions were ice-covered during the last ice age; the uplift, then, is commonly interpreted as "rebound" of the crust after the ice load has melted off some 10,000 years ago. The Baltic area has been particularly carefully studied; classic investigations had already been published by Kukkamäki[219,220] and Gutenberg[221]. The older results have generally been confirmed[222,223]. The picture is similar in the area around

216 Gopwani, M. V., Scheidegger, A. E.: Ann. Geofis. (Roma) 24(1):1 (1971)
217 Pavoni, N., Green, R. (eds.): Proc. Symp. Recent Crustal Movements, Tectonophysics 29:1 (1975)
218 Vyskocil, P.: Tectonophysics 38:49 (1977)
219 Kukkamäki, T.: Veroeff. Finn. Geodaet. Inst. No. 26:120 (1939)
220 Kääriäinen, E.: Veroeff. Finn. Geodaet. Inst. No. 42 (1953)
221 Gutenberg, B.: Bull. Geol. Soc. Am. 52:750 (1941)
222 Balling, N.: Geoskrifter (Aarhus) 10:199 (1978)
223 Mörner, N. A.: In: Earth Rheology, Isostasy and Eustasy (ed. Mörner). New York: Wiley, Pap. 138, 1979

Hudson Bay[224-226]. Figure 26 shows a compilation of the uplift rates in the formerly ice-covered areas of the world[227].

Large uplift rates do not only occur in evident glacial rebound areas[228,229]. They are also commonly observed in high mountain ranges: Indeed, inasmuch as the observed erosion rates[230] are of the order of mm/year, the uplift rates must be of the same order of magnitude; otherwise the mountain ranges would disappear very rapidly. Thus, uplift rates of the order of mm/year have actually been found in the Swiss[231-233] and Austrian Alps[234,235], in the Carpathians[236], in the Ponto-Caspian[237] orogenic region, and in Southern California[238,239].

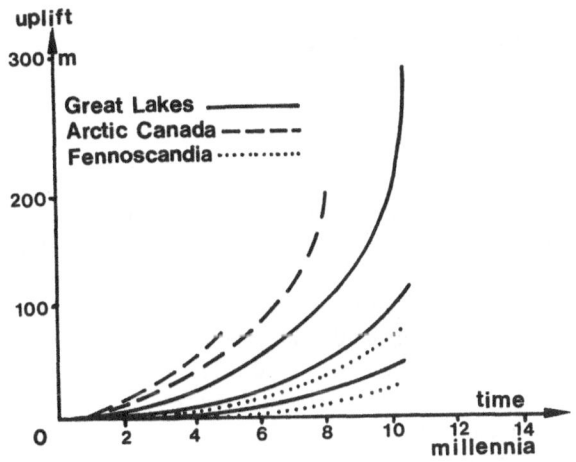

Fig. 26. Postglacial uplift curves. Begin of deglaciation is at zero; this varies for each locality. (After Schofield[227])

224 Walcott, R. I.: Rev. Geophys. Space Phys. 10(4):849 (1972)
225 Hillaire-Marcel, C., Fairbridge, R. W.: Geology 6:117 (1978)
226 Lambert, A., Vanicek, P.: Can. J. Earth Sci. 16:647 (1979)
227 Schofield, J. C.: N. Z. J. Geol. Geophys. 7:359 (1964)
228 Vyskocil, P.: Tectonophysics 38:49 (1977)
229 Schofield, J. C.: N. Z. J. Geol. Geophys. 7:359 (1964)
230 Corbel, J.: Z. Geomorphol. 3:1 (1959)
231 Jeanrichard, F.: Tectonophysics 29:289 (1975)
232 Gubler, E.: Schweiz. Mineral. Petrol. Mitt. 56:675 (1976)
233 Schaer, J. P.: Eclogae Geol. Helv. 72:263 (1979)
234 Senftl, E.: Verh. Geol. Bundesanst. (Austria) 1973:209 (1973)
235 Höggerl, N.: Rock Mech. Suppl. 9:201 (1980)
236 Kvitkovic, J.: Tectonophysics 29:369 (1979)
237 Blagovolin, N. S., Lilienberg, D. A., Pobedonostsev, S. V.: Tectonophysics 29:395 (1975)
238 Yeats, R. S.: Science 196:295 (1977)
239 Vanicek, P., Elliott, M. R., Castle, R. O.: Tectonophysics 52:287 (1979)
240 Mohr, P. A., Girnius, A., Cherniack, J. R., Gaposchkin, E. M., Latimer, J.: Tectonophysics 29:461 (1975)
241 Reilinger, R. E., York, J. E.: Geology 7:139 (1979)

Californian basins[242]. It cannot be generally stated, however, that rifts always show subsidence. Since the commonly assumed model for such areas is that material is upwelling from below, the vertical motions can be either way. Thus, uplifts have been found in Iceland[243] and in the Gulf of Aden[244].

The data for continental cratons are also equivocal: One may find subsidence or uplift. Thus, for Africa, an uplift (at least in the Tertiary) has been claimed relative to the Americas, Australia, and Europe[245]. For the Rhenisch shield[246], a quaternary uplift has been found. On the other hand, subsidences were found in Czechoslovakia[247] (1.5 mm/year) and in France (26 mm/year)[248].

In summary, it can be stated, therefore, that large uplifts occur generally in active land-based mountain ranges and in glacial rebound areas. On cratons and in rift zones, the vertical displacements may be either way.

1.7.6.3 Horizontal Displacements

1. Types of Displacements. A measurement of relative horizontal displacements between points is somewhat more difficult than the measurement of vertical displacements, because a vector rather than a scalar has to be determined.

Again, the scales of the motions of interest vary over a great range. First, there are attempts to measure local differential displacements of points on the Earth's crust across, say, an active fault or before and after an earthquake. Such measurements are relatively easy to perform by making repeated surveys of markers. Second, there are attempts at finding the displacements of areas of a continental scale. Third, one can try to determine, by direct measurements, the possible motions of continents with regard to each other.

2. Local Displacements. The problem of determining the relative shift of two points on the Earth's surface is one of making repeated geodetic surveys. The accuracy of newer equipment is to an error about 10^{-6} of the measured distance.

The most localized applications of the indicated measurements is to the determination of the displacements on "unstable terrains", such as scree slopes, soil creep regions, etc. These displacements are generally induced by gravitational effects.

242 Yeats, R. S.: Geology 6:456 (1978)
243 Schäfer, K.: Tectonophysics 29:223 (1975)
244 Faure, H.: Tectonophysics 29:479 (1975)
245 Bond, G.: J. Geol. 86:47 (1978)
246 Illies, J. H., Prodehl, C., Schmincke, H. U., Semmel, A.: Tectonophysics 61:197 (1979)
247 Vyskocil, P.: Tectonophysics 29:349 (1975)
248 Gopwani, M. V., Scheidegger, A. E.: Ann. Geofis. Roma 24(1):1 (1971)

Somewhat larger-scale motions, such as those connected with mountain fractures and valley closures, may already be partially conditioned by the tectonic stress field[249-252].

The strain accumulation across faults has been monitored for a long time, particularly in California[253-260], and Alaska[261,262]. For the vicinity of the San Andreas fault, shear strain rates of about 10^{-6} per year have been found. In terms of displacements, one obtains 10 mm/year for the San Andreas Fault[263], up to 20 mm/year for other faults in California[264] with the Alpine Fault in New Zealand[265] attaining a maximum of 70 mm/year.

Studies of surface displacements associated with earthquakes are very numerous. Large earthquakes ($M \geqslant 7$) are connected with substantial, obvious shifts on the Earth. A comprehensive study of such effects in Japan has been made by Sato[266]. Similar studies for Alaska have been reported by Parkin[267], for Iceland by Tryggvason[268].

3. Regional Studies. The next level of scale is to make investigations of regional rather than local strains. The most elegant idea proposed to do this is based upon a comparison of geodetic nets at subsequent times. This method has been applied successfully by Thurm et al.[269] to a region of Saxony where two very accurate surveys, lying about 80 years apart, were available. In principle, the shift of the coordinates of "fixed" points leads directly to the strain tensor built up between the two surveys. Thus, the principal strain directions could be calculated (at least in plan), which are represented as strain trajec-

249 Carniel, P., Hauswirth, E. K., Roch, K. H., Scheidegger, A. E.: Verh. Geol. Bundesanst. (Austria) 1975:305 (1975)
250 Hauswirth, E. K., Scheidegger, A. E.: Riv. Ital. Geofis. Sci. Aff. 3:85 (1976)
251 Hauswirth, E. K., Pirkl, H., Roch, K. H., Scheidegger, A. E.: Verh. Geol. Bundesanst. (Austria) 1979:51 (1971)
252 Hauswirth, E. K., Scheidegger, A. E.: Interpraevent 1980, 1:159 (1980)
253 Scholz, C. H., Fitch, T. J.: J. Geophys. Res. 74(27):6649 (1969)
254 Scholz, C. H., Fitch, T. J.: J. Geophys. Res. 75(23):4447 (1970)
255 Savage, J. C., Burford, R. O.: J. Geophys. Res. 78(5):832 (1973)
256 Prescott, W. H., Savage, J. C.: J. Geophys. Res. 81(26):4901 (1976)
257 Huggett, G. R., Slater, L. E., Langbein, J.: J. Geophys. Res. 82(23):3361 (1977)
258 Harsh, P. W., Pavoni, N.: Bull. Seismol. Soc. Am. 68(4):1191 (1978)
259 Thatcher, W.: J. Geophys. Res. 84:2283 (1979)
260 Burford, R. O., Harsh, P. W.: Bull. Seismol. Soc. Am. 70(4):1233 (1980)
261 Brogan, G. W., Cluff, L. S., Koringa, M. K., Slemmons, D. B.: Tectonophysics 29:73 (1975)
262 Savage, J. C.: J. Geophys. Res. 80(26):3786 (1975)
263 Lensen, G. L.: Bull. R. Soc. N. Z. 9:97 (1971)
264 Scholz, C. H., Fitch, T. J.: J. Geophys. Res. 75:4447 (1970)
265 Pavoni, N.: Bull. R. Soc. N. Z. 9:7 (1971)
266 Sato, H.: Bull. Geogr. Surv. Inst. Jpn. 19(61):89 (1973)
267 Parkin, E. J.: Surv. Mapping 27(3):423 (1967)
268 Tryggvason, E.: J. Geophys. Res. 75(23):4407 (1970)
269 Thurm, H., Bankwitz, P., Bankwitz, E., Harnisch, G.: Peterm. Geogr. Mitt. 1977(4):281 (1977)

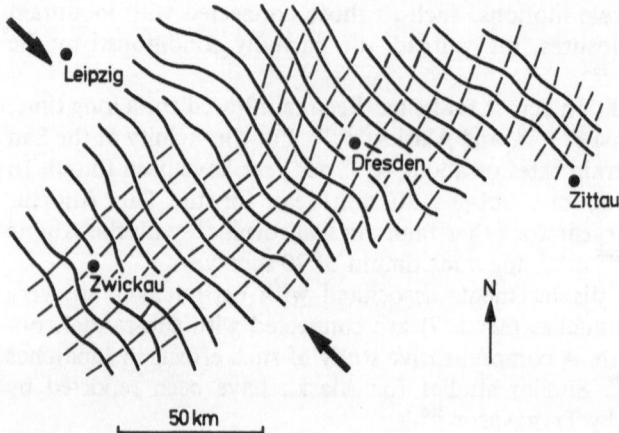

Fig. 27. Strain trajectories in Eastern Germany from the comparison of geodetic nets. Heavy lines: compression. (After Thurm et al.[269])

tories in Fig. 27. It turned out that there is a maximum compression in the NW direction. A similar study was made for Japan[270].

4. Global Studies. Of greatest interest is a confirmation (or otherwise) of the existence of continental drift motions as postulated in the "new global tectonics". The rates in question are around 30 – 40 mm/year[271]. Techniques to obtain such a confirmation that are based on a statistical analysis of geodetic measurements have not yet been quite sufficient to yield unequivocal results[272]. It is to be hoped that particularly satellite geodesy will give the required accuracy in the near future. However, other observations have generally supported continental drift. Thus, Proverbio and Quesada[273,274] have shown that the secular variations in latitudes and longitudes fit exactly the values of relative continental displacements postulated from plate tectonic theory and from other geophysical (such as magnetic) observations.

270 Harada, T., Shimura, M.: Tectonophysics 52:469 (1979)
271 Garland, G. D.: In: Continental drift, secular motion of the pole, and rotation of the Earth. Int. Astron. Union, Symp. No. 32, Stresa, Italy, 1967. Berlin-Heidelberg-New York: Springer, p. 19, 1968
272 Van Mierlo, J.: Tectonophysics 52:457 (1979)
273 Proverbio, E., Quesada, V.: J. Geophys. Res. 79(32):4941 (1974)
274 Proverbio, E., Quesada, V.: Bull. Geodesique 1974(112):187 (1974)

2. Geophysical Data Regarding the Earth

2.1 Results from Seismic Wave Propagation Studies

2.1.1 Earthquakes and Seismic Waves

Seismology, the study or earthquakes, has yielded some very pertinent information about the structure of the Earth. Earthquakes are shocks that occur within the Earth. It appears that these shocks originate each in a region which is small compared with the Earth as a whole; this region can be regarded as a point for most purposes and is referred to as the *focus* of the earthquake. The point directly above the focus on the Earth's surface is termed the *epicenter* of the earthquake.

After an earthquake has occurred, one observes effects at seismic stations throughout the world. Such stations are equipped with *seismographs*, instruments designed to amplify and register any tremors of the Earth's surface in their vicinity. The seismograph writes a *seismogram*, a line related to the motion of the Earth in any one chosen direction. Any change of amplitude or frequency in the seismogram of an earthquake is called a *phase*. The principal phases in the seismogram of an earthquake have been called P, S, and L and it has been established that they represent the first onsets of compressional, transverse bodily and surface waves (cf. Sect. 3.2.1), respectively. The characteristic periods increase from P to S waves; the very largest periods correspond to the eigenoscillations of the Earth.

The various phases of a seismogram have, even to the eye, characteristic traits. Thus, it is often possible to a trained observer to recognize at a glance not only the type of phase, but even the epicenter region of an earthquake from its record written at a station. A catalog of typical seismogram phases has been published by Simon[1].

One of the principal outcomes of observational seismology has been the recognition that it is possible to regard "phases" as travelling along "rays". It is thus possible to construct traveltime tables for the various phases and to trace their paths through the interior of the Earth. The traveltime of a phase does not depend appreciably on the location of the epicenter and the station, but only on the epicentral distance and the depth of the focus.

1 Simon, R. B.: Earthquake Interpretations. Golden: Colorado School of Mines, 1969

From the above description it is obvious that seismology yields data about the Earth in two ways: first, one can analyze the effect that the interior of the Earth has upon elastic wave transmission, and second, one can analyze the occurrence and the mechanism of the earthquakes themselves.

We shall turn our attention first to the facts that may be gleaned from the transmission of seismic waves through the Earth; this yields much information about the Earth's interior. It is clear that it is immaterial whether the source of the seismic wave energy is natural or artificial (conventional or nuclear explosions). The discussion of the nature of seismic foci will be relegated to Sect. 2.2.

Many treatises exist on the various aspects of seismology. We shall be concerned here only with those aspects that are pertinent with regard to geodynamics. For further details, the reader is referred to the literature[2-11].

2.1.2 The Basic Division of the Earth into Layers

The study of the records written by natural or artificial earthquakes at seismic stations forms a subject of vast complexity. The distortion of the various seismic phases at discontinuities follows complex laws and many theoretical investigations of these laws have been undertaken. These theoretical investigations have been reviewed in the books on seismology mentioned in Sect. 2.1.1, and the reader is referred thither for the details.

The main result of the studies of elastic wave transmission within the Earth is that there are two fundamental discontinuities which divide the interior of the Earth into three principal layers called the *crust*, the *mantle*, and the *core*.

The most shallow of these discontinuities has been termed *Mohorovičić discontinuity* (after its discoverer[12]); it lies at depths of from 5 to 60 km beneath the surface of the Earth. On it, the velocity of longitudinal elastic waves jumps from some low value to a fairly uniform value of about 8.1 km/s. This discontinuity divides the crust from the mantle of the Earth.

2 Byerly, P.: Seismology. New York: Prentice-Hall 1942

3 Macelwane, J. B.: When the Earth Quakes. Milwaukee: Bruce 1947

4 Leet, L. D.: Earth Waves. Cambridge (Mass.): Havard Univ. Press 1950

5 Savarenskiy, E. F., Kirnos, D. P.: Elementy seysmologiy i seysmometriy. Moscow: Gos. Iz-vo Tekh.-Teoret. Lit. 1955

6 Richter, C. F.: Elementary Seismology. San Francisco: Freeman & Co. 1958

7 Bullen, K. E.: An Introduction to the Theory of Seismology, 2nd edn. London: Cambridge Univ. Press 1959

8 Bath, M.: Mathematical Aspects of Seismology. Amsterdam: Elsevier 1968

9 Müller, S.: Erdbeben. Zürich: Mitt. No. 217 aus dem Inst. f. Geophysik der ETH, 1978

10 Bath, M.: Introduction to Seismology, 2nd edn. Basle: Birkhäuser 1979

11 Ben-Menahem, A., Singh, S. J.: Seismic waves and Sources. Berlin-Heidelberg-New York: Springer 1980

12 Mohorovičić, A.: Jahrb. Met. Obs. Zagreb 1909 9, Pt. 4; 1 (1910)

The second main discontinuity is the core boundary at a depth of 2900 km beneath the surface of the Earth (discovered by Oldham[13] and Gutenberg[14]). On it, transverse elastic waves disappear, from which it has been inferred that the Earth's core is liquid. However, it has come to light that there may be an *inner core* within the core which could again be solid (cf. Sect. 2.1.5).

Furthermore, it is possible to determine the velocity distribution in the Earth from the P and S traveltime curves of seismic body waves outside the discontinuities. In principle, this is achieved by an inversion of the time – distance function into a velocity – depth function; the calculations are, in practice, very complicated. We present here in Fig. 28 a recent result due to Haddon and Bullen[15].

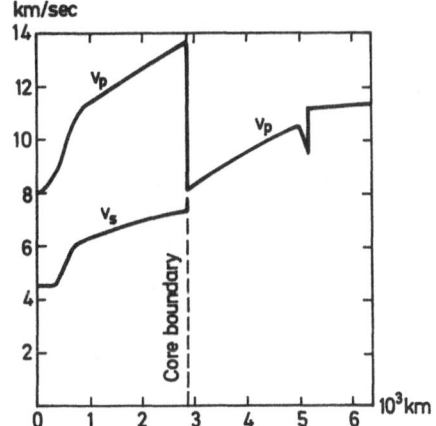

Fig. 28. Velocity distribution of P and S waves in the Earth. (After Haddon and Bullen[15])

In connection with the values given above, it should be noted that they represent averages. In the upper mantle of oceanic regions, there is a significant horizontal anisotropy (about 5%) with the highest velocity perpendicular to the direction of local magnetic anomalies[16].

The values for the P and S wave velocities can be used to estimate the density distribution within the Earth. The equation of equilibrium at the distance r from the center of the Earth requires

$$dp/dr = -g\varrho = -\varkappa m\varrho/r^2 \tag{2.1.2-1}$$

where p is the pressure, g the local gravitational pull, ϱ the density, m the mass of the matter inside a sphere of radius r, and $\varkappa$ the gravitational constant. Assuming adiabatic conditions in a homogeneous substance, we have

13 Oldham, R. D.: Q. J. Geol. Soc. 62:456 (1906)
14 Gutenberg, B.: Nachr. Ges. Wiss. Goettingen, Math.-Phys. Kl. 1914, 1:125 (1914)
15 Haddon, R., Bullen, K. E.: Phys. Earth Planet. Inter. 2:35 (1969)
16 Bisbee, L. D., Shor, G. G.: Geophys. Res. Let. 3:639 – 642 (1976)

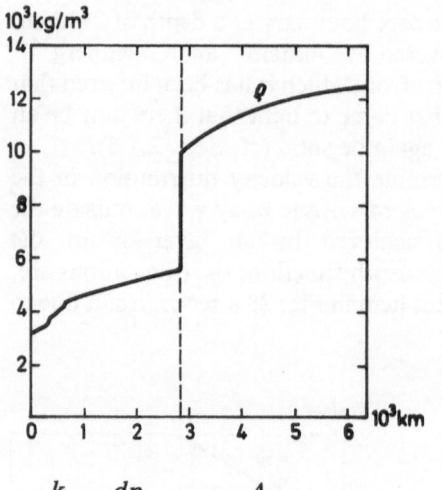

Fig. 29. Density distribution in the
Earth. (After Haddon and Bullen)

$$\frac{k}{\varrho} = \frac{dp}{d\varrho} = v_p^2 - \frac{4}{3} v_s^2 \qquad (2.1.2-2)$$

where k is the (adiabatic) incompressibility, and v_p, v_s denote the local P and S velocities, respectively. This result is a consequence of infinitesimal elasticity theory [cf. Eq. $(3.2.1-17/18)$ and $(3.2.1-6c)$]. Hence

$$\frac{d\varrho}{dr} = - \frac{\varkappa m \varrho}{r^2 \left(v_p^2 - \frac{4}{3} v_s^2 \right)}. \qquad (2.1.2-3)$$

This equation can be evaluated numerically, but there is some difficulty in determining the analytical continuation across the surfaces of discontinuity. Fortunately, it turns out that various reasonable assumptions that can be made do not influence the result very much; thus, the density variation is probably pretty close to that shown in Fig. 29, as calculated by Haddon and Bullen [16a].

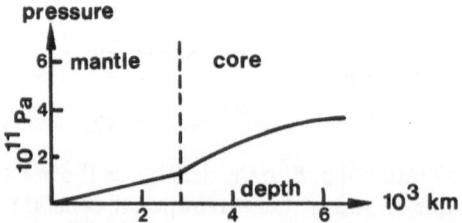

Fig. 30. Pressure distribution in the Earth [16a]

Once the density distribution is known, one can calculate the pressure distribution from the law of gravitation; the result of such a calculation is shown in Fig. 30.

16a Haddon, R., Bullen, K. E.: Phys. Earth Planet. Inter. 2:35 (1969)

Fig. 31. Bullen's principal Earth model[18]

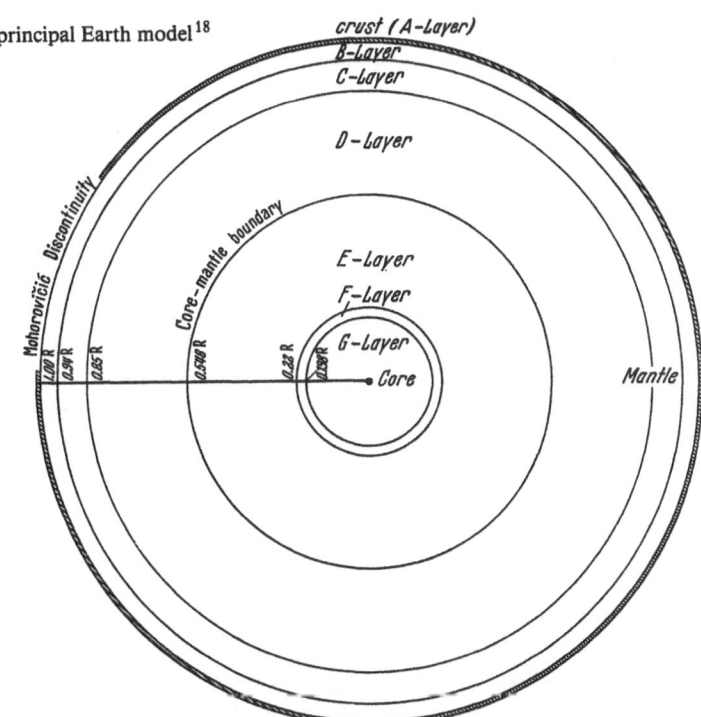

If one collects all the information presented above, one arrives at a *model* of the Earth[17]. In essence, the above discussion refers to Bullen's[18] model; other models have been suggested by Molodenskiy. In constructing Earth models, the Earth is split into various layers with homogeneous composition and with more or less homogeneous physical properties. Bullen's (principal) model[17] shown in Fig. 31.

The fact, Bullen's model[18] as shown in Fig. 31 exhibits more details in the Earth's structure than have been discussed thus far. It is, therefore, necessary to consider the various regions called "crust", "mantle", and "core" in greater detail.

2.1.3 The Crust

2.1.3.1 General Remarks

As noted above, our knowledge of the various characteristic layers of the Earth stems from the analysis of the propagation of seismic waves. In general,

17 Bolt, B. A.: Sci. Am. 228(3):24 (1973)
18 Cf. e. g., Bullen, K. E.: An Introduction to the Theory of Seismology. London: Cambridge Univ. Press 1959

these seismic waves originate from earthquakes. In the case of crustal studies, waves from *artificial* "earthquake" sources, i.e., from conventional and atomic explosions, have also proven to be very useful. In an artificial source, the origin time and place of the shock are a priori known, which makes the evaluation of seismograph records relatively simple.

The methods for interpretation are basically those that have been developed by exploration geophysicists[19]. In principle, one wishes to trace the course of the Mohorovičić discontinuity, so as to determine the "thickness" of the crust. Subsequently, one also wishes to determine the internal structure of the crust. The velocity contrast between the crust and the mantle is much greater than the contrasts within the crust, so that the prevailing characteristic case is that of a two-layer problem. With regard to P waves, this case is characterized by a traveltime curve consisting of two linear segments. The slope of each segment is inversely proportional to the velocity in the corresponding layer; the intercept time (obtained by continuing the straight segment beyond the "kink") indicates the depth of the layer. The traveltime curves for S waves are similar, but they are harder to identify on a seismogram, so that most studies have been made with P waves. If additional layers are present within the crust, one has a "multilayer"-case: each layer is represented by a segment of the traveltime curve, characterized by a slope and an intercept, from which the depths and thicknesses and the wave velocities can be deduced. If the velocity changes continuously with depth, or if velocity reversals are present, the problem of interpretation of the seismograms becomes very complicated. For the details, the reader is referred to the pertinent literature on exploration geophysics; here we shall state only those results which are of importance to geodynamics.

For the determination of the thickness of the crust, one can also use surface waves. Such waves show the phenomenon of velocity *dispersion*: The phase (or group) velocity of the wave is a function of the frequency. In the two-layer case, this function depends on the velocity and thickness of the upper and lower layers. In multilayer cases, this dependence becomes more complicated. Correspondingly, it is possible to determine the layer thicknesses from the dispersion curves deduced from the seismograms of an earthquake. These curves refer to the entire wave path of the surface waves, so that only, so to speak, a picture of an "averaged" crustal structure (referring to the entire wave path) can be obtained[20].

The main result of the analyses of the type indicated above is that the P wave velocity at the Mohorovičić discontinuity is around 8.1 km/s, as had been mentioned earlier. In addition, it turns out that there are basically two types of crustal structure: A continental one and oceanic one. Under continents, the Mohorovičić discontinuity lies usually at a depth of some

---

19 See e.g., Dobrin, M. B.: Introduction to Geophysical Prospecting, 3rd edn. New York: McGraw-Hill 1976

20 For a review, see e.g., Kovach, R. L.: Rev. Geophys. Space Phys. 16(1):1 (1978)

35 km, but may descend to 60–70 km beneath moutain ranges. Under oceans, the Mohorovičić discontinuity lies usually at a depth of 5 km (beneath the ocean bottom); under mid-ocean ridges, it many disappear altogether.

We shall discuss the details of the crustal structure in the two types of regions as well as in their transition region in the paragraphs following.

2.1.3.2 Continents

We have noted that the Mohorovičić discontinuity lies commonly at a depth of 35 km beneath continents. In fact, this value applies to most plains areas. In mountainous areas, it has been found that the Mohorovičić discontinuity is depressed so that the corresponding mountains have *roots*.

Within the crust, several additional seismic discontinuities may exist. The most important of these is the *Conrad*[21] *discontinuity* where the seismic velocity jumps from approximately 6.1 to 6.4–6.7 km/s. It exists at varying depths in continental areas and is supposed to separate a "granitic" from a "basaltic" (or "intermediate") layer; not too much chemical significance, however, should be attached to these designations. In the light of the existence of such a Conrad discontinuity, the problem of mountain roots takes on a dif-

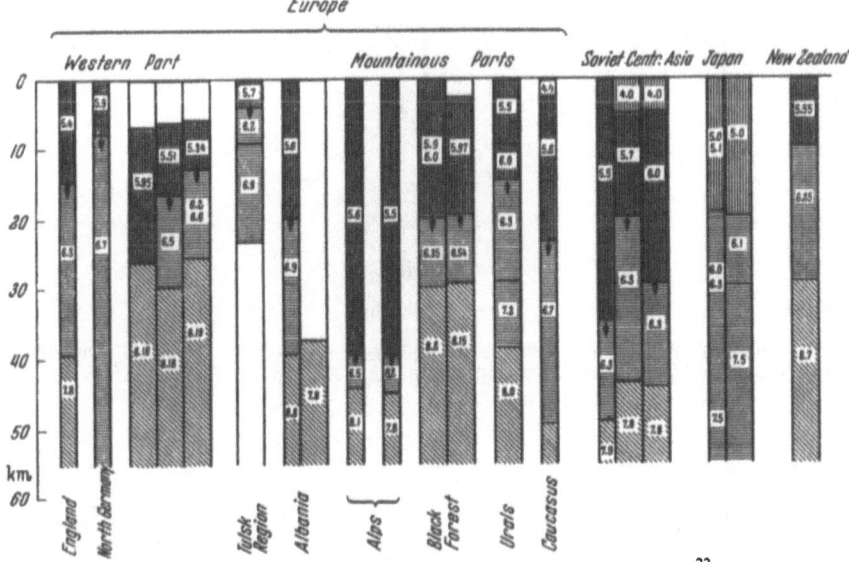

Fig. 32. Typical continental seismic sections. (After Savarenskiy and Kirnos[22])

---

21 Conrad, V.: Laufzeitkurven des Tauernbebens vom 28. Nov. 1923. Wien: Mitt. Erdb. Komm. No. 59 (1925)

22 Savarenskiy, E. F., Kirnos, D. P.: Elementy seysmologiy i seysmometriy. Moscow: Gos. Iz-vo Tekh.-Teoret. Lit. 1955

ferent slant. It is conceivable that mountains may have roots in the Conrad discontinuity rather than in the Mohorovičić discontinuity. Instances where this has been claimed to the case have been reported from Central Asia[23] and from California[24]. The "mountain roots" may therefore possibly be roots in the "granitic" layer, roots in the "basaltic" layer, or may in certain instances not exist at all. A series of typical continental seismic velocity profiles is shown in Fig. 32.

In the discussion given above, only the broadest features of continental structure have been presented. In fact, there may be velocity reversals (low-velocity layers lying underneath high-velocity ones) present in the lithosphere[25,26]. In Fig. 33 we show typical velocity graphs for various regions of the world.

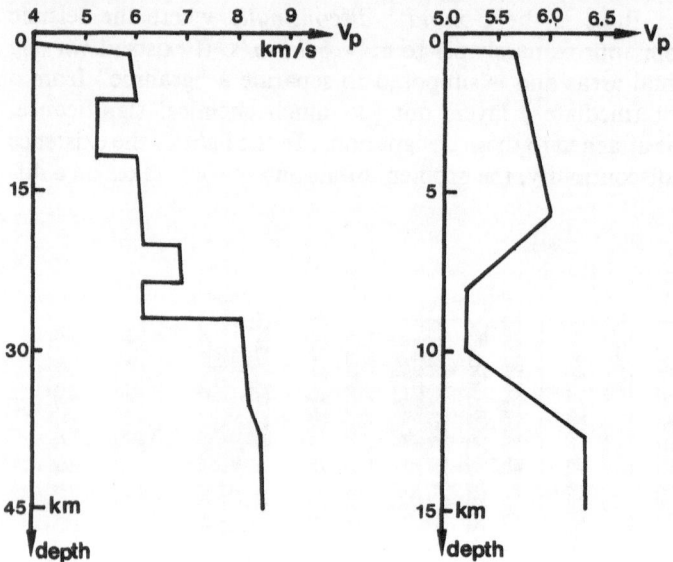

Fig. 33. Fine structure of the *P*-wave velocity in the continental crust (*left* Jura mountains between France and Switzerland; *right* midwestern North America). (After Müller[25])

2.1.3.3 Oceans

The crustal structure beneath oceanic plains can be inferred from seismic and gravity investigations. A typical result of such investigations is shown in

23 Kosminskaya, I. P., Mikhota, G. G., Tulina, Yu. V.: Bull. (Izv.) Akad. Nauk SSSR Ser. Geofiz. 1958, No. 10, 1162 (1958)
24 Gutenberg, B.: Geol. Rundsch. 46:30 (1957)
25 Müller, S.: Geophys. Monogr. Am. Geophys. Union 20:289 (1977)
26 Meissner, R. O., Flüh, E. R.: J. Geophys. 45:349 (1979)

Fig. 34. Two typical sections of the Earth's crust in oceanic areas; (*A*) in an abyssal plain. (*B*) in area where an archipelagic apron is present. (After Gaskell[27])

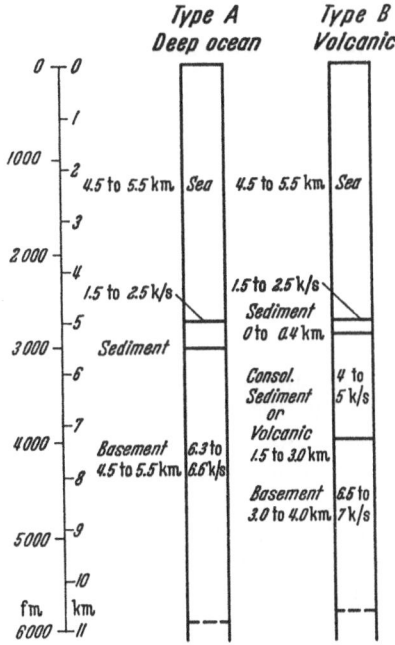

Fig. 34 (after Gaskell[27]). It is a remarkable result that the Mohorovičić discontinuity is, beneath oceans, at a depth of only about 5 km beneath the solid surface.

Regarding the crustal structure beneath mid-ocean ridges, the most important fact is that the Mohorovičić discontinuity becomes lost there: The *P*-wave velocity rises only gradually with depth from 7.1 km/s to the values found in the mantle (cf. Sect. 2.1.4). This above fact, together with the observation that rocks on the ocean bottom are rather young, gives rise to the hypothesis that mantle material wells up in the vicinity of the ridges, gradually moving as a slab toward the nearest continental margins and becoming subducted there. This corresponds to the "plate-tectonic cycle", of rifting, drifting, and subduction (cf. Sect. 1.5.3).

2.1.3.4 Transition and Transformation Between Continental and Oceanic Crust

The basically different velocity structure of the continental and oceanic crusts naturally poses the question regarding the possibility of a transformation between the two. In fact, Mueller[28] thought that such a transformation is part

27 Gaskell, T. F.: Proc. R. Soc. London Ser. A 222:341 (1954)
28 Müller, S.: In: Tectonic and Geophysics of Continental Rifts (eds. I. B. Ramberg, E.-R. Neumann) Dordrecht: Reidel pp. 11 − 28 (1978)

of a cyclic evolutionary process. Necking in rifting zones would initiate the mechanism of crustal attenuation. As the rifting process proceeds, the broken parts would move away from each other. In the space left in between, new oceanic crust would be produced by upwelling of mantle material, presenting the picture of the plate-tectonic cycle.

2.1.4 The Mantle

2.1.4.1 General Remarks

As noted, the mantle comprises the region between the Mohorovičić discontinuity and the core – mantle boundary. In the mantle, various regions, as indicated in Fig. 34 have been delineated. First of all, there is the question regarding the nature (in the seismic sense) of the Mohorovičić discontinuity and of the boundary velocity itself. Next, there is the so-called upper mantle which is taken to reach to a depth of about 600 – 800 km (deepest earthquakes). Finally, there is the lower mantle.

The results on these various regions have been obtained from the analysis of seismograms, caused by natural earthquakes and by explosions. The method is generally that of constructing synthetic seismograms (by computer) based on hypothetical models, and varying the models until a fit with the *form* of the actually observed seismograms has been obtained.

We shall discuss these regions in their turn below.

2.1.4.2 The Mohorovičić Discontinuity

The Mohorovičić discontinuity has, until recently, usually been regarded as a first-order seismic discontinuity separating layers of the lower crust from the upper mantle. Based upon this assumption, the *P*-wave velocity in the uppermost mantle (boundary velocity) has been found to be about 8.1 km/s.

Fig. 35. *P*-wave velocities at the Mohorovičić Discontinuity (km/s) in the United States. (After Herrin and Taggart[29])

29 Herrin, E., Taggart, J.: Bull. Seismol. Soc. Am. 52:1037 (1962)

However, there are regional variations. The most celebrated study of boundary velocities has been made by Herrin and Taggart[29] for the United States; its results are summarized in Fig. 35. It is seen that boundary velocities are highest in the stable mid-continent areas and lowest in the Rocky Mountain regions.

Recently, however, it has become likely that the model of an abrupt velocity change at the Mohorovičić discontinuity may be oversimplified. Davydova[30] has summarized the possibilities as indicated in Fig. 36. Of the three types envisaged, Type 1 represents a first-order discontinuity, Type 2 discontinuous or continuous transition zones whose thickness is less than two wavelengths, and Type 3 laminated transition zones. The latter are much

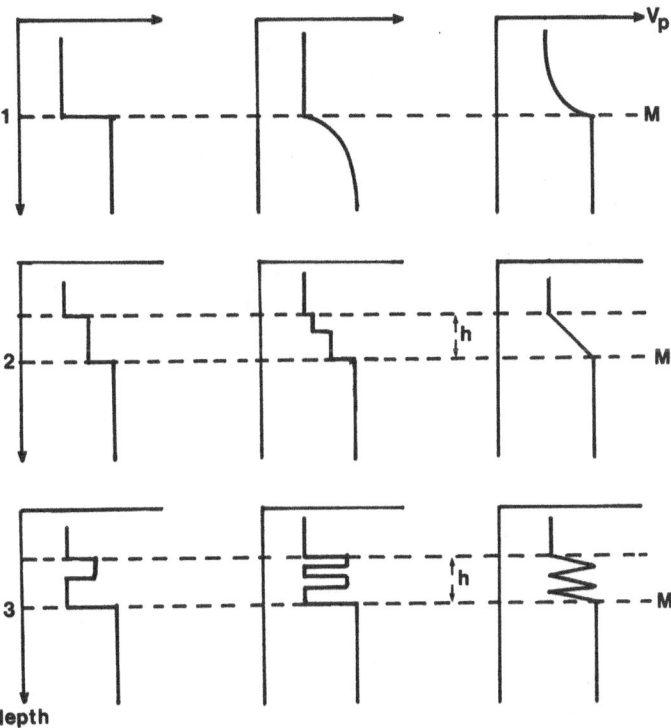

Fig. 36. Types of P velocity – depth functions for the crust – mantle transition zone M: Type 1, first-order discontinuities; Type 2, continuous or discontinuous transition zones with thickness h less than two wavelengths; and Type 3, laminated transition zones consisting of thin alternating high- and low-velocity layers. (After Davydova[30])

30 Davydova, N. I.: Possibilities of the DSS technique in studying properties of deep seated seismic interface, in Seismic Properties of the Mohorovičić Discontinuity (in Russian). (ed. N. I. Davydova). Moscow: Izdatel'stvo Nauka 1972

favored today[31]. The implications of the above observations with regard to the nature of the Mohorovičić discontinuity will be discussed in Sect. 2.8.

2.1.4.3 The Upper Mantle

As one proceeds downward from the Mohorovičić discontinuity, the most interesting feature is a velocity *decrease*; this refers to P as well as to S waves. Deeper down, the velocities increase again. One is, thus, faced with the presence of a low-velocity layer (region B in Bullen's model shown in Fig. 31); this low-velocity layer is often called the *asthenosphere* channel. Gutenberg[32] was probably the first to postulate this channel as early as in 1926. In broad terms, one can state that the velocity begins to decrease near the Mohorovičić discontinuity and stays below the value immediately below the Mohorovičić discontinuity to a depth of almost 200 km. This "low-velocity channel" has commonly been regarded as the bottom of the tectonic plates (cf. Sect. 1.4.2).

The problem of velocities has later been studied by many people. Again, it turned out that conditions are not as simple as they appeard at first[33]: There are many oscillations back and forth in the course of the velocity with depth. Furthermore, there are regional variations in the velocity distribution. Ansorge[34] has recently made a compilation of available data; Fig. 37 shows some typical results for the uppermost mantle.

The region below the B-layer in Bullen's model (Fig. 31), the so-called C-layer, has also been called "transition layer" as it is the interval of relatively rapid velocity increase below ca. 600 km. In this region at 900 km, Birch[34a] has postulated a discontinuity upon chemical grounds. It will be discussed more fully in Sect. 2.8.

The "transition" (C-) layer separates the "upper" from the "lower" mantle.

2.1.4.4 The Lower Mantle

The lower mantle is a region of relative uniform P- and S-wave velocity increase. Nevertheless, here, too, lateral velocity variations have been found[35]. This can best be expressed by giving the per-cent r.m.s. deviation of the velocity in random blocks of size $10° \times 10°$ and 500 km thickness. Figure 38 shows a typical result of such studies. There appears to be no correlation between surface tectonics and lower-mantle structure.

31 Cf. Müller, S.: Geophys. Monogr. Am. Geophys. Union 20:289 (1977)
32 Gutenberg, B.: Z. Geophys. 2:24 (1926)
33 Mayer-Rosa, D., Müller, S.: Z. Geophys. 39:395 (1973)
34 Ansorge, J.: Die Feinstruktur des obersten Erdmantels unter Europa und dem mittleren Nordamerika. Diss., Karlsruhe 1975
34a Birch, F.: Trans. Am. Geophys. Un. 32:533 (1951)
35 Sengupta, M. K., Toksöz, M. N.: Geophys. Res. Lett. 3(2):84 (1976)

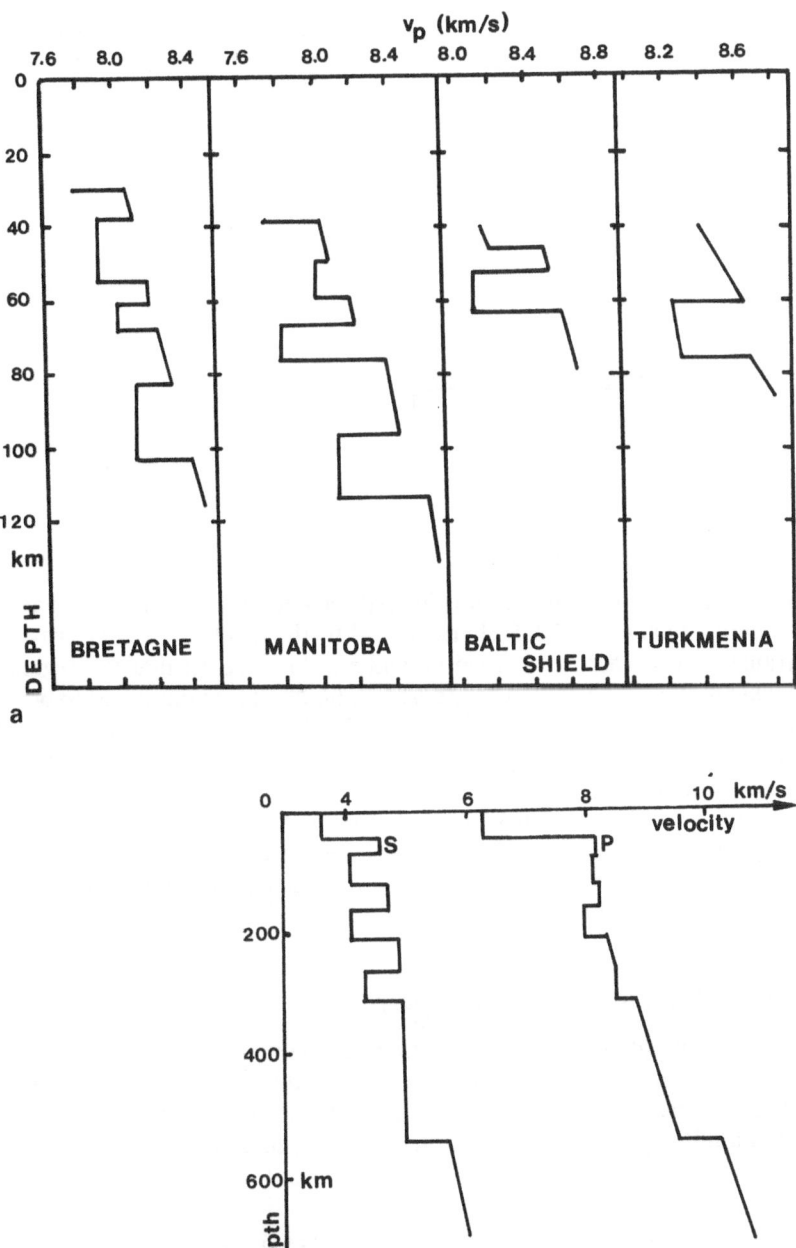

Fig. 37. a *P*-wave velocity in the uppermost mantle in various regions of the world. (After Ansorge[34]). **b** Gross velocity-depth distribution of *P*- and *S*-waves in the upper mantle of Europe. (After Mayer-Rosa and Müller[33])

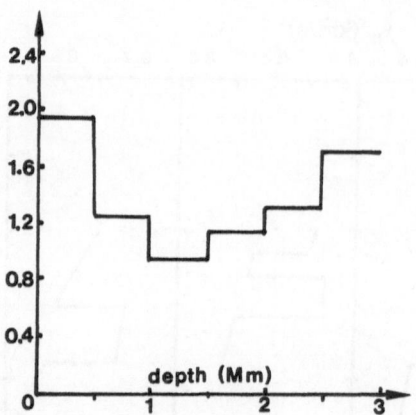

Fig. 38. r.m.s. relative perturbations (in %) of *P*-wave velocity for each 500 km depth interval in the mantle. (After Sengupta and Toksöz[35])

2.1.5 The Core

2.1.5.1 General Remarks

As has already been noted, a principal seismic discontinuity exists at a depth of about 2900 km which is characterized by the fact that no shear waves can be transmitted across it. This led Gutenberg[36] to suggest that the region below 2900 km, which he named "the core", is liquid. This notion had to be changed after the discovery of waves[37] that evidently are refracted once more within the core (so-called *PKIKP* waves), so that an additional boundary at about 5000 km depth had to be introduced[37]. Hence, the classic model of the core became one which consists of two layers: a fluid outer core and a solid inner core with a transition layer above the inner core boundary[38,39]. However, the details of the velocities in the various regions and around the discontinuities are not yet entirely certain.

2.1.5.2 The Core – Mantle Boundary

Data on the nature of the core – mantle boundary are gleaned from a comparison of the wave form of *P* waves reflected from the core (*PcP* waves) with that of direct *P* waves. The evidence is highly equivocal. Thus, Bullen and Haddon[40] suggested a sharp boundary, Dorman et al.[41] as well as Phinney and Alexander[42] suggested the existence of a soft layer, characterized by a low

36 Gutenberg, B.: Nachr. Ges. Wiss. Goettingen 1914:166 – 218 (1914)
37 Lehmann, I.: Publ. Bur. Cent. Int. Seismol. Trav. Sci. 14:87 (1936)
38 Jeffreys, H.: Mon. Not. R. Astron. Soc. Geophys. Suppl. 4:548 (1939)
39 Gutenberg, B, Richter, C. F.: Mon. Not. R. Astron. Soc. Geophys. Suppl. 4:363 (1939)
40 Bullen, K. E., Haddon, R. A. W.: Geophys. J. 17:179 (1969)
41 Dorman, J., Ewing, J., Alsop, L. E.: Proc. Natl. Acad. Sci. 54:364 (1965)
42 Phinney, R. A., Alexander, S. S.: J. Geophys. Res. 71:5959 (1966)

velocity, at the bottom of the mantle. Finally, the existence of a thin high-impedance liquid layer of several km in thickness embedded between the mantle and the core has been suggested[43]. Similarly, Ibraham[44] proposed a multi-layered core – mantle boundary.

2.1.5.3 The Outer Core

The structure of the Earth's outer core is also not yet entirely ascertained. Thus, based on the analysis of shear waves (inside the core: P waves; this type is denoted as *SKS*), Kind and Müller[45] proposed a model which is not as uniform as had been thought heretofore[46]. Figure 39 shows the course of the velocity of P waves in the outer core as proposed by the above-mentioned authors. Jacobs[47] gave a possible physical explanation of a non-smooth velocity curve by proposing that the temperature may cross the solidus – liquidus boundary several times.

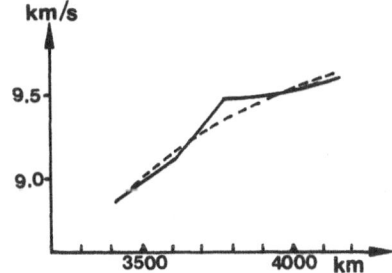

Fig. 39. Velocity of P waves versus depth in part of the outer core. *Solid curve* after Kind and Müller[51], *dashed curve* after Gilbert and Dziewonski[52]

2.1.5.4 The Inner Core Boundary

The inner core boundary, at roughly 5000 km depth, has been posing some questions for quite some time. A sharp discontinuity on the bottom of the outer core is supported by the fact that waves reflected from such a boundary were found on records. However, there may be more than one such discontinuity; thus, Bolt[48] supposes another discontinuity inside the core 450 km above the main inner-core discontinuity; this discontinuity was modified somewhat by Ruprechtova[49]. Finally Bullen[50] introduced a velocity reversal. The various possibilities are illustrated in Fig. 40.

43 Buchbinder, G., Poupinet, G.: Bull. Seismol. Soc. Am. 63:2047 (1973)
44 Ibrahim, A. K.: Pure Appl. Geophys. 91:95 (1971)
45 Kind, R. Müller, G.: Bull. Seismol. Soc. Am. 67:1541 (1977)
46 E. g., Gilbert, F., Dziewonski, A. M.: Philos. Trans. R. Soc. London Ser. A 178:187 (1975)
47 Jacobs, J. A.: J. Geophys. 44:675 (1978)
48 Bolt, B. A.: Bull. Seismol. Soc. Am. 54:191 (1964)
49 Ruprechtova, L.: Z. Geophys. 38:441 (1972)
50 See e. g., Haddon, R. A., Bullen, K. E.: Phys. Earth Planet. Inter. 2:35 (1969)
51 Kind, R., Müller, G.: Bull. Seismol. Soc. Am. 67:1541 (1977)
52 Gilbert, F., Dziewonski, A. M.: Philos. Trans. R. Soc. London Ser. A 178:187 (1975)

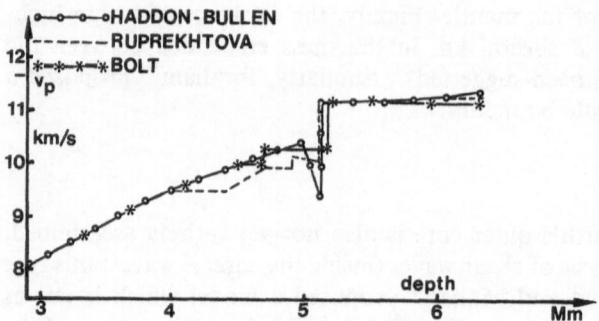

2.1.5.5 The Inner Core

Turning finally to the inner core, we note that the latter is generally again assumed as solid. The evidence for this stems from the fact that two types of converted waves, interpreted as longitudinal and shear waves, are found to pass through the inner core (these waves, of course, travel as longitudinal waves through the outer core). This question, however, has also not yet been entirely settled [53].

2.2 Studies of the Earthquake Focus

2.2.1 General Remarks

We shall now turn our attention to data related to the occurrence and the mechanism of the earthquakes themselves.

An earthquake manifests itself as a "shock", which occurs at a certain "place" and at a certain "time". Actually, this is an oversimplification inasmuch as the shock occupies an extended time and place interval. Nevertheless, one usually can define at least the time and space of the *beginning* of the shock. One calls in this the "origin-time" and the "hypocentral (or focal) coordinates" of the earthquake. Evidently, the last must be given by three numbers: the latitude and longitude of the geographical location and the depth of the focus. The point on the Earth's surface directly above the focus is often called the "epicenter" of the earthquake.

Focal coordinates and origin time are the two principal data on an earthquake source. They can be determined from records at seismic observatories with the help of the traveltime curves of seismic phases.

---

53 Bukowinski, M., Knopoff, L.: Is the core liquid after all?, paper presented at the 23rd Technical Conference of the Institute of Geophysics and Planetary Physics, Univ. Calif., May 21, 1973

In principle, the epicentral distance of an earthquake from a station can be determined from the difference in *P*- and *S*-wave arrival times. If various stations are used, the focus and origin time can be found accurately.

The investigation of the geographical location of earthquake foci leads to seismicity studies, the investigation of the focal energy leads to magnitude studies and to fault plane studies. In addition, a number of further focal parameters have recently been defined, such as stress drop and seismic moment.

Finally, not only the distribution of earthquake foci in space, but also in time is of great interest. Studies of such type lead to the investigation of the recurrence time of earthquakes in a region and to such problems as the description of aftershock series.

We shall discuss the mentioned aspect of earthquake foci in their turn below.

2.2.2 Seismicity Studies

Seismicity studies have as their objective a definition of the geographical distribution of earthquakes. In order to achieve this aim, it is necessary to make determinations of the focal coordinates of as many shocks as is possible and to plot the latter on a map and on cross-sections. It is obvious that these shocks will refer to a certain time interval, so that the density of hypocenters in a given region also represents in some way the *temporal frequency* of the shocks. However, specific questions regarding the temporal sequence of the shocks will be relegated to Sect. 2.2.6.

One of the first comprehensive studies of the seismicity of the Earth was made by Gutenberg and Richter[54]. This study referred to the shocks that occurred since the introduction of seismographs around 1900 until around 1945. The general character of the seismicity picture of the world has been confirmed by more accurate later studies, such as that, for instance, of Barazangi and Dorman[55] for the period of 1961 – 1967: The relative density of foci in any regions remains, for periods of several years, approximately constant. The zones of the globe in which large and frequent earthquakes occur are shown (shaded) in Fig. 41.

One of the outcomes of seismicity studies was the recognition that earthquakes may occur at various levels, to a depth of about 700 km. It is therefore convenient to separate earthquakes according to their depth of focus; a common classification of earthquakes is into *shallow* ones (depth of focus less than 65 km), *intermediate* ones (depth of focus between 65 and 300 km), and *deep* ones (depth of focus more than 300 km). Gutenberg and Richter found that the distribution of foci beneath orogenetic (mountain and island) arcs is

54 Gutenberg, B., Richter, C. F.: Seismicity of the Earth and Associated Phenomena. Princeton: Princeton Univ. Press 1949
55 Barazangi, M., Dorman, J.: Bull. Seismol. Soc. Am. 59:369 (1969)

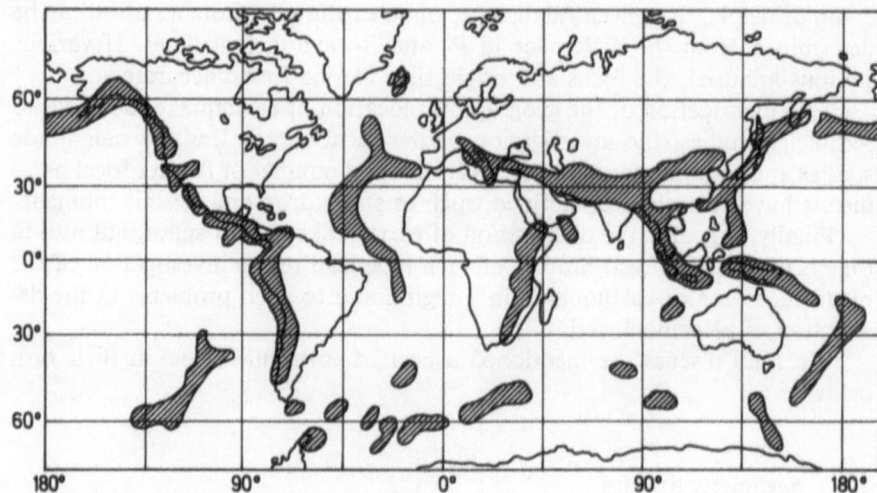

Fig. 41. The high-seismicity zones of the Earth

characteristic of the latter. Accordingly, earthquakes occur in localized zones which are almost planar, dipping at some intermediate angle (30° to 60°) into the Earth (see Fig. 42).

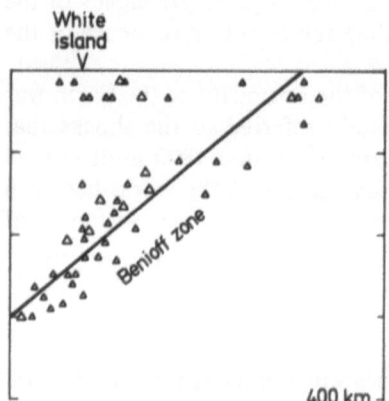

Fig. 42. Distribution of earthquake foci beneath the Bonin Island arc at about lat. 31° N and long. 138 – 141° E. (After Carr et al.[56])

The characteristic patterns of dipping seismic zones beneath many mountains and island arcs were studied closely and linked to orogenesis by Benioff[57]; hence these zones are now often referred to as "Benioff zones". An inspection of the figures mentioned above indeed indicates a very close relationship between seismicity and orogenetic activity. Evidently, the seismicity

56 Carr, M., Stoiber, R., Drake C. L.: Bull. Geol. Soc. Am. 84:2917 (1973)
57 Benioff, H.: Bull. Geol. Soc. Am. 65:384 (1954)

is concentrated geographically along the margins of the "plates" referred to in Sect. 1.4.2. The dipping Benioff zones may be thought of as connected with *subducting* plates.

2.2.3 Magnitude and Energy Studies

After an earthquake occurs, it is a natural desire to try to ascribe to it a certain value representative of its "strength". Here, one has to distinguish at once between two physically totally different attempts to do this. To the layman experiencing an earthquake the fact of importance is the "strength" of the "earthquake" *at the point of his position*. This leads to the concept of "intensity" of the ground motion at any given point. The original intensity scales were set up by investigating the destructiveness of earthquakes upon human structures. It is obvious, however, that a classification achieved in this fashion does not only depend on the intrinsic severity of the ground motion, but also on the type and number of man-made structures that happen to be in the area. Modifications of the intensity scale have been attempted by specifying more exactly the types of structures affected, the type of ground they stand on, etc. In spite of these efforts, intensity scales arrived at in this manner remain at best mostly descriptive. Any attempts to assign a value to the severity of the ground motion, therefore, must be based upon an analysis of instrumental records. Then, the velocity of acceleration *spectra* of the ground motion can be determined. This, then, can be used for earthquake-resistant design purposes.

For a scientific study of the phenomenon "earthquake" as such, the above procedure is unsatisfactory. It will be noted that the "intensity" referred to above is that of the ground motion *at a specific point*. For a given earthquake, the intensity varies from place to place as it depends on the epicentral distance. For an investigation of the earthquake phenomena itself, a property descriptive of the *focal mechanism* is required. Such a property is the earthquake *magnitude*, which has been defined empirically by Richter[58]. Accordingly, the relation between the magnitude M_L of two earthquakes at a standard epicentral distance of 100 km, and for the maximum recorder trace amplitude B (mm) is

$$M_{L_1} - M_{L_2} = \log_{10}B_1 - \log_{10}B_2 \qquad (2.2.3-1)$$

where the seismogram is supposed to have been written on a standard torsion seismometer of free period 0.8 s, static instrumental magnification of 2800 and damping ratio of 50:1. The zero of the scale is defined by setting $M = 3$ for $B = 1$ mm.

The above definition of earthquake magnitude was devised for the analysis of local shocks in southern California. In order to assign magnitudes to more

---

58 Richter, C. F.: Bull. Seismol. Soc. Am. 25:1 (1935)

distant shocks and to shocks observed elsewhere, various different magnitude scales have been devised, all based upon measurements of certain "standard" amplitudes in conjunction with an equation of the type shown in Eq. (2.2.3 – 1). Thus, a surface wave magnitude M_S can be determined, in addition to the Richter magnitude M_L introduced initially, from the amplitudes and periods of shallow teleseisms. From the analysis of many records, Bath and Duda [59] found

$$M_S = \log_{10}\left(\frac{A}{T}\right) + 1.66\log_{10}\Delta + 3.3 \qquad (2.2.3-2)$$

where A is the ground amplitude of surface waves in microns and T the wave period in seconds (one chooses the maximum value of A/T); Δ is the epicentral distance in degrees. In analogous fashion, a body wave magnitude m_b can be introduced. The various magnitude scales, based on Eq. (2.2.3 – 1) were originally adjusted to coincide at a magnitude of 7, but it was subsequently found impossible to have them coincide everywhere, without introducing a "calibrating function", $f(\Delta, h)$ depending on epicentral distance and focal depth [60]:

$$m_b = \log\left(\frac{A}{T}\right) + f(\Delta, h). \qquad (2.2.3-3)$$

An evaluation of data by statistical regression gives the following relation between surface-wave (M_s) and body-wave (m_b) magnitude [61]:

$$m_b = 0.56\,M_s + 2.9. \qquad (2.2.3-4)$$

The ambiguity encountered in the attempts at defining a "magnitude" is, of course, a reflection of the fact that analogous parts of the seismograms of two earthquakes are not proportional, not even on identical instruments. In view of this it must be conceded that the term "magnitude" does not yield an absolute characterization of an earthquake; rather it indicates the amplitude of a particular part of a seismogram which that earthquake has produced on a particular instrument.

An unambiguous indication of the intensity of an earthquake would be obtained if it were possible to give the amount of energy released by it. Unfortunately it is very difficult to do this. In fact, the only indication of the energy released in an earthquake is obtained by an inference from the above-mentioned, incompletely defined magnitude scales.

---

59 Bath, M., Duda, S. J.: Some aspects of global seismicity. Seismol. Inst. Uppsala Rep. 1 – 79:41 (1979)
60 Gutenberg, B., Richter, C. F.: Ann. Geofis. (Roma) 9:1 (1956)
61 Bath, M.: Ann. Geofis. (Roma) 30:299 (1977)

Table 3. Relation between energy E, magnitudes M_s and m_b and volume V. (Modified after Bath[64])

K	E_0 (joules)	M_s	m_b	$\log V$ (m³)
14	10^7	1.2	3.3	5.34
15	10^8	1.9	3.7	6.37
16	10^9	2.6	4.1	7.40
17	10^{10}	3.3	4.6	8.46
18	10^{11}	4.0	5.0	9.49
19	10^{12}	4.7	5.5	10.52
20	10^{13}	5.4	5.9	11.55
21	10^{14}	6.1	6.3	12.58
22	10^{15}	6.8	6.8	13.60
23	10^{16}	7.5	7.2	14.60
24	10^{17}	8.2	7.7	15.63
25	10^{18}	8.9	8.1	16.66

Attempts to correlate magnitude with earthquake energy E started in 1942 when Gutenberg and Richter[62] correlated the amplitude expectable in a particular seismogram with the total energy released as estimated from the total energy flux going through one station, calculated from the seismic trace by using elasticity theory. The originally proposed connection between magnitude and energy has since been modified several times; it also depends on the radiation model and the magnitude scale chosen. Thus, there are a number of such relations current which, however, all have the form[63]

$$K \equiv \log_{10} E = \alpha + \beta M \qquad (2.2.3 - 5)$$

where α and β are coefficients. For convenience's sake, as indicated above, one often introduces the "energy class" $K = 7 + \log_{10} E$, if E is in joules. The commonly used relations between K, E, M_s and m_b are shown in Table 3; these were calculated with $\alpha = 12.32$ and $\beta = 1.42$ for M_s[64]. The range of values quoted for these coefficients is $6.1 < \alpha < 13.5$ and $1.2 < \beta < 2.0$. The largest earthquakes occurring in nature have a magnitude between 8 and 9, corresponding to an energy release of about 10^{18} joules.

2.2.4 Fault Plane Studies

Investigations into the mechanism at the focus of an earthquake yielded the result that the latter can be regarded as a faulting process. Within the focus, a mechanical event must take place such as the sudden occurrence of failure of the material. Such a phenomenon would of necessity be connected with displacements within the focal region which, in turn, would cause seismic waves.

62 Gutenberg, B., Richter, C. F.: Bull. Seismol. Soc. Am. 32:163 (1942)
63 King, D. Y. Knopoff, L.: Bull. Seismol. Soc. Am. 59:269 (1969)
64 Bath, M.: Phys. Chem. Earth 7:115 (1966)

The nature of the seismic waves originating from the focal region, therefore, should be an indication of the displacements occurring therein. In this instance, it cannot be expected that it will be possible to infer every detail of the focal mechanism from seismic waves, but rather the general behavior of the focal region as a whole. As a first approximation, therefore, it seems reasonable to enclose the focal region within a sphere (termed "focal sphere")[65] and to study the motion of the surface of this sphere as it may be inferred from seismic evidence. The size of the focal sphere must be such that it encloses the whole region of the focus in which mechanical deformation may have taken place during the earthquake, but that it is small compared with the size of the whole Earth.

The various phases of a seismic disturbance, as has also already been pointed out, may be thought to have traveled along curved paths. The direction by which these phases leave the focal region can be calculated from traveltime tables. Thus, to each phase observed at any one seismic station corresponds a point on the focal sphere, viz. the point where the direction which the phase traveled when starting out from the focus intersects the focal sphere. If the motion near the focal sphere corresponding to each phase can be calculated from the observed motion of the station, the geometrical change of the surface of the focal sphere during the occurrence of the earthquake considered can be inferred. Finally, it is a problem of mechanics to determine the possible failure patterns inside the focal sphere that might produce the observed displacements on the surface.

The problem is thus to determine, from seismic evidence, the displacement pattern on the focal sphere. This can be achieved by means of the hypothesis of conservation of phase signs[66]. This hypothesis can best be demonstrated by referring the reader to Fig. 43 representing the focal sphere. The vector of the displacement at each point of the surface of the focal sphere can be split into its radial and tangential (to the surface of the sphere) component. It is easy to see that these components, if the mode of dislocation of the focal sphere is assumed to correspond to that of an orange which is sliced down the middle with one part being shifted over the other, are as depicted in Fig. 43. In this figure, a radial component "out" from the center of the sphere is denoted by 0, a radial component "in" by Δ. The tangential components are shown as *arrows*. The tangential components may further be split into a component lying in the plane of the ray through the point in question (called SV) and another one at right angles to it (called SH). The hypothesis of conservation of phase signs can then be stated as follows: "the directions (signs) of the phases P, SH and SV arriving at a seismological observatory are identical to the directions on the focal sphere in that point through which the ray has passed".

65 Scheidegger, A. E.: Trans. R. Soc. Can. 49, Ser. III, Sect. 4:65 (1955)
66 Scheidegger, A. E.: Bull. Seismol. Soc. Am. 47:89 (1957)

Fig. 43. The focal sphere

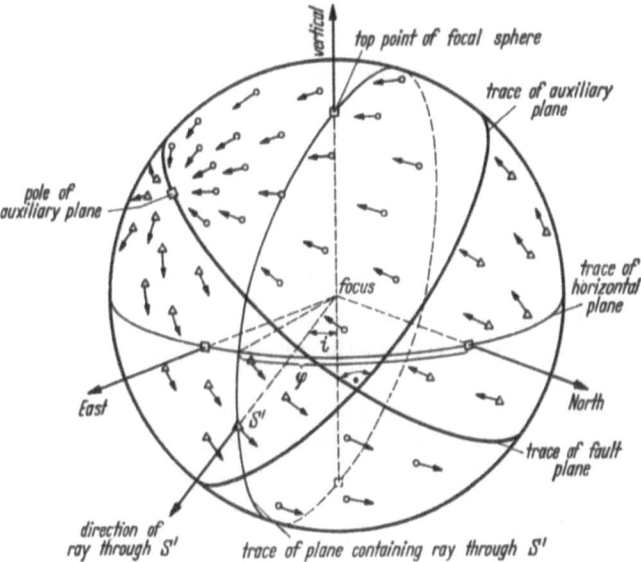

It is thus evident that the radial displacements of the focal sphere can be determined from *P* phases, whereas for the tangential displacements an analysis of *S* waves is necessary. It so happens that it is much easier to read *P* phases from seismograms than *S* phases, and therefore the direction of the radial displacement of the focal sphere can be obtained for many more points than the direction of the tangential displacement. In general, one has thus to be content with the knowlewdge of the radial displacements only. From these, however, it cannot be decided which of the solidly drawn two planes (in Fig. 43) is the true "fault" plane, and which is the plane orthogonal to it (the latter is often termed "auxiliary plane" of the fault). "Fault-plane solutions" obtained from *P* phases only, are therefore ambiguous to this extent. However, the quadrants moving "in" and those moving "out" can be deter-

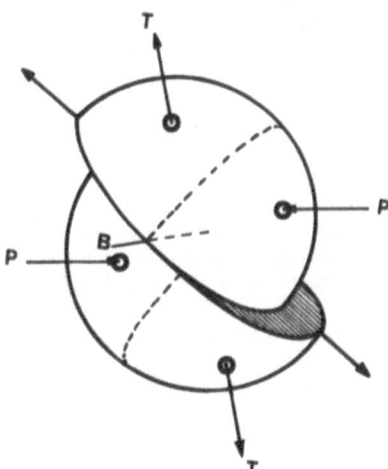

Fig. 44. Simple model of an earthquake source with corresponding principal stresses

mined unequivocally. The bisectrices of these quadrants would represent the maximum (P) and minimum (T) compressions, respectively, acting in the focal region (Fig. 44).

The most common procedure for the determination of fault-plane solutions is still to plot the direction of the first onsets on some representation of a sphere surrounding the focus; the position where to plot the direction is found by following the ray from the recording station back to the focal region. Then, the best-fitting nodal planes are constructed "by eye". However, computer procedures have now been developed[67-69], based on optimal strategies for finding best-fitting solutions. Computer-calculated fault-plane solutions of larger earthquakes, as they occur, are also available from the International Seismological Centre in Edinburgh.

Fault-plane solutions can be made more reliable if onsets of S waves are also considered[70-73]. Computer procedures for S waves have also been developed[74-78]. Similarly, surface waves have been employed. More sophisticated techniques make use of amplitude-spectra analysis[79-82]. Langston[83] even developed a method to obtain a complete fault-plane solution from body wave forms from a *single* station.

The number of actually available solutions is legion. Inasmuch as the source mechanism is a manifestation of the geophysical stress field, the latter can be reconstructed from a series of fault-plane solutions in an area. We shall discuss these matters further when we come to a description of the methods for determining the geophysical stress field (Sect. 2.4.4).

2.2.5 Seismic Source Parameters

2.2.5.1 General Remarks

In addition to the earthquake parameters discussed heretofore, a number of additional ones have been introduced. These are divided into two broad cate-

67 Kasahara, K.: Bull. Seismol. Soc. Am. 53:1 (1963)
68 Guinn, S., Long, L. T.: Earthquake Notes 48(4):21 (1977)
69 Julian, B. R.: Tectonophysics 49:223 (1978)
70 Ferraes, S. G.: Geofis. Anal. (México) 9:33 (1963)
71 Kasahara, K.: Bull. Seismol. Soc. Am. 53:643 (1963)
72 Otsuka, M.: Kyoto Univ. Geophys. Inst. Spec. Contrib. 4:37 (1964)
73 Stefanson, R.: Tectonophysics 3:35 (1966)
74 Udias, A.: Bull. Seismol. Soc. Am. 54:2037 (1964)
75 Udias, A., Stauder, W.: Bull. Seismol. Soc. Am. 54(6):2049 (1964)
76 Stauder, W., Bollinger, G. A.: Bull. Seismol. Soc. Am. 59:503 (1969)
77 Udias, A., Baumann, D.: Bull. Seismol. Soc. Am. 59:503 (1969)
78 Chandra, U.: J. Geophys. Res. 75(17):3411 (1970)
79 Kishimoto, Y.: Bull. Disaster Prov. Res. Inst. Kyoto 67:1 (1964)
80 Khattri, K.: Bull. Seimol. Soc. Am. 67:1355 (1977)
81 Chandra, U.: Bull. Seismol. Soc. Am. 60:1221 (1970)
82 Schick, R.: Z. Geophys. 36:205 (1970)
83 Langston, C. A.: Geophys. Res. Lett. 6(1):41 (1979)

gories, viz. kinematic parameters and dynamic parameters. The more important ones of them will now be discussed individually.

2.2.5.2 Kinematic Parameters

1. Earthquake Volume and Strain. An important parameter of an earthquake is its volume V. This is the region around the focus in which the elasticity conditions no longer apply. In earthquake models assuming a singularity, this volume would evidently be zero, which is a physically unrealistic assumption. Thus, Bath and Duda[84] identified the volume of a large earthquake with the total volume occupied by its aftershocks (cf. Sect. 2.2.6). By this procedure one finds a relation between volume and magnitude M_s which has the following form

$$\log V = \alpha + \beta M_s . \qquad (2.2.5-1)$$

The coefficients are empirical; Bath and Duda found $\alpha = 9.58$, $\beta = 1.47$. Inasmuch as there is a relation between magnitude and energy, one obtains the relation shown in Table 3 between the energy parameter K and the earthquake volume V.

In a strained elastic body, the strain is proportional to the square root of the elastic energy stored. It is therefore possible to interpret the energy released in an earthquake in terms of a corresponding *strain* (ε_{ij}) *release*[85]. The complete relation is

$$2W = \lambda \theta^2 + 2\mu \varepsilon_{ij}^2 \qquad (2.2.5-2)$$

where W is the elastic energy per unit volume, θ is the cubical dilatation and λ, μ are Lamé's parameters. The last equation implies that the total elastic strain energy E of the focal region of volume V is

$$E = VW . \qquad (2.2.5-3)$$

Inasmuch as V increases, according to Eq. (2.2.5 – 1), with magnitude in the same manner as E, it follows that W, and therewith the strain, *is independent of magnitude.* This is simply the "critical" strain which the material can support before it breaks. Alternatively, if the existence of a universal critical strain in earthquake focus is assumed a priori, one has a means of calculating earthquake volumes[86,87].

2. Fault Length, Width and Area. Inasmuch as an earthquake must be regarded as some type of faulting process, it stands to reason that the geometrical parameters, viz. the length L and width w the "fault" can be deter-

84 Bath, M., Duda, S.: Ann. Geofis. (Roma) 17:353 (1964)
85 Benioff, H.: Geol. Soc. Am. Spec. Pap. 62:61 (1955)
86 Duda, S.: Bull. Seismol. Soc. Am. 60:1479 (1970)
87 Dzhibladze, E. A.: Bull. (Izv.) Acad. Sci. USSR Earth Phys. 1971(5):89 (1971)

mined. This, of course, is a formidable problem. However, one can grossly assume that the linear fault dimensions L are related to the volume V as

$$L^3 \sim V \qquad (2.2.5-4)$$

Inasmuch as (see above) V is related to M_s by Eq. (2.2.5 – 1), we have approximately ($\beta \approx 1.5$) [88,89]

$$M_s = 2 \log_{10} L + \text{const} . \qquad (2.2.5-5)$$

Relations of this type, viz.

$$M_s = c \log w + d \qquad (2.2.5-6)$$

were also applied to the width w of the fault, as well as to the relative movement D [90-94]

$$M = h \log D + f . \qquad (2.2.5-7)$$

In the above equations, c, d, h, and f are constants.

However, inasmuch as it turned out that the regression of earthquake magnitude on the logarithm of the fault length is not symmetric, equations of the type of Eq. (2.2.5 – 5) have been questioned [95] on principle. Indeed, King and Knopoff [96] had already proposed a more complicated relation earlier:

$$M = c_1 + c_2 \log L D^n \qquad (2.2.5-8)$$

where c_1, c_2 and n are again constants, L is a linear dimension, and D the relative movement in the focus.

2.2.5.3 Dynamic Parameters

1. Seismic Efficiency. Of the total strain energy in an earthquake focus only a part is radiated in the seismic waves. The total energy radiated is the seismic energy; the fraction of the total strain energy this represents has often been called "seismic efficiency". Attempts at determining this quantity are based on the search for similarity laws. Generally, values for moderate to large earthquakes range over two orders of magnitude which represents a considerable uncertainty in this parameter [97].

88 Schick, R.: Z. Geophys. 34:267 (1968)
89 Drimmel, J.: Tectonophysics 55:T1 (1979)
90 Tocher, D.: Bull. Seismol. Soc. Am. 48:147 (1958)
91 Iida, K.: J. Earth Sci. Nagoya Univ. 13:115 (1959)
92 Wyss, M., Brune, J. N.: J. Geophys. Res. 73:468 (1968)
93 Wyss, M., Molnar, P.: J. Geophys. Res. 77:1433 (1972)
94 Randall, M. J.: J. Geophys. Res. 77:969 (1972)
95 Bolt, B. A.: Geology 6:233 (1978)
96 King, C. Y., Knopoff, L.: Bull. Seismol. Soc. Am. 58:249 (1968)
97 Sibson, R. H.: Tectonophysics 51:T39 (1978)

2. Rupture Velocity. If an earthquake is envisaged as a faulting process, the latter must expand at a finite rate across the fault plane.

3. Stress Drop. In view of the fact that the limiting strain in an earthquake appears to be constant, there also must be a limiting stress, since there is a single relation between stress and strain in elasticity theory. When an earthquake occurs, the stress *drops* at the fault; this stress drop corresponds to the shearing strength. Chinnery[98,99] estimated this stress drop in shallow earthquakes by measuring the horizontal displacement of the fault at the surface assuming that the vertical plane of the fault can be treated as the slip patch wich is displaced by a dislocation. Stress drops can also be estimated from strain estimates of an earthquake assuming suitable elastic constants. Values of $1-10$ MPa were obtained in this fashion, with $3-6$ MPa common values in the magnitude range from $6.8-8.3$. Extremes may range from 1 to several times 10 MPa[100]. There does not appear to be a correlation between stress drop and magnitude of an earthquake. No systematic variations of stress drops with the location of an earthquake relative to the plate boundaries (interior, margin) were found either[101].

4. Seismic Moment. The seismic moment M_0 is the mechanical moment of the point source equivalent to the faulting motion:

$$M_0 = \mu \int_S dS \cdot D(x, y) = \mu \bar{D} S \qquad (2.2.5-9)$$

where μ is the rigidity, D the displacement of the fault, S its total area, and $\bar{D}$ the "average" slip. It was for the first time determined by Aki[102] for the Niigata earthquake of 1964. As defined above, the seismic moment is a scalar. Inasmuch as it depends on the orientation of the fault plane as well, it should actually be defined as a tensor (this is particularly important in curved fault surfaces), viz. as the volume integral of the stress-drop tensor[103]. The theory of stress-moment tensors and equivalent body forces has been described by Backus and Mulcahy[104].

The seismic moment of an earthquake can be determined directly from very long period seismograms, because there exists a simple linear relation between the amplitude of free oscillations excited by a point source and the seismic moment tensor. A similar linear relation exists for surface waves[105,106], but the inversion requires rather sophisticated instrumentation.

98 Chinnery, M. A.: J. Geophys. Res. 69:2085 (1964)
99 Chinnery, M. A.: Publ. Dom. Obs. Ottawa 37(7):211 (1969)
100 Trifunac, M. D.: Bull. Seismol. Soc. Am. 62:1283 (1972)
101 Richardson, R. M., Solomon, S. C.: Pure Appl. Geophys. 115:317 (1977)
102 Aki, K.: Bull. Earthquake Res. Inst. Tokyo Univ. 44:73 (1966)
103 Gilbert, F.: Geophys. J. R. Astron. Soc. 22:223 (1970)
104 Backus, G., Mulcahy, M.: Geophys. J. R. Astron. Soc. 46:341 (1976); 47:301 (1976)
105 McCowan, D. W.: Geophys. J. R. Astron. Soc. 44:595 (1976)
106 Mendiguren, J. A.: J. Geophys. Res. 82:889 (1977)

Nevertheless, catalogues[107] of seismic moments of major earthquakes are now available. The range[108] of observed seismic moments extends from 10^8 to 10^{23} Nm (joules). The relationship between the logarithm of the seismic moment and the source radius is not linear.

The seismic moment is an extremely important quantity, because it represents the average slip in an earthquake. Seen world-wide, one chould obtain a measure of the seismic slip of the world. Relative to the time in which this slip occurs, one obtains from it the "seismic flow" of the Earth's crust[109].

2.2.6 Temporal Sequence of Earthquakes

2.2.6.1 Introduction

Not only the spacial, but also the temporal distribution of earthquakes is of importance. There appears to be, in fact, a rather simple relation between the magnitude and the frequency of earthquakes in any one area. Inasmuch as magnitude can be interpreted in terms of energy, strain, and seismic moment, analogous relations can also be determined for these focal parameters. Finally, one may try to characterize the occurrence of earthquakes as a time series. It is, however, well known that earthquakes have a tendency to cluster in groups (mostly as aftershocks), so that a characterization of their occurrence as a time series is not very easy to achieve.

We shall now analyze these questions in detail.

2.2.6.2 Magnitude-Frequency Relations

It was recognized fairly early that the number N of earthquakes occurring during a given time interval in a given area decreases with their magnitude M. Gutenberg and Richter[110] postulated the following relation:

$$\log N(M) = a - bM \tag{2.2.6-1}$$

where N is the number of earthquakes in a given region in a certain magnitude interval. The relation can be applied to the world as a whole as "region"; Bath and Duda[111] have made a study for all the earthquakes known from 1905 to 1977 and found

$$\log N = 9.51 - 1.00 M \tag{2.2.6-2}$$

where M intervals are 0.1 unit and N refers to the number of earthquakes

107 Aki, K., Patton, H.: Tectonophysics 49:213 (1978)
108 Gibowicz, S. J.: Acta Geophys. Pol. 25(2):119 (1977)
109 Riznichenko, Yu. V.: Bull. (Izv.) Acad. Sci. USSR 13(10):698 (1977)
110 Gutenberg, B., Richter C. F.: Seismicity of the Earth and Associated Phenomena. Princeton: Univ. Press 1949
111 Bath, M., Duda, S. J.: Tectonophysics 54:T1 (1979)

occurring in one year. Formulas of the type of Eq. (2.2.6–1) are, of course, valid only for $M \leqslant M_{max}$ for an area.

The above relation is valid for the Earth as a whole. There are, however substantial differences form place to place. Thus, Miyamura[112] collected the available evidence and found that very low b values (0.4–0.6) were reported for old shields, low values ($b = 0.6-0.7$) for continental rift zones, medium values ($b = 0.7-1.0$) for the Circum-Pacific and Alpine orogenic belts, and high values ($b = 1.0-1.8$) for oceanic regions. The basic form of the magnitude–frequency relation [Eq. (2.2.6–1)] was found to be valid in all regions of the world, except perhaps in the New Hebrides-Solomon Islands-New Guinea Region[113]. Furthermore, in the Ukraine[114] and in the vicinity of New Zealand, the coefficient b was found to decrease with depth[115].

In terms of energy rather than magnitude – using Eq. (2.2.3–5) – the formula equivalent to Eq. (2.2.6–1) becomes

$$\log N = \log A - \gamma(K - K_0) \qquad (2.2.6-3)$$

where K_0 is some conventionally fixed energy-class value; A is then the activity and γ the "tectonic" parameter. The fact that the coefficients characterizing the magnitude–frequency relation appear to vary from place to place has been ascribed to tectonic effects[116,117]. In addition to spatial variations there are also *temporal* variations in the magnitude–frequency relationship (cf. Sect. 2.2.6.5).

Inasmuch as the magnitude–frequency relations are only valid up to a maximum magnitude M_{max} (or energy-class K_{max}), the latter represents an independent parameter for any one area. Riznichenko[118,119] developed a correlation for K_{max}

$$\log A = \log \alpha + \beta(K_{max} - K_\alpha) \qquad (2.2.6-4)$$

where the coefficients (assuming $K_0 = 10$, time unit = 1 year, area unit = 1,000 km^2) $\log \alpha = 2.84$, $\beta = 0.21$, $K_\alpha = 15$ were determined by a least squares procedure from observations for Central Asia. For the Crimea Riznichenko et al. used $\log \alpha = 2.63$; the other parameters had the same values as mentioned above.

112 Miyamura, S.: Proc. Jpn. Acad. Sci. 38(1):27 (1962)
113 Ranalli, G.: Veroeff. Zentrl. Inst. Phys. d. Erde, Potsdam 31:163 (1975)
114 Butovskaya, E. M., Kuznetsova, K. I.: Bull. (Izv.) Acad. Sci. USSR Earth Phys. 1971(2):11 (1972)
115 Gibowicz, S. J.: Tectonophysics 23:283 (1974)
116 Molnar, P.: Bull. Seismol. Soc. Am. 69:115 (1979)
117 Rikitake, T.: Tectonophysics 35:335 (1976)
118 Riznichenko, Yu. V.: Proc. (Dokl.) Acad. Sci. USSR 157(6) (1964)
119 Riznichenko, Yu. V., Bune, V. I., Zadzharova, A. I., Seyduzova, S. S.: Bull. (Izv.) Acad. Sci. USSR 1969(8):3 (1969)

However, the uncertainty regarding the value of M_{max} is large, and some investigators[120-122] have proposed different maximum magnitudes for each region. There is, thus, a fundamental problem with the range of applicability of Eq. (2.2.6−1), especially for high magnitudes, which has given rise to criticisms of its basic form. Therefore, a modification in the form (A, B, α are constants)

$$\log N(M) = A - B \exp(\alpha M) \qquad (2.2.6-5)$$

has been suggested[123] which approximates (2.2.6−1) asymptotically for small magnitudes. The form (2.2.6−5) can be justified on the basis of a simple stochastic model.

2.2.6.3 Aftershock Sequences

Large earthquakes seldom occur as single events. They are usually followed by a series of smaller earthquakes, called "aftershocks". A comprehensive statistical study of the phenomenology of aftershock sequences has been made by Ranalli[124], and by Ranalli and Scheidegger[125]. Accordingly, the statistical laws postulated by earlier writers have been fully confirmed by a detailed analysis of the analysis of 15 aftershock sequences. These laws are the following:

First, there is Omori's[126] law which states that the frequency $n(t)$ of the number of aftershocks in a series per unit time decrease hyperbolically with time t ($t = 0$ for the origin-time of the main shock):

$$n(t) = a t^{-\beta} \qquad (2.2.6-6)$$

where β is a parameter close to 1. Second, there is a law of magnitude stability due to Lomnitz[127]: The sliding mean magnitude (over, say, ten earthquakes) in one and the same aftershock series is constant. Naturally, there is some scattering, but this scattering is random. The third law due to Gutenberg and Richter[128] states that the frequency distribution of earthquakes in an aftershock series as a function of magnitude has an exponential form:

$$N(M) = K e^{-b(M - M^*)} \qquad (2.2.6-7)$$

120 Cornell, C. A., Vanmarcke, E.-H.: Proc. World Conf. Earthquake Eng. 4th, Santiago, 1(A-I):69 (1969)
121 Yegulalp , T. M., Kuo, J. T.: Bull. Seismol. Soc. Am. 64:393 (1974)
122 Smith, S. W.: Geophys. Res. Lett. 3:351 (1976)
123 Lomnitz-Adler, J., Lomnitz, C.: Bull. Seismol. Soc. Am. 69(4):1209 (1979)
124 Ranalli, G.: Ann. Geofis. (Roma) 22(4):359 (1969)
125 Ranalli, G., Scheidegger, A. E.: Ann. Geofis. (Roma) 22:3 (1969)
126 Omori, F.: Proc. 3. Coll. Sci. Imp. Univ. Tokyo 7:III (1894)
127 Lomnitz, C.: Bull. Seismol. Soc. Am. 56:247 (1966)
128 Gutenberg, B., Richter, C. F.: Seismicity of the Earth and Associated Phenomena. Princeton: Univ. Press 1949

here, $N(M)$ is the number of earthquakes in the aftershock sequence with a magnitude greater or equal to M. M^* is the minimum magnitude that could be registered, and K and b are constants. The parameter b is found to change for different aftershock sequences; in the 15 sequences studied by Ranalli[129], it was found to range from 0.44 to 1.36. Whether these differences are significant is, as for earthquakes in general, still an unsolved question. Finally, it may be added that aftershocks are generally crustal events[130]; deep aftershocks occur relatively rarely. The deepest sequence observed[131] (Dzhurm region, 14 March 1965) had a depth of about 240 km. Although the statistical properties of various aftershock sequences are similar, the absolute *level* of aftershock activity depends very much on the geological conditions of an area. (Maximum probability of aftershocks in mountainous areas.)[132]

2.2.6.4 Energy and Strain Release

The seismic energy release in a region is evidently a characteristic of the local conditions. Thus, Bath[133] has used this quantity for describing the seismicity

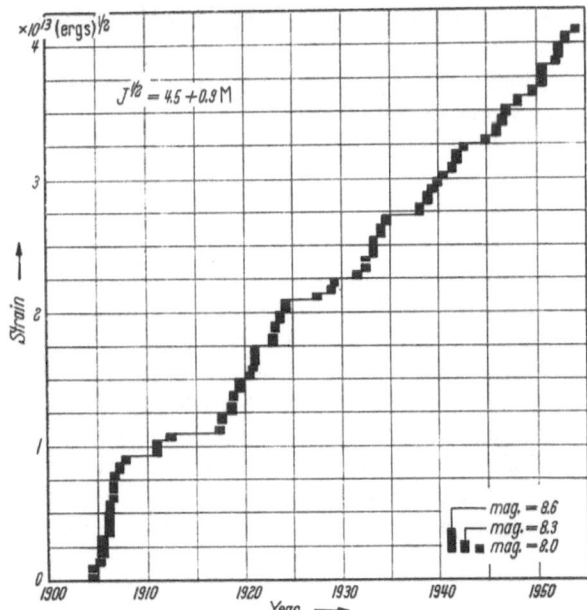

Fig. 45. Strain release in the world's shallow earthquakes. (After Benioff[135])

129 Ranalli, G.: Ann. Geofis. (Roma) 22(4):359 (1969)
130 Page, R.: J. Geophys. Res. 73:3897 (1968)
131 Lukk, A. A.: Bull. (Izv.) Acad. Sci. USSR Earth Phys. 1968(5):83 (1968)
132 Olsson, R.: On the occurrence of aftershock sequences. Rept. No. 4-79, Seismological Institute Uppsala (1979)
133 Bath, M.: Gerl. Beitr. Geophys. 63:173 (1953)

of an area. More common than the use of energy release has been the use of strain release for the same purpose. As was noted in Sect. 2.2.5.2, the square root of the energy radiated in an earthquake is proportional to the strain released in that earthquake. This idea was used for the first time by Benioff[134] to plot curves of cumulative strain release for a given period for a given area (which may be the whole world). A typical curve, obtained by Benioff[135] for the world's shallow earthquakes, is shown in Fig. 45.

The time differential of the cumulative strain release has been called *tectonic flux* μ of an area[136]. Accordingly, it can be defined as follows

$$\mu = \frac{\omega}{A} \lim_{\langle t, t_2 \rangle \to 0} \frac{I}{\langle t_1 t_2 \rangle} \int\int_{A\langle t_1 t_2 \rangle} E^{1/2} dA \, dt \qquad (2.2.6-8)$$

where ω is an arbitrary constant to arrive at a reasonable scale, A is the area in question, and $\langle t_1 t_2 \rangle$ is the time interval, thought to be very small. In practice, the seismic energy E is calculated by one of the magnitude – frequency relationships.

Characterizations of seismicity based on strain and energy release have been reported for Southern California[137], Canada[138], and the Western United States[139].

Finally, the seismic-moment release has also been used to characterize frequency relationships[140,141].

2.2.6.5 Variability of Release Rates

On several occasions it has already been indicated that the various frequencies of earthquake parameters may vary not only with location, but also in time[142]. An inspection of Benioff's picture of strain release in the world's shallow earthquakes shown in Fig. 45 reveals that, while there is a constant release rate over large time intervals, there are, in fact, large fluctuations in detail. Thus, there was a "lull" in earthquake activity around 1915 and around 1925, increased activity around 1905.

Similar fluctuations are found also if simply the *number* of earthquakes per time interval is considered[143]. One of the principal problems is the search for suspected periodicities in earthquake occurrences. Man such cases have

134 Benioff, H.: Bull. Geol. Soc. Am. 62:331 (1951)
135 Benioff, H.: Geol. Soc. Am. Spec. Pap. 62:61 (1955)
136 St. Amant, P.: Bull. Seismol. Soc. Am. 46:1 (1956)
137 Allen, C. R. et al.: Bull. Seismol. Soc. Am. 55:753 (1965)
138 Milne, W. G.: Can. J. Earth Sci. 4:797 (1967)
139 Prescott, W. H., Savage, J. C., Kinoshita, W. T.: J. Geophys. Res. 84:5423 (1979)
140 Kanamori, H.: Nature (London) 271:411 (1978)
141 Mogi, K.: Tectonophysics 57:T43 (1979)
142 Bath, M., Duda, S. J.: Tectonophysics 54:T1 (1979)
143 Kanamori, H.: J. Geophys. Res. 82:2981 (1977)

been reported; thus, among many, Dubourdieu[144,145] claimed to have found a seismic period of 4 years, Sadeh and Meidav[146] a period of about 24 h. Of the possibilities suggested, the most reasonable ones are correlations[147-149] with the solid tides of the Earth and with the Chandler wobble[150]. However, a very careful analysis of the available data has usually yielded inconclusive results. Hunter[151] has reviewed the evidence and found no support for the claim of the existence of any periodicities. Similarly, Bath[152] had a close look at suspected periodicities of earthquakes in Sweden and found none. Thus, the claims to periodic "triggering effects" for earthquakes must be dismissed.

The time series of earthquake events obviously contains clusters, because of the occurrence of aftershocks. Thus, most tests of time series of earthquake occurrences of low magnitude yielded the result that they do not fit a Poisson process very well[153-156]. However, if the obvious clusters are removed, the time series is that corresponding to a Poisson process. For the latter, the probability $p(j; t)$ of j earthquakes occurring in the time interval t is given by

$$p(j; t) = \frac{(kt)^j}{j!} e^{-kt}. \tag{2.2.6-9}$$

The recurrence time, then, is exponentially distributed and the mean and variance of the number of events per unit time are both equal to k. Ben-Menahem[157] concluded this already some time ago; his result was later confirmed[158]. Based on the assumption of a Poisson process, earthquake sequences were simulated numerically by a computer[159].

The above analysis does not obviate the possibility of the existence of (aperiodic) mechanical triggering effects of earthquakes. Indeed, individual cases such triggering effects may exist, as will be discussed in Sect. 7.3.4.

144 Dubourdieu, G.: Sur la période sismique de 4 ans. Neubourg; Imprimerie Neubourg 1972
145 Dubourdieu, G.: Mise en évidence de la période sismique de 4 ans an 19e siècle. Paris: Imprimerie Laboureur 1980
146 Sadeh, D. S., Meidav, M.: J. Geophys. Res. 78:7709 (1973)
147 Mauk, F. J., Kienle, J.: Science 182:386 (1973)
148 Klein, F. W.: Geophys. J. R. Astron. Soc. 45:245 (1976)
149 Young, D., Zürn, W.: J. Geophys. 45:171 (1979)
150 Kanamori, H.: J. Geophys. Res. 82:2981 (1977)
151 Hunter, R. N.: Earthquake Inf. Bull. U.S.C.S. 8(5):4 (1976)
152 Bath, M.: Tectonophysics 51:T55 (1978)
153 Vere-Jones, D., Turnovsky, S., Eiby, G. A.: N. Z. J. Geol. Geophys. 7:722 (1964)
154 Knopoff, L.: Bull. Seismol. Soc. Am. 54:1871 (1964)
155 Isacks, B. L., Sykes, R., Oliver, J.: Bull. Seismol. Soc. Am. 57:935 (1967)
156 Ferraes, S. G.: J. Geophys. Res. 72:3741 (1967)
157 Ben-Menahem, A.: Gerlands Beitr. Geophys. 69:68 (1960)
158 Gardner, J. K., Knopoff, L.: Bull. Seismol. Soc. Am. 64:1363 (1974)
159 Rundle, J. B., Jackson, D. D.: Bull. Seismol. Soc. Am. 67:1363 (1977)

2.3 Gravity Data

2.3.1 Gravity and Gravity Anomalies

We are turning now our attention towards the *gravity field* of the Earth.
The average value of the Earth's gravity acceleration

$$g = 9.8 \text{ m/s}^2 \tag{2.3.1-1}$$

yields a means to estimate the Earth's mass and hence its mean density. The
latter turns out to be about

$$\varrho_{\text{mean}} = 5600 \text{ kg/m}^3 . \tag{2.3.1-2}$$

It is very significant that this is much higher than the density of the rocks
in any of the accessible parts of the Earth. The above value, therefore, gives
rise to much speculation about the composition of the Earth's interior.

It turns out that the gravity value upon the Earth's surface is not constant,
but depends on the geographical location of the point under consideration.
There are two effects which a priori may be assumed to affect gravity: The
Earth's rotation, and the altitude at which gravity is being measured. The
effects of the *Earth's rotation* will be discussed more fully in Sect. 4.2.2 where
the theoretical equilibrium figure of the Earth will be given. The gradient of
the potential function W given there yields the theoretical value of gravity at
sea level as a function of the latitude φ. The formula for the "normal" gravity
value currently internationally adopted is the following[160]:

$$g = 9.7803185 \, (1 + 0.00530233 \sin^2 \varphi - 0.00000589 \sin^2 2\varphi) \text{ ms}^{-2} . \tag{2.3.1-3}$$

The second effect, that of *altitude*, can be taken care of as follows. For a
spherical mass M, the attraction (neglecting the centrifugal force) is at a
distance a from its center

$$g = \varkappa M/a^2 \tag{2.3.1-4}$$

where $\varkappa$ is the gravitational constant. Hence the change of g for a change of
altitude is

$$\partial g/\partial a = -2g/a = -0.3086 \text{ milligal/m} \tag{2.3.1-5}$$

where 1 gal equals 0.01 m/s^2.

The above formula is valid for a change in altitude if this change occurs in
free air. Hence gravity "anomalies" calculated by means of Eq. (2.3.1−5) [as
referred to the "theoretical" value obtained from Eq. (2.3.1−3)] are called
free-air anomalies.

---

160 Heiskanen, W. A., Vening Meinesz, F. A.: The Earth and its gravity field. New York:
McGraw-Hill 1958

However, in most cases, if gravity is measured on the surface of the Earth, it would appear as more logical to make a correction for the attraction of the rocks between the gravity station and the level of reference. If these rocks are approximated by an infinite horizontal slab, the induced gravity change is

$$\Delta g = 2\pi \varkappa \varrho h \qquad (2.3.1-6)$$

where ϱ is the density of the slab and h its thickness. For an average value of the density (2670 kg/m^3) one obtains

$$\Delta g/h = 0.1119 \text{ milligal/m} . \qquad (2.3.1-7)$$

Combination of the corrections (2.3.1 – 5) and (2.3.1 – 7) yields, if applied to any measured gravity value, the latter's *Bouguer reduction*. Further corrections can be applied to account for the topographical shape of the surrounding terrain[161].

The reduced gravity values, compared with the "normal" values, can be represented in a map of gravity anomalies. These anomalies are able to give some indications regarding the mass distribution in the Earth's crust.

It should be noted that, semantically, all "reductions" of gravity values depend on a model that is chosen for the ground. The "anomalies", then, which result, are nothing but an indication of the *deviation* of the true structure from the assumed one[162].

2.3.2 Distribution of Gravity Anomalies

2.3.2.1 Continental Areas

If the Bouguer anomalies are calculated for various land areas of the world, then it becomes at once obvious that they tend toward large negative values with increasing elevations. In fact, it appears that the free air reduction leads to smaller anomalies than the Bouguer reduction. One can interpret this in terms of *isostasy*. By this term is meant that the mountain ranges are supported by "roots" of low density from below, as floating masses upon a denser substratum. There are two hypotheses how this effect could be achieved. According to Pratt[163], the assumed mass deficiency would consist of an anomaly in density extending to constant depth, whereas according to Airy[164], the density of the floating matter would be assumed as constant and the latter would therefore have to extend to a depth varying with surface elevation. It is now generally the Airy hypothesis which is favored. Based upon, it, Heiskanen[165] has published tables for isostatic reduction of gravity

161 Garland, G. D.: Handbuch der Physik, Bd. 47, S. 202. Berlin-Göttingen-Heidelberg: Springer 1956
162 Ervin, C. P.: Geophysics 42:1468 (1977)
163 Pratt, J. H.: Philos. Trans. R. Soc. London 149:745 (1859)
164 Airy, G. B.: Philos. Trans. R. Soc. London 145:101 (1855)
165 Heiskanen, W.: Publ. Isostatic Inst. 2 (1938)

values which is supposed to reduce to zero all such values as correspond to proper isostatic adjustment. Gravity anomalies obtained by isostatic reduction are called *isostatic anomalies*. It must be again noted, however, that these anomalies simply indicate the deviation of the true crustal structure from that which was assumed to make the gravity reductions.

It is quite certain that the isostatic theory gives a proper interpretation of many gravity observations. There are, however, notable exceptions. These show that isostatic adjustment can at best be expected to be achieved on a broad scale; it certainly does not hold for small-scale features.

Of large-scale deviations from isostasy one should mention the Fennoscandian shield. An explanation suggested is that isostatic adjustment has not yet been achieved since the melting of the ice cover which was present during the most recent (Pleistocene) ice age.

Characteristic gravity anomalies are also found in the vicinity of continental rift systems, such as the East African Rift System. Across its axis, a Bouguer anomaly of about 50 mgal is observed which indicates that the rifts are tension cracks in the crust. However, there is in addition a long-wavelength negative anomaly which is more than 1000 km wide and extends for several thousand kilometers over the whole rift system. The gradients are only of the order of 0.3 mgal/km and indicate the existence of a low-density contrast at considerable depth, i.e., at the base of the crust or in the upper mantle[166]. A possible interpretation is obtained if a thinning of the lithosphere is assumed, the "hole" being filled by somewhat lighter material from the asthenosphere.

Typical anomaly patterns occur commonly also at the margins of continents. At *passive* (non-subducting) margins, one observes an increased isostatic anomaly on the continental side and a decreased one on the shelf side. Some examples are shown in Fig. 46. These patterns can be explained in terms of basement elevation[167].

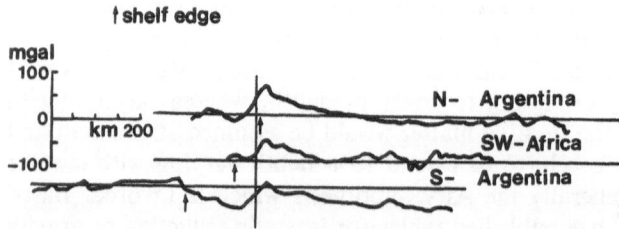

Fig. 46. Isostatic gravity anomalies on passive Argentine and South African continental margins. (After Rabinowitz and Labrecque[167])

166 Girdler, R. W.: Mem. Bur. Rech. Geol. Min. (Paris) 91:19 (1977)
167 Rabinowitz, P. D., Labrecque, J. L.: Earth Planet. Sci. Lett. 35:145 (1977)

In subduction zones and island arcs, another characteristic gravity pattern is found. A typical gravity profile across an island arc is shown in Fig. 47. It thus appears that the actual "active" (volcanic) part of an island arc is connected with positive isostatic anomalies, whereas the foredeeps show large negative ones. Tsuboi[168] has developed a method for the direct calculation of the depth of the interface between "crust" and "mantle" (corresponding to a density difference of 400 kg/m³) which, according to the Airy hypothesis, is supposed to cause the gravity anomalies. He found that this assumed interface (which may perhaps be identified with the Mohorovičić discontinuity) must dip rather sharply at the margin of continents, at an angle ϑ which is remarkably constant from place to place so that

$$\tan \vartheta = 0.1 . \tag{2.3.2-1}$$

Gravity anomalies other than those associated with typical large-scale figures on land are mostly simply the results of very local conditions. In fact, the interpretation of gravity anomalies has been much used as a local geophysical exploration tool. In this, it must be remembered, however, that the source (mass distribution) producing a given gravity pattern can never be determined *uniquely*: There is always an infinity of possible mass distributions that produce the same gravitational anomaly pattern.

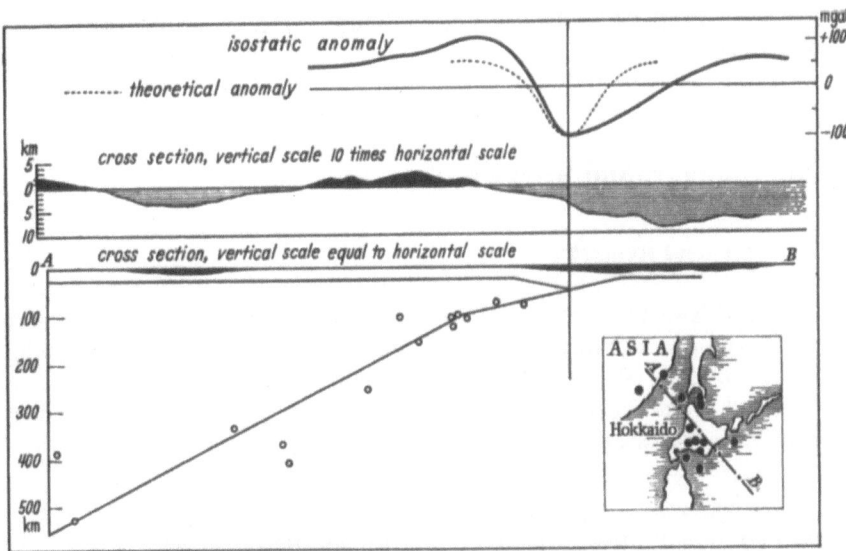

Fig. 47. Gravity anomalies and vertical distribution of the foci of deep earthquakes. (After Gutenberg and Richter[169])

168 Tsuboi, C.: Vening Meinesz Aniversary Volume (publ. by K. Ned. Geol. Mijnb. Genot.), p. 287 (1957)
169 Gutenberg, B., Richter, C. F.: Seismicity of the Earth and Associated Phenomena. Princeton: Princeton Univ. Press. 1949 (2nd edn. 1954)

2.3.2.2 Oceanic Regions

Above oceans, the Earth's gravity field seems to be in reasonable correlation with the elevation of the ocean bottom. Thus, mid-ocean ridges are consistently characterized by a gravity maximum, the trenches by a minimum[170-172]. Attempts at explaining the observed anomalies in terms of Earth models assume stiff plates overlying a weaker substratum, such that deviations from isostasy can be accounted for[173,174]. Further explanations are based on temperature effects[175].

2.3.2.3 Global Scale

Finally, it should be mentioned that large-scale gravity anomalies exist on a global scale. On such a scale, positive anomalies were found in South America, Europe, the SW Indian Ocean and the W Pacific. The corresponding negative anomalies lie in the NW and central Atlantic, in the E Indian and S Pacific oceans, as well as in the northernmost Pacific. These anomalies reach ±20 mgal and are uncorrelated with the distribution of continents and lithospheric plates[176].

Thus, the origin of the disturbing masses producing such global anomalies must be sought at great depths, e. g., in undulations of the core – mantle interface. Undulations or at least inhomogeneities in suspected boundaries in the upper mantle (such as the Birch discontinuity, cf. Sect. 2.1.4.3) have also been invoked[177,178].

2.4 Underground Stresses

2.4.1 General Remarks

The Earth's surface shows a varied topography. Hence, it cannot represent an equilibrium state; forces must be present which maintain disequilibrium. Since the material of the Earth is continuous, one cannot actually speak of "forces" in the Earth; the proper physical entities involved are force per unit area, i. e., "stresses".

170 Cochran, J. R., Talwani, M.: Geophys. J. R. Astron. Soc. 50:495 (1977)
171 Cochran, J. R., Talwani, M.: J. Geophys. Res. 83(B10):4907 (1978)
172 Cochran, J. R., Talwani, M.: Revs. Geophys. Space Phys. 17(6):1387 (1979)
173 Cochran, J. R.: J. Geophys. Res. 84:4713 (1979)
174 Watts, A. B.: J. Geophys. Res. 83(B12):5989 (1978)
175 Kahle, H. G., Werner, D.: Tectonophysics 29:487 (1975)
176 Kahle, H.: Verh. Schweiz. Naturforsch. Ges. 1975:134 (1975)
177 Lambeck, K.: J. Geophys. Res. 81(35):6333 (1976)
178 Jordan, S. K.: J. Geophys. Res. 83:1816 (1978)

Stress is physically a tensorial quantity (cf. also Sect. 3.1.2). A stress tensor has three principal axes; these are the axes in whose direction solely compressional or tensional stresses are present and shearing stresses are absent. In this, the stresses of interest are not those that are simply caused by the hydrostatic effect of the overburden, but the stresses deviating from the hydrostatic ones.

The stress tensor varies from point to point within the Earth. Therefore, the stresses form a *field* in the Earth; this is the *geophysical stress field*. This field has varied throughout geological history; hence, in discussion of the geophysical stress field, its time dependence must be taken into consideration.

Geological materials are often *porous*; in their *pores* they contain a *pore fluid* (water, petroleum, gas, etc.). As will be proved in Sect. 3.1.3, the total stress present at a point in the material can be considered in such cases as the sum of the fluid stress (pressure) and the "effective" stress which is the difference between the total and fluid stress.

As noted, the stress at any point within the Earth will primarily be determined by the weight of the overburden. The overburden pressure, in a first approximation, will simply be equal to the hydrostatic pressure caused by the superjacent layers. This pressure, equal to $\rho g h$ (where ρ is the density, g the gravity acceleration and h the depth), is approximately 23 kPa per m depth. However, in a second approximation, several corrections should be made to the hydrostatic pressure.

First, it should be noted that the Earth materials are *solid*. In an isotropic elastic solid, the lateral pressure is given by Poisson's relationship (Sect. 3.2.1); this yields, assuming that no lateral displacement can occur

$$\sigma_{\text{lateral}} = \frac{m}{1 - m} \sigma_{\text{vertical}} \tag{2.4.1 - 1}$$

where m is Poisson's number. For $m = 0.25$, one has

$$\sigma_{\text{lateral}} = \frac{1}{3} \sigma_{\text{vertical}} \tag{2.4.1 - 2}$$

(see e.g., Isaacson[179]).

Second, the vertical shear components of stress should also be taken into consideration[180]. Near the surface these are generally negligible, but at greater depths any lateral tectonic stress will also create vertical shear components. These are then automatically superposed upon the overburden pressure and cannot be neglected.

The geological stress state within the Earth's crust cannot be entirely arbitrary. It is obvious that the cannot be a pressure or a tension normal to the Earth's surface and that there cannot be a shearing force parallel to it. The

---

179 Isaacson, E.: Rock Pressure in Mines. London: Mining Publ., 260 p. 1962
180 Howard, J. H.: Bull. Geol. Soc. Am. 77:657 (1966)

latter condition implies that the normal to the surface is one of the principal directions of stress at or near the surface (cf. Sect. 7.2.2). Thus, except in strongly folded areas, one principal direction of stress is nearly vertical, the other two are horizontal. The relative magnitudes of the principal stresses characterize the possible standard geological stress states. These states are described in detail in Sect. 7.2.2.

The present-day stress field manifests itself in a variety of phenomena. First of all, there are direct effects in tunnels and mines. Secondly, such tectonic effects as earthquakes are manifestly due to the action of tectonic stresses. Thirdly, there are geological effects: Faults and joints are caused by stresses, as well as other phenomena in rock texture. Fourthly, many geomorphic features, such as the directions of valley trends are thought to be due to the action of the geotectonic stress field. Finally, certain types of petrofabrics (such as stylolites) are caused by the stresses. All of the listed phenomena are due to the acting stresses which, in turn, can be inferred and deduced from a study and interpretation of these phenomena. It is the aim of the present Section to give a review of the current state of the art of the study of the phenomena mentioned.

2.4.2 Direct In Situ Stress Determinations

2.4.2.1 Introduction

The determination of the present-day stress field in various localities is based on the effects of such stresses. It is possible to make "experiments", usually involving displacements, in which the present stress state will affect the outcome. General reviews of the methods available to measure present-day stresses underground have been given by Hast[181] and in the monograph of Obert and Duvall[182].

We shall now discuss the various possibilities in detail.

2.4.2.2 Stress Relief Methods

The most common methods for the in-situ determination of present day stresses are *stress relief* methods. In these methods the differential strains experienced by a sample of material (rock) when it is removed from its natural surroundings (where it is stressed) into an unstressed state, are measured. A knowledge of the stress – strain behavior of the sample (usually assumed as elastic) enables one to calculate the stresses from the observed strains.

---

181 Hast, N.: The measurement of rock pressure in mines. Stockholm: Sveriges Geologiska Under. 1958
182 Obert, L., Duvall, W. I.: Rock Mechnics and the Design of Structure in Rock. New York: Wiley 1967

The most common stress-relief method is the so-called doorstopper method: A two-dimensional strain gauge (doorstopper) is attached to the polished end of a borehole; overcoring is then performed and the core with strain gauge attached is withdrawn. From the strains undergone by the core upon destressing, the stresses acting before the destressing are calculated. The principle of the method was probably first used by Hast[181]. It was later improved by Leeman[183-185] in South Africa and became known as the "South African Doorstopper Method". Practical procedures, with interpretative charts, for the application of the doorstopper method have been given by Van Heerdern[186].

Other versions of the method use photoelastic (rather than piezoelectric) gauges[187]. In principle, a determination of the complete stress tensor in the vicinity of a locality requires three holes, since each hole yields only two components. The usual theory assumes the rock to be an isotropic elastic medium, the modifications necessary to allow for anisotropy have been discussed by Becker and Hooken[188,189]. In a similar vein, the modifications required owing to anelastic behavior in rock have been discussed by Barla and Wane[190] and by Skilton[191].

Since drilling is expensive, attempts have been made to determine more than two components of the stress tensor in a *single* borehole. This has been achieved by the introduction of *triaxial* stress gauges[192-197]. These gauges measure deformation not only radially to the borehole, but also longitudinally. Unfortunately, the requirements on the stability of the rock material are rather high.

As noted above, the stress-relief techniques are based on a measurement of deformations. In order to calculate the stresses, a knowledge of the elastic properties of the core material is required. The most obvious means for attaining this is by applying loads to the core in the laboratory and measuring the resulting displacements. However, it is also possible to attempt a determina-

183 Leeman, E.R.: Mine Quarry Eng. 30(6):250 (1964)
184 Leeman, E. R.: J. S. Afr. Inst. Min. Metall. 65:45 (1964)
185 Leeman, E. R.: Proc. Sixth Symp. on Rock Mech. Univ. Missouri, 407 (1964)
186 Van Heerdern, W. L.: Proc. Symp. Explor. Rock Eng. Johannesburg 189 (1976)
187 Hawkes, I., Moxon, S.: Int. J. Rock Mech. Min. Sci. 2(4):405 (1965)
188 Becker, R. M., Hooker, V. E.: U. S. Bur. Mines Rept. Inv. 6965, 23 p. (1967)
189 Becker, R. M., Hooker, V. E.: Trans Soc. Min. Eng. AIME 244(4):436 (1969)
190 Barla, G., Wane, M. T.: Int. J. Rock Mech. Min. Sci. 5(2):187 (1968)
191 Skilton, D.: Int. J. Rock Mech. Min. Sci. 8(4):283 (1971)
192 Leeman, E. R., Hayes, D. J.: Proc. Ist. Int. Soc. Rock Mech. Congr. Lisbon, 2:17 (1966)
193 Crouch, S. L., Fairhurst, C. A.: Int. J. Rock Mech. Min. Sci. 4(2):209 (1967)
194 Merrill, R. H.: Three-component borehole deformation gage for determining the stress in rock. U. S. Bur. Mines Rept. Inv. 7015, 38 p. (1967)
195 Oka, Y.: Int. J. Rock Mech. Min. Sci. 7:503 (1970)
196 Leeman, E. R.: Rock Mech. 3(1):25 (1971)
197 Herget, G.: Int. J. Rock Mech. Min. Sci. 10:509 (1973)

tion of the elastic properties of the rock material in situ by applying loads and measuring displacements[198,199].

2.4.2.3 Restoration Methods

Methods opposite to those described in the last section are determinations of the geophysical stresses by the restoration of the virgin state after the latter has been disturbed. The most common procedure of this type involves making a slot in a rock wall which will cause a displacement of the material forming the edges of the slot. Then a flatjack is inserted into the slot and the force is measured which has to be applied to the jack to restore the original state, i. e., to reverse the displacements caused by making the slot. The method seems to have been developed primarily in France[200,201]. It was, however, widely applied elsewhere as well[202-207].

Rather than making a slot, attempts have been made at using simply a borehole which has a simpler geometry, to restore the original virgin state. This led to the development of borehole deformation strain cells and corresponding loading devices[208-215].

2.4.2.4 Wave Fields and Stresses

The wave field of a seismic wave is affected by the prevailing stresses. In a small measure, the rock density and the character of minute cracks are

198 Rotter, D., Stoll R., Thon, H.: Bergakademie 1967(10):575 (1967)
199 Lötgers, G., Voort, H.: Rock Mech. 6:65 (1974)
200 Habib, P., Marchand, R.: Ann. Inst. Tech. Bat. Trav. Publ. Ser. Sols Fond. 58:966 (1952)
201 Tincelin, M. E.: Ann. Inst. Tech. Bat. Trav. Publ. Ser. Sols Fond. 58:972 (1952)
202 Panek, L. A., Stock, J. A.: Development of a rock stress monitoring station based on the flat slot method of measurement. U. S. Bur. Min. Rep. Inv. 6537 (1964)
203 Hoskins, E. R.: Int. Rock Mech. Min. Sci. 3(4):249 (1966)
204 Rocha, M.: New techniques for the determination of the deformability and state of stress in rock masses. Lab. Nac. Eng. Civil Mem. 328:1 (1969)
205 Rocha, M., Baptista Lobes, J., Neves da Silva, J.: Proc. 1st. Int. Soc. Rock Mech. Congr. Lisbon 2:57 (1966)
206 Rocha, M., Baptista Lopes, J., Neves da Silva, J.: A new technique for applying the method of the flat jack in the determination of stress inside rock masses. Lab. Nac. Eng. Civil Mem. 324:1 (1969)
207 Merrill, R. H., Williamson, J. V., Ropchan, D. M., Kruse, G. H.: Stress determination by flatjack and borehole-deformation methods. U. S. Bur. Mines Rep. Inv. 6400 (1964)
208 Panek, L. A.: Trans. Am. Inst. Min. Metall. Eng. 220:287 (1961)
209 Merrill, R. H., Peterson, J. R.: Deformation of a borehole in rock. U. S. Bur. Mines Rep. Inv. 5881 (1961)
210 Obert, L., Merrill, R. H., Morgan, T. A.: Borehole deformation gage for determining the stress in mine rock. U. S. Bur. Mines Rep. Inv. 5978 (1962)
211 Suzuki, K.: Proc. 1st. Int. Soc. Rock Mech. Congr. Lisbon 2:35 (1966)
212 Wöhlbier, H., Natau, O.: Proc. 1st Int. Soc. Rock Mech. Congr. Lisbon 2:25 (1966)
213 Dutta, S. K., Singh, B.: Metals Minerals Rev. 7(7):17 (1968)
214 Hiramatsu, Y., Oka, Y.: Int. J. Rock Mech. Min. Sci. 5(4):337 (1968)
215 Stephenson, B. R., Murray, K. J.: Int. J. Rock Mech. Min. Sci. 7(1):1 (1970)

affected by the stress and therewith also the wave velocity. In addition, the rupture process around an explosion is also affected by the preexisting stresses so that the radiation pattern of that explosion can be used to determine the stresses.

The effect of stresses on seismic wave propagation near a well has been studied by Plokhotnikov and Dzeban[216] who showed that the normal stress in the direction of wave propagation is of prime inportance. However, the effect is small. An experimental study involving the continuous monitoring of the seismic velocity in a granite quarry by Reasenberg and Aki[217] showed that the tidal variations of the stresses can be inferred. However, velocity measurements have to be made with a precision better than $1:10^5$. The cited authors attributed the velocity variations to the opening and closing of minute cracks. These authors also gave a historical review of attempts to monitor stresses by monitoring seismic velocity.

Finally, Archambeau and Sammis[218] investigated the radiation field from an explosion in a prestressed medium and its possible usage for determining the geophysical stress field.

As a conclusion to this section on wave fields, it should be mentioned that not only seismic waves, but also electromagnetic waves have been used for attempting the determination of the stress field. Indeed, the pressure dependence of conductivity and dielectric constant ought to have the result that electromagnetic waves are affected by the prevailing stresses. In an investigation of this possibility, Kaspar and Dokoupil[219] claim that it is realizable.

2.4.2.5 Results

There are now a number of in-situ stress measurements available so that one can attempt to construct a global picture. The available results have been collected by Ranalli and Chandler[220]. When inspecting such a collection of data, it is seen that there is a great uncertainty and scatter with regard to the absolute stress values. These seem to be influenced not only by the amount of overburden, but also by the local topography. However, one quantity which does show some consistency is the azimuth of the maximum principal compression.

Particularly accurate studies have been made in Europe. Thus, a catalogue of measurements from central Europe has been published by Illies and Greiner[221,222]. Further measurements have been reported from France[223],

216 Plokhotnikov, A. N., Dzeban, I. P.: Bull. (Izv.) Acad. Sci. USSR Earth Phys. 1974(1):101 (1974)
217 Reasenberg, P., Aki, K.: J. Geophys. Res. 79:399 (1974)
218 Archambeau, C., Sammis, C.: Rev. Geophys. Space Phys. 8(3):473 (1970)
219 Kaspar, M., Dokoupil, S.: Proc. 1st Int. Soc. Rock Mech. Congr. 1st, Lisbon, 1:31 (1966)
220 Ranalli, G., Chandler, T. E.: Geol. Rundsch. 64:643 (1975)
221 Illies, J. H., Greiner, G.: Bull. Geol. Soc. Am. 89:770 (1978)
222 Greiner, G., Lohr, J.: Rock Mech. Suppl. 9:5 (1980)
223 Paquin, C., Froidevaux, C., Souriau, M.: Bull. Soc. Geol. Fr. 20(5):135 (1978)

Italy[224,225] and Austria[226-228]. If one confines himself to the azimuths of the
maximum compressions listed in the above references, it turns out that the
mean in Central Europe (excepting Italy) is N 145° E[229]. Figure 48 shows a
diagram of the regional values[229].

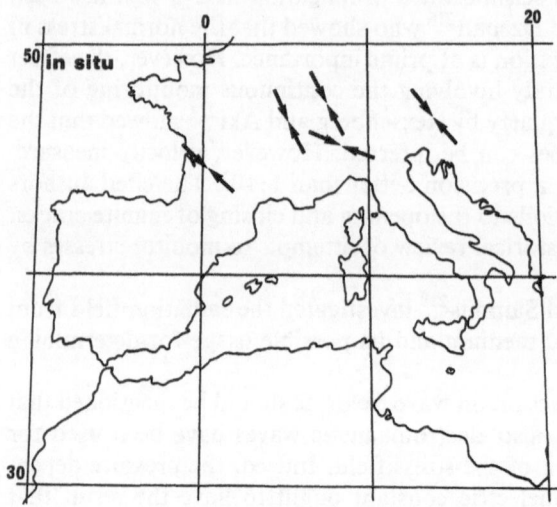

Fig. 48. Maximum principal
pressure directions in regions
of West-Central Europe.
(After Scheidegger[229])

A very careful study has also been made for Iceland[230]. It was again found
that the maximum compression had the direction NW – SE which is particularly surprising as this is *normal* and across the generally observed rifting
in Iceland.

In regions of the world[231] other than Europe, the results of direct in-situ
stress measurements are mainly characterized by a ("non"-)pattern of extreme
diversity. It is, therefore, generally not possible to arrive at even regional
pictures of the course of stress trajectories.

However, it has been possible to establish some general rules regarding the
variations of the stresses with depth. Thus, Herget[232] analyzed the variation of
the principal stresses with depth in a Canadian mine. The smallest compres-

224 Martinetti, S., Ribacchi, R.: Proc. 3d Int. Congr. Rock Mech. Denver 2A:458 (1974)
225 Martinetti, S., Ribacchi, R.: Rock Mech. Suppl. 9:31 (1980)
226 Brückl, E., Roch, K. H., Scheidegger, A. E.: Tectonophysics 29:315 (1975)
227 Carniel, P., Roch, K. H.: Riv. Ital. Geofis. Sci. Aff. 3:233 (1976)
228 Kohlbeck, F., Roch, K. H., Scheidegger, A. E.: Rock Mech. Suppl. 9:21 (1980)
229 Scheidegger, A. E.: Geophys. Surv. (Reidel) 4:233 (1981)
230 Schäfer, K., Keil, S.: Messtech. Briefe (Hottinger-Baldwin) 15(2):35 (1979)
231 A good catalog exists for North America: Lindner, E. N., Halpern, J. A.: Int. J. Rock
    Mech. Min. Sci. 15:183 (1978). For other regions one has to dig out the individual study
    reports if the work is newer than what had been collected by Ranalli and Chandler, loc. cit.
    1974
232 Herget, G.: Int. J. Rock Mech. Min. Sci. 10:37 (1973)

sion was always vertical (σ_v), the largest horizontal (σ_h). With his data, as well as those from other sources, he arrived at the following regression equations

$$\sigma_v = (1.9 \pm 1.26) + (0.027 \pm 0.003)\, H \,\text{MPa}$$

$$\bar{\sigma}_h = \frac{\sigma_{h1} + \sigma_{h2}}{2} = (8.3 \pm 0.5) + (0.041 \pm 0.002)\, H \,\text{MPa}$$

where H is depth in meters. The form of the regression equation, evidently, depends on the locality. A comparison of various regions has been made by Brown and Hoek[233].

2.4.3 Stresses from Hydraulic Fracturing

2.4.3.1 The Method

Inasmuch as the regional stresses affect the character of the fractures that occur if a well is pressured up (see Sect. 7.4.3), the observation of the inducement of fractures, in turn, can be used for a determination of the regional stresses. This method of stress determination was first suggested by Scheidegger[234,235] who based his analysis on pressure and flow records from routine oil well fracturing operations. As model of a "well", a cylindrical pressure center was used corresponding to the simple theory of Hubbert and Willis referred to in Sect. 7.4.3.3. Later, other types of pressure centers (spherical, etc.) were tried. Various theoretical investigations of the possible models were made by Morgenstern[236] and Kehle[237]. Further interpretations gained from oil field operations were reported by Pulpan and Scheidegger[238].

The above results were often not satisfactory because the records from routine oil well fracturing operations (bottom hole pressure measurements) were often inaccurate. In addition, it is not immediately possible to ascertain the orientation of the fractures at the well bottom which is all-important for the determination of the orientation of the principal stress directions. Special types of packers have to be lowered into the wells which take an impression of the well periphery so that the orientation of the fractures can be determined. From the latter, following the theory outlined in Sect. 7.4.3, the principal stress directions can be found: the largest compression is parallel to the fracture surface. Thus, hydraulic fracturing operations have to be especially designed with the express purpose of determining the stress field. The

233 Brown, E. T., Hoek, E.: Int. J. Rock Mech. Min. Sci. 15:211 (1978)
234 Scheidegger, A. E.: Pure Appl. Geophys. 46:66 (1960)
235 Scheidegger, A. E.: Geologie Bauwes. 27:45 (1962)
236 Morgenstern, N.: Pure Appl. Geophys. 52:104 (1962)
237 Kehle, R. O.: J. Geophys. Res. 69:259 (1964)
238 Pulpan, H., Scheidegger, A. E.: J. Inst. Petrol. 51:169 (1965)

technique was developed particularly by Haimson and his coworkers[239–242]; later it was followed up by many groups, especially in North America.

2.4.3.2 Results

The method of stress determination by means of specially designed hydraulic fracturing operations has almost exclusively been applied in North America. Global results can, therefore, not be presented, and we must confine ourselves to some remarks on specific regions. As with other in-situ stress results, it is essentially the *direction* of the maximum principal (horizontal) stress which is best defined. A compilation of the available results has been made by Haimson[243]; the result is shown in Fig. 49. It is seen that there is basically a constant NE – SW direction for the maximum compression throughout the continental U.S.A.

Fig. 49. Maximum horizontal principal stress directions in the continental United States based on the orientation of vertical hydro-fractures. (After Haimson[243])

Another set of data refer to well breakouts in the Province of Alberta in Canada[244]. These occur in the direction of the smallest compressive stress as this is the site of the largest shearing stress. The maximum horizontal compression found is much in conformity with that encountered to the South in the U.S.A.

Finally, it should be remarked that Haimson and Voight[245] also measured stresses by hydrafacturing two wells in Iceland. The results were entirely in

239 Haimson, B. C.: Hydraulic fracturing in porous and nonporous rock and its potential for determining in-situ stresses at great depth. Technical Report 4-68, Missouri River Division Corps of Eng., also Ph. D. thesis, Univ. Minnesota (1968)
240 Haimson, B. C.: A simple method for estimating in situ stresses at great depth, field testing and instrumentation of rock, STP 554, Am. Soc. for Testing of Materials, Philadelphia, pp. 156 – 182 (1974)
241 Haimson, B., Fairhurst, C.: Proc. 2nd. Symp. Rock Mech. Berkeley: 559 (1969)
242 Haimson, B. C., LaComb, J., Green, S. J., Jones, A. H.: Proc. 3rd Int. Congr. Rock Mech. Denver 2:557 – 562 (1974)
243 Haimson, B. C.: Int. J. Rock Mech. Min. Sci. 15:167 (1978)
244 Bell, J. S., Gough, D. I.: Earth Planet. Sci. Lett. 45(2):475 (1979)
245 Haimson, B. C., Voight, B.: Pure Appl. Geophys. 115:153 (1977)

conformity with the in-situ measurements mentioned in Sect. 2.4.2: The maximum compression is NW – SE, corresponding to the "European" orientation.

2.4.4 Stress Determination from Seismic Effects

2.4.4.1 Introduction

The analysis of seismic events yields a means to obtain an idea of the present-day stress field which exists somewhat below the surface of the Earth. Two types of investigations are of importance: The triggering of earthquakes and fault-plane solutions.

2.4.4.2 Triggering of Seismic Events

The number of seismic events taking place in a particular area may be an indication of the stresses that are present there. Upon this basis, attempts have been made, for instance, to predict the likelihood of a major rock burst occurring in a mine. Unfortunately, the number of microseismic events is indicative of the rate at which strains are being released rather than of the rate at which stresses are being built up. Nevertheless, for a material of known rheological properties, the two rates may be proportional to each other.

With regard to rock bursts in mines, the possibility of their prediction by monitoring microseismic events has been mentioned in many textbooks on mining engineering. A specific study of the problem has been made by Kunsdorf and Rotter[246].

Earthquakes may also be artifically released by the impounding of large water reservoirs. In many instances, it has been found that the filling-up of such reservoirs has caused a series of earthquakes[246a]. The manner in which this occurs may also be indicative of the preexisting stresses in the area. However, whilst there is a fair amount of literature in which the problem of artificial triggering of earthquakes is treated[247], there do not seem to be any instances where the occurrence of such artificial earthquakes would have been used to determine the tectonic stress field; – although this should be possible.

2.4.4.3 Stresses from Fault-plane Solutions of Earthquakes

The main procedure for determining present-day stresses from seismic events is by means of an analysis of fault-plane solutions of earthquakes. In principle, as was illustrated in Fig. 47, every fault-plane solution of an earthquake yields the direction of the maximum (P) and minimum (T) com-

246 Kunsdorf, W., Rotter, D.: Freiberg Forschungsh. C120:1 (1961)
246a Gupta, H. K., Rastogi, B. K.: Dams and Earthquakes. Amsterdam: Elsevier (1976)
247 E. g., Milne, W. G.: Induced Seismicity. Spec. Issue Eng. Geol. 10(2 – 4):81 (1976)

pression. However, for an evaluation of the regional stress field, many such solutions pertaining to an area have to be "averaged". The problem, thus, is to average a series of *axes* which requires some sophisticated procedures. Fara and Scheidegger[248] have shown that the problem, given some statistical assumptions, can be solved by an eigenvalue calculation of a certain matrix, best carried out on a computer.

Thus, regional principal stress directions can be determined by averaging principal stress directions obtained from individual earthquakes. Studies of individual areas are legion. Thus, for instance, the best fitting regional P direction in Switzerland has an azimuth of N 142° E[249] which is very close to the value found for Central Europe from in-situ stress measurements (N 145° E). There is, of course, the intrinsic possibility that the stresses at the depths of earthquake foci differ from those close to the surface, but this seems not to be the case. One must say, thus, that there is an excellent correspondence between the stresses determined in Europe by seismic and by in-situ methods.

Investigating earthquake fault-plane solutions, one was able to observe a global pattern very early. Thus, Scheidegger[250] determined the stress patterns in the main tectonic regions (the Circum–Pacific belt, the Alpine– Himalayan belt and the Mid-Atlantic rift); the global pattern found was later confirmed by Isacks et al.[251] and by Balakina et al.[252], using more complete

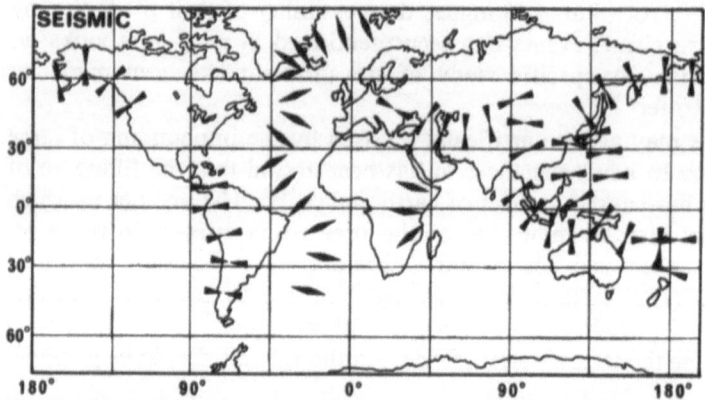

Fig. 50. The global seismotectonic stress field. (After Scheidegger[252a] as drawn from data given by Balakina et al.[252])

248 Fara, H. D., Scheidegger, A. E.: Bull. Seismol. Soc. Am. 53:811 (1963)
249 Scheidegger, A. E.: Ann. Geofis. (Roma) 30:259 (1977)
250 Scheidegger, A. E.: Z. Geophys. 31(6):300 (1965)
251 Isacks, B., Oliver, J., Sykes, L. R.: J. Geophys. Res. 73:5855 (1968)
252 Balakina, L. M., Misharina, L. A., Shirokova, E. I., Vvedenskaya, A. V.: Publ. Dom. Obs. Ottawa 37(7):194 (1969)
252a Scheidegger, A. E.: Rock Mech. Suppl. 6:55 (1978)

data. It was found that subducting plate boundaries show generally a compressive, rifting plate margins generally a tensional stress pattern in their vicinity (Fig. 50).

2.4.5 Stress Determination from Faults and Joints

2.4.5.1 Introduction

The geological features termed faults and joints must have been produced by the action of stresses. Hence, it must be possible to infer the field of acting stresses in turn from an analysis of the phenomenology of faults and joints. If is the aim of the present section to demonstrate how to do this.

2.4.5.2 Faults

The relation between fault planes and principal stress directions, according to Anderson's theory, is discussed in Sect. 7.2.2.2. A discussion of the numerical relationships (depending on the fracture angle δ) has been given by Thrasher and Scheidegger[253]. Hence, a mapping of a series of faults in an area can be immediately interpreted in terms of the stresses that produced them. In case of *old* faults, this is, naturally, not the present-day stress system, but the stresses that acted at the time the fault was created. In an older formation, the dating of the faulting event may pose some problem. Upon the above basis, paleo-stresses were determined by Kakimi et al.[254] in Japan, by Bullock and Jones[255] in Labrador, and by Chappel[256] in New Guinea.

Stresses can also be reconstructed from an analysis of a single large fault or a population of faults. The theory has been developed by Angelier[257]; it is based on the same idea as the determination of P and T directions in fault-plane solutions of earthquakes (the normal to the "slickensides" corresponds to the "auxiliary" plane!). Thus, Eisbacher[258] investigated the stress field along a fault in Nova Scotia. Similarly, the investigation of swarms of dikes leads to a determination of at least the minimum principal direction[258-260].

2.4.5.3 Joints

More convenient for the determination of the tectonic stress field than faults are the ubiquitous joints in rock outcrops. In Sect. 1.6.3 we have adduced

253 Thrasher, J. E., Scheidegger, A. E.: Z. Geophys. 29:309 (1963)
254 Kakimi, T., Hirayama, J., Kageyama, K.: J. Geol. Soc. Jpn. 72(10):1 (1966)
255 Bullock, W. D., Jones, D.: Can. Min. J. 89(7):43 (1968)
256 Chappell, J.: J. Geol. 81:705 (1973)
257 Angelier, J.: Tectonophysics 56:T17 (1979)
258 Eisbacher, G. H.: Can. J. Earth Sci. 6(5):1095 (1969)
259 Gilliland, W. N.: Geol. Mag. 100(5):425 (1963)
260 Harms, J. C.: Bull. Geol. Soc. Am. 76:981 (1965)

some phenomenological facts about joints. In particular, we have noted that joints occur generally in three sets at an outcrop; two thereof are near-vertical, one near-horizontal.

The preferred orientations of the joint sets can be ascertained by a statistical computation procedure; they form usually a more or less rectangular system of three planes. The vertical joints have been termed tectonic joints. Of great importance is the fact that they seem to be of *very recent* origin: Evidently older, filled, "inactive" joints are cut through at acute angles, pebbles are sheared off, and there is absolutely *no* relation between the orientation of joints in fresh outcrops and features related to the genesis of the rocks[261] (such as small folds).

It has generally been assumed that the tectonic joints (as defined above) are Mohr-type fracture surfaces (cf. Sect. 7.2.2), but the commonly observed large angles between conjugate joint sets may, in fact, indicate that the fractures are not of the Mohr type at all. The joint sets align themselves very closely with the planes of maximum shear in a stress field which may indicate that they are the result of some ductile or plastic slippage process. Thus, an obvious phenomenological explanation of joints would be that they are the response to an instantaneous creep process induced in the horizontal plane by the momentarily acting tectonic stresses.

On the basis of the above remarks, it is possible to deduce the orientation of the principal tectonic stress directions from a measurement of joint orientations. From the position of the two density maxima, the principal stress directions can then be calculated as the bisectrices of the two "preferred" joint planes. In the case of Mohr-type fractures, the bisectrix of the *smaller* angle should be the greatest compression. However, inasmuch as, as noted, the angle between steeply dipping conjugate joint sets is usually close to 90°, it is often not possible to distinguish reliably between the largest and smallest principal stress direction. For the sake of visualization, it is often useful to draw the principal stress directions into pole density diagrams of a joint set (Fig. 16, Sect. 1.6.3). However, for routine evaluations, the entire procedure is naturally fully computerized.

Results of the interpretation of the orientation of joints in terms of the neotectonic stress field are now available from many parts of the world. In general, a good correspondence has been observed between the maximum compression calculated from joints and that postulated from plate tectonic theory (e. g., in the Himalaya[262], the Caribbean[263], Eastern Canada[264], and México[265]).

---

261 Scheidegger, A. E.: Rock Mech. 8:23 (1976)
262 Scheidegger, A. E.: Arch. Meteorol. Geophys. Bioklimatol. Ser. A 28:89 (1979)
263 Scheidegger, A. E.: Geofís. Int. (México) 18:219 (1979)
264 Scheidegger, A. E.: Arch. Meteorol. Geophys. Bioklimatol. Ser. A 27:375 (1978)
265 Scheidegger, A. E.: Geofís. Int. (México) 18:329 (1979)

As an illustration, the results from Europe are shown here in somewhat more detail[266] (Fig. 51). In this figure, it should be kept in mind that the identification of P and T is not certain. From an inspection of Fig. 51 it becomes obvious that there is evidently a large-scale European neotectonic stress system present which is homogeneous from the Alps north to Norway. It agrees closely with the stresses determined from in-situ (Fig. 48), seismological (Fig. 50), and geodetic (Fig. 27) investigations. Significant changes occur only in the Dinarides to the East.

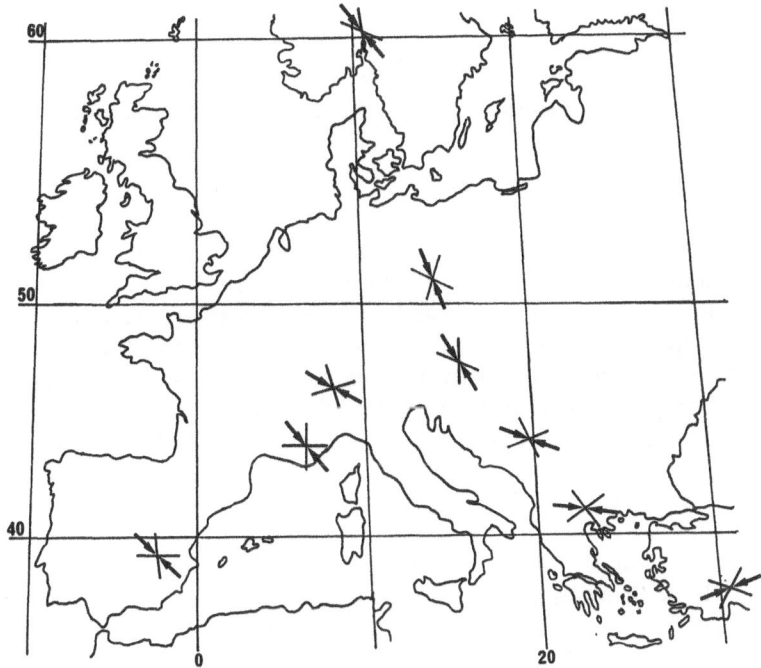

Fig. 51. Principal stress directions determined from joints in Europe. (After Scheidegger[266])

2.4.6 Geomorphology and Stresses

The interaction of the stresses with the Earth's *surface* leaves traces that can be used, in turn, for a reconstruction of the stresses that produced them.

A study of this possibility has been made by Gerber and Scheidegger[267,268], particularly in connection with the geomorphological appearance of mountain massifs. However, other geomorphic features, such as the direction of valleys and the course of shorelines, also might be usable.

266 Scheidegger, A. E.: Rock Mech. Suppl. 9:109 (1980)
267 Gerber, E., Scheidegger, A. E.: Riv. Ital. Geofis. 2(1):47 (1975)
268 Gerber, E.: Rock Mech. Suppl. 9:93 (1980)

Thus, we have already shown in Sect. 1.6.4, that the orientation structure of valley trends is not random. Often, one finds two preferred orientations in an area. As is the case with joints, the bisectrices of them can be taken as the principal stress directions under the assumption that the valleys are lines of maximum shear. If the valleys are supposed to be Mohr-type fractures, the greatest pressure would be contained in the smaller quadrant. However, it is generally uncertain which is the latter because the angle between them is usually close to 90° and the identification as to which of the principal stress directions is the greatest and which is the smallest pressure, is, thus, not assured statistically.

The procedure with valleys is therefore the same as that with joints. One obtains as predicted stress directions the bisectrices of the preferred valley trends. Referring to the example of Austria in Sect. 1.6.4, one obtains

σ_1: N 42° E

σ_2: N 132° E .

This is very close indeed to the "European" principal stress directions found from other sources in the NW – SE quadrants.

In other areas the correspondence between valley trends and other neotectonic stress features is not always so good as in Austria. Sometimes, the valley trends have more than two maxima so that the possibility exists that stress systems of different ages are expressed therein.

A peculiar type of geomorphic features that have been invoked as possible indicators of neotectonic stresses is that of volcanoes[269]. The underlying concept is that magmatic dikes tend to propagate in a direction normal to the minimum regional stress and thus indicate the strike of the maximum regional compression. The latter direction, obtained from the surface study (particularly flank eruptions) of Alaskan volcanoes coincides well with the direction of convergence of the Pacific and American tectonic plates postulated from other evidence[269].

2.4.7 Petrofabrics and Tectonic Stresses

Petrofabric analyses have been generally successful for attempts at reconstructing the stress field that produced them. Inasmuch as the rock texture records the effects of *all* past tectonic events, the attempt at unraveling a deformation history becomes a formidable venture.

Thus, it has been claimed that the residual stress, which is stress at a point in a body subject to zero external tractions, is an indicator of ancient stress fields in the Earth's crust[270,271]. Analyses of this "frozen-in" stress are com-

269 Nakamura, K., Jacob, K. H., Davies, J. D.: Pure Appl. Geophys. 115:87 (1977)
270 Friedman, M.: Tectonophysics 15:297 (1972)
271 Varnes, D. J., Lee, F. T.: Bull. Geol. Soc. Am. 83:2863 (1972)

monly based on x-ray diffraction techniques. However, it has been noted that
one cannot generally infer the directions of the principal tectonic stresses that
might have been responsible for the residual stresses in a body without
knowing the sources of such stresses beforehand[272].

Nevertheless, supposing that one starts from an isotropically oriented
assemblage of objects, it is possible to infer the strain (and therewith at least
the principal stress directions) that has been active in rendering these objects
anisotropic. Some of the theoretical considerations involved in doing this will
be discussed in Sect. 7.2.4. Applications of these ideas to a determination of
paleostrain have been reported to Vermont[273] and to Pembrokeshire[274]. As
noted, the main difficulty is to date the time when the deformations occurred.

The most successful application of petrofabric analyses to investigations
of the tectonic stress field has been by the study of stylolites (cf. Sect. 1.6.6).

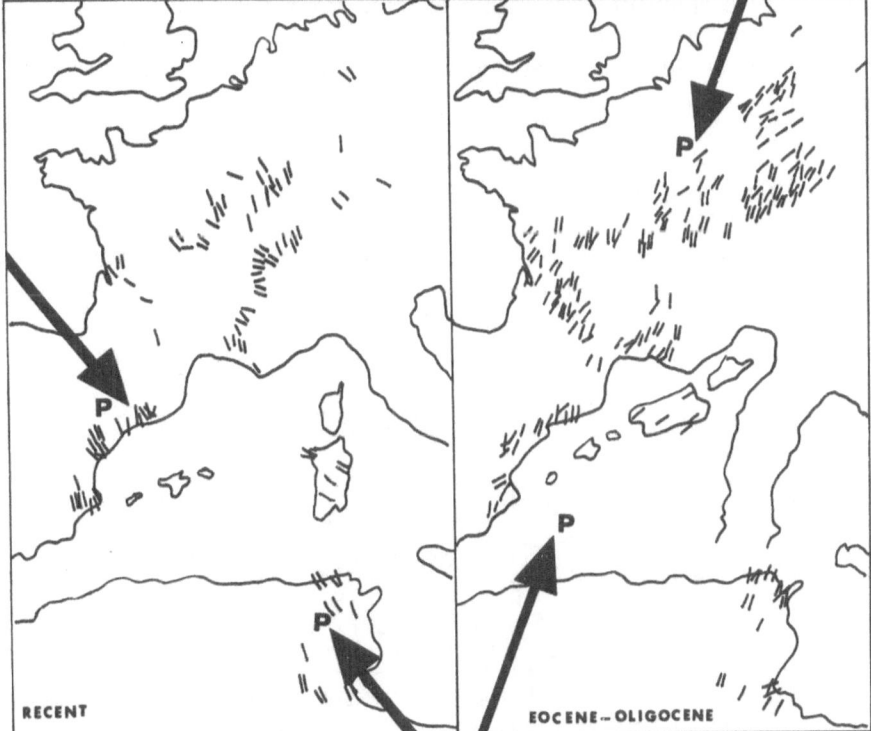

Fig. 52. Tectonic stresses (*P*) from stylolites. (After Schäfer[275])

272 Holzhausen, G. R., Johnson, A. M.: Tectonophysics 58:237 (1979)
273 Detenbeck, J., Stanley, R. S.: Bull. Geol. Soc. Am. II 90:1377 (1979)
274 Hanna, S. S., Fry, N.: J. Struct. Geol. 1(2):155 (1979)
275 Schäfer, K. H.: Fridericiana (Karlsruhe) 23:30 (1978)

Inasmuch as these solution-precipitation features grow in the direction of the maximum compression present during their genesis, a means is found to determine this maximum compression direction directly in the field.

Thus, a study has been reported by Schäfer[275a] of stylolites in Europe and North Africa. Figure 52 (after Schäfer) shows the principal compression directions obtained from stylolites in the western part of the Alpine-Mediterranean region, referring to the present-day stress field. It may be interesting to note that, inasmuch as stylolites preserve the orientation in which they were formed, an indication of the Paleogene stress field can also be obtained. The corresponding picture, again after Schäfer, is also shown in Fig. 52. It is seen, thus, that a certain reorientation of the principal tectonic stress directions occurred in the Miocene. It is remarkable to note how well the neotectonic stress direction fits with that (NW − SE) deduced from the other sources discussed in this chapter. The method, in the Alpine-Mediterranean area, was applied to a study of even earlier tectonic stress fields by Letouzey and Trémolières[276] who were able to determine the evolution of the plate-tectonic mechanism in the Mediterranean area from the Late Cretaceous to the present.

2.4.8 Global Results

The foregoing discussion should have made it clear that the present-day stress field is a single geophysical field whose comportment can be inferred from the various sources mentioned. Apparently, it is constant over the whole depth of the lithospheric plates and over large "intraplate" regions. It is now possible

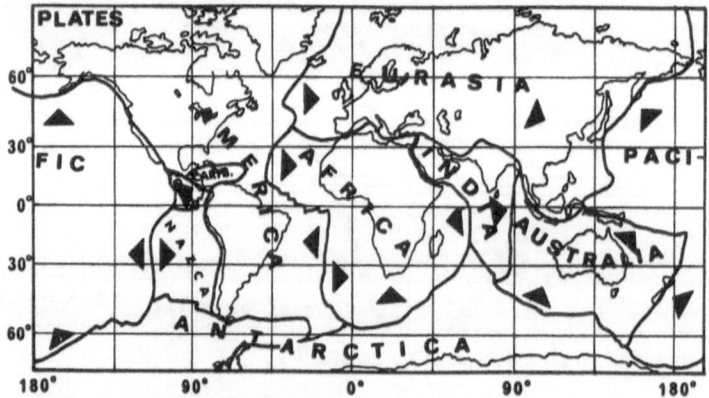

Fig. 53. Direction of intraplate stresses

275a Schäfer, K. H.: Fridericiana (Karlsruhe) 23:30 (1978)
276 Letouzey, J., Trémolières, P.: Rock Mech. Suppl. 9:173 (1980)

to collect the available data regarding this field from many sources[277,278]; the result is as shown in Fig. 53.

2.5 Data from Age Determinations

2.5.1 General Remarks

The geologic time scale discussed in Sect. 1.2.2 begins with the Cambrian, owing to the absence of fossils in earlier strata. Originally, the relative ages of the various geological formations were established by geologists solely upon the fossil record in accordance with the supposed evolution of life forms. Naturally, no absolute ages could be assigned in this way; for doing this, the advent of radioactive dating methods had to be awaited.

Although the usual geologic time scale begins with the Cambrian, it is inconceivable that the Earth would not have existed for a very long time before the advent of animals that could be fossilized. It is also most remarkable that Cambrian strata already contain remants of fishes which belong to the highest phylon of animal life. It stands to reason, therefore, that not only in the geological evolution, but also in the evolution of animal species a very long time passed before the Cambrian epoch.

It is thus fortunate that the physics of radioactive decay has enabled one to extend the geologic time scale backward to almost 4 billion years ago.

2.5.2 Methods

The methods of physical age determination are based on the radioactive decay of atoms. Radioactive decay of fissionable material occurs spontaneously. The number of atoms decaying in any one (short) time interval is thus proportional to the total number of atoms present and to a decay constant λ. Thus the decay of, e. g., rubidium can be written as follows:

$$-\frac{d}{dt} \text{Rb}^{87} = \lambda \text{Rb}^{87} \qquad (2.5.2-1)$$

where Rb^{87} indicates the number of atoms of rubidium-87 present in any quantity under consideration. Integrating, one obtains

$$\text{Rb}^{87} = \text{Rb}_0^{87} e^{-\lambda t}. \qquad (2.5.2-2)$$

Rubidium-87 decays into strontium-87; the number of radiogenic strontium-

277 Richardson, R. M., Solomon, S. C., Sleep, N. H.: Rev. Geophys. Space Phys. 17(5):981 (1979)
278 Bergman, E. A., Solomon, S. C.: J. Geophys. Res. 85:5389 (1980)

87 atoms contained in, say, a rubidium crystal must be equal to the number of rubidium atoms that have decayed (barring any losses) since the formation of that crystal. Thus

$$Sr^{87} = Rb_0^{87} - Rb^{87} = Rb_0^{87}(1 - e^{-\lambda t}) \qquad\qquad (2.5.2-3)$$

or

$$Sr^{87} = Rb^{87}(e^{\lambda t} - 1). \qquad\qquad (2.5.2-4)$$

Thus, by measuring the ratio of Sr^{87}/Rb^{87} it is possible to determine the time of formation of the crystal. This is the basis of radioactive age determinations. The actual technique of carrying out the measurements is rather involved; the required isotope ratios are obtained by means of mass spectrometry. The materials, therefore, have to be changed chemically into such a state that they can be introduced into a mass spectrometer which, in itself, is not a simple piece of equipment.

The strontium decay scheme mentioned above is only one of several nuclear decay reactions occurring in nature which are suitable for radioactive age determinations. For a naturally decaying material to be suitable for such age determinations it is necessary, firstly, that its half-life be comparable with the age of the Earth, and secondly, that the material be reasonably abundant in rocks. The decay schemes which are thus usable for the present purpose are shown in Table 4. A detailed discussion of the techniques involved in using these schemes may be found in a book by York and Farquhar[279] and by Faure[280], and in an article of O'Nions et al.[281].

Table 4. Radioactive decay schemes usable for age determinations

Isotope	End product	Half-life
K^{40}	Ca^{40} and A^{40}	1.3×10^9 years
Rb^{87}	Sr^{87}	6.1×10^{10} years
Re^{187}	Os^{187}	4.5×10^{10} years
Sm^{147}	Nd^{143}	1.1×10^{11} years
Th^{232}	Pb^{208}	1.4×10^{10} years
U^{235}	Pb^{207}	7.1×10^8 years
U^{238}	Pb^{206}	4.5×10^9 years

A somewhat different method of dating rocks is based on the idea of isochrons. These are curves of particular ratios of isotopes obtained in lead-containing minerals. As noted in Table 4, Pb^{207} is the daugther of U^{235} and Pb^{206} the daugther of U^{238}, with corresponding decay constants λ_{235} and λ_{238},

279 York, D., Farquhar, R. M.: The Earth's Age and Geochronology. London: Pergamon, 1972
280 Faure, G.: Principles of Isotope Geology. New York: Wiley, 464 pp., 1977
281 O'Nions, R. K., Hamilton, P. J., Evensen, N. M.: Sci. Am. 242(5):120 (1980)

respectively. One has at the time t after the beginning of the decay (atom symbols denoting *numbers* of atoms, as above)

$$U_t^{235} = U_0^{235} e^{-\lambda_{235}t} = U_0^{235} - Pb_t^{207} \tag{2.5.2-5}$$

and

$$U_t^{238} = U_0^{238} e^{-\lambda_{238}t} = U_0^{238} - Pb_t^{206} \tag{2.5.2-6}$$

or

$$\frac{Pb_t^{207}}{Pb_t^{206}} = \frac{U_0^{235}(1 - e^{-\lambda_{235}t})}{U_0^{238}(1 - e^{-\lambda_{238}t})}. \tag{2.5.2-7}$$

If the zero time is taken as that of the formation of the Earth, then U_0^{238}/U_0^{235} has a fixed value which was found as equal to 3.33. If a lead-containing mineral was removed from the decay scheme at time t after formation of the Earth, then the ratio Pb^{206}/Pb^{207} in it is fixed. Conversely, the time of the formation of the mineral can be inferred from this ratio alone. In practice, one measures the ratios of the number of atoms Pb^{206} and Pb^{207} to the number of non-radiogenic lead atoms Pb^{204}; one then plots Pb^{207}/Pb^{204} along the ordinate and Pb^{206}/Pb^{204} along the abscissa of a graph. Then all lead minerals of a given age lie on a straight line which is called *isochron*. The scheme, in fact, is somewhat complicated by the fact that not all Pb^{206} and Pb^{207} atoms are radiogenic. The original abundance at time 0 of such atoms, however, is a certain constant so that the isochron, although it does not pass through the origin of the graph, still has a *slope* indicative of the age of the mineral.

2.5.3 Results from Radioactive Dating

The results of radioactive dating methods lie at the root of the geologic time table (Table 1) shown in Sect. 1.2.2. It is now possible, however, to extend this table backward. A scale compounded from various sources is shown in Table 5.

Table 5. Extended geological time scale

Events	Time (in billion years)	Rock types
Present day	0	
	0.08	Cenozoic group
	0.23	Mesozoic group
First index fossil	0.6	Paleozoic group
	1.0	Grenville-Huronian types
Change in orogenesis	2.0	
Oldest known rocks	4.0	Keewatin-Temiskaming types
Age of the Earth's crust	4.6	No rocks preserved
Origin of Earth	4.8 – 5.0	

A most interesting observation in considering this table is that the so-called Precambrian time takes up about five-sixth of the Earth's history, the rest only one-sixth. Thus, conventional geology usually is concerned with only the latter sixth of the life of the Earth.

Concurrently with the extension of the geologic time scale, attempts have been made to elucidate the tectonic history of the Precambrian time. It is known that old shields have the appearance of worn-down mountain roots rather than that of homogeneous blocks. It many thus be conjectured that orogenetic activity has been going on for a long time before the appearance of the first fossilizable animals upon Earth.

Originally, the Precambrian areas had been divided into only two "ages", Proterozoic and Archean, the former being assumed as the younger. This division came about because of the presence of two types of Precambrian rocks, a sedimentary type resting upon a highly altered type. The former was simply called Proterozoic, the latter Archean. It was observed, however, soon after radioactive age determinations had been undertaken, that the classification into Proterozoic and Archean rocks is not a chronological one at all. Although in any one region Proterozoic rocks rest upon and are younger than Archean type of rocks, Proterozoic rocks are not at all of the same age and may be older in one area than the Archean-type rocks in another.

In fact, the arrangement of ages in the Canadian Shield (which has been most thoroughly investigated) seems to suggest a phenomenon of *continental growth* (see Fig. 54). The Canadian Shield may be divided into "provinces", i.e., regions of roughly uniform ages. The oldest province is found around James Bay, younger ones progressively following on its side. The above picture is due to Wilson et al.[282] and was established in 1956. Manwhile, newer

Fig. 54. Age provinces in America. *Black:* >1700 million years; *hatched:* 800–1700 million years; *dotted:* <800 million years. (After Hurley[284])

282 Wilson, J. T., Russell, R. D.,Farquhar, R. M.: Can. Min. Metall. Bull. 49:550 (1956)

investigations[283,284], have fully confirmed the general structural character of the Canadian Shield; it is even possible to make a finer division into "provinces".

Accordingly, the oldest province in a shield would form a *continental nucleus*. There are indications that orogenesis in the continental nuclei occurred in a manner different from that occurring today[285]: Mountain building processes taking place in continental nuclei produced many small sinuous belts characterized by poorly differentiated sediments and by a high proportion of basic volcanic materials. The changeover to present-day type of orogenesis (as described in Sect. 1.4) occurred about 2000 million years ago. An analysis of rock ages and tectonic patterns recognizable since that time yields that about 10 "modern-type" orogenetic cycles occurred up to the present (as mentioned already earlier; cf. Sect. 1.4.2).

The existence of orogenic cycles leads to a continual rejuvenation of continental rocks. This has as consequence an exponential distribution with age of continental areas (and volumes)[286].

The above age determinations refer to continental areas. Rocks dredged from the oceans are all much younger than Precambrian. We have already mentioned that, at least on the ridges, no rocks seem to be older than Tertiary. This had been a puzzle for a long time, but it now fits very well with the notion of oceanic plate tectonics, according to which oceanic material is constantly supplied at the ridges and moves outward therefrom. Thus, the ocean bottoms consist only of recent materials.

2.6 Thermal Data

2.6.1 General Remarks

It has been postulated for some time that the Earth may essentially be a heat engine. This means that the energy causing geodynamic effects may stem from thermal phenomena. Unfortunately, the thermal history of the Earth is only very imperfectly known as it is closely linked with problems of the origin of the Earth and with the chemistry of the Earth's interior. These topics are beyond the scope of the present study and we shall, therefore, only present here a survey of those investigations and speculations that are of importance with regard to problems of geodynamics.

283 Lowdon, J. A.: Geol. Surv. Can. Pap. 61–17 (1961)
284 Hurley, P. M.: Earth Planet. Sci. Lett. 8:189 (1970)
285 Embleton, J. J., Schmidt, P. W.: Nature (London) 282:705 (1979)
286 Veizer, J., Jansen, S. L.: J. Geol. 87:341 (1979)

2.6.2 Surface Heat Flow Measurements

The most tangible information regarding thermal properties of the Earth is obtained from surface heat-flow measurements. The fact that heat flows everywhere from the Earth's interior into outer space is evident from the well-known observation that the temperature increases with depth in any borehole or shaft; the thermal gradient varies[287] from 25°C to 40°C per kilometer. Consequently, a thermal steady state can exist only if heat flows from the interior of the Earth into space. General reviews of such problems have been given by Lee[288], Kappelmeyer and Haenel[289], and by Adam[290].

Heat-flow measurements require the determination of the thermal gradient in a mine, borehole or such like, and a determination of the thermal conductivity of the rock strata. There are now a large number of heat-flow measurements in many parts of the world which cannot be mentioned individually within the scope of this book. Rather, let us note a few general conclusions.

The heat flow data available in the world up to about 1975 have been represented on a map by the World Data Center A[291]. On this map, every single measurement has been represented by a colored dot indicating the value of the heat flow. Inasmuch as there is considerable scattering, the data become meaningful only if some sort of smoothing, like the fitting of spherical harmonics, is applied. On this basis, the world heat-flow data have been looked at recently by Chapman and Pollack[292]. Accordingly, the average heat flow HF through the surface of the Earth is

$$HF = 59 \, mW/m^2 \, .$$

It should be noted that, until the recent introduction of the SI system of units, it used to be common to give heat flows in "heat-flow units", understood as $\mu cal \, s^{-1} \, cm^{-2}$. One has as relation

$$1 \, mW/m^2 = 0.0239 \, \mu cal \, m^{-2} \, s^{-1}$$

$$1 \, \mu cal \, cm^{-2} \, s^{-1} = 41.84 \, mW/m^2 \, .$$

The first observation that one makes in looking at heat-flow measurements throughout the world is that there is no essential difference between continental and oceanic heat flows. This observation is most significant. It suggests that each unit of area of the Earth's surface is underlain by the same

287 Schubert, G., Anderson, O. L.: Phys. Today 1974(3):28 (1974)

288 Lee, H. K. (ed.): Terrestrial Heat Flow. Washington: Am. Geophys. Union Monogr. No. 8 (1965)

289 Kappelmeyer, O., Haenel, R.: Geothermics. Berlin: Borntraeger 1976

290 Adam, A. (ed.): Geoelectric and geothermal studies. Budapest: Publ. House Hung. Acad. Sci. 1977

291 Anonymous: Terrestrial Heat Flow Data. Map published by World Data Center A, Boulder, Colorado (1976)

292 Chapman, D. S., Pollack, H. N.: Earth Planet. Sci. Lett. 28:23 (1975)

amount of heat-generating radioactivity. This is very surprising in view of the fact that continents and oceans have entirely different structures; thus it is known that continental rocks contain much more radioactive material near the surface than the oceanic rocks. Somehow, this difference must be compensated at depth.

There are, naturally, deviations from the mean heat flow in various areas of the Earth. The lowest heat flow values are found in old shields like Africa (50 mW/m²) and in the abyssal plains of the Atlantic Ocean. The heat flow rises to 63 mW/m² in orogenetic zones, and reaches 75 mW/m² in the rifts on mid-oceanic ridges. The highest values (250 – 290 mW/m²) are attained in volcanic areas. Because of the scatter of the results and the uneven distribution of measurement locations, a significant result for the *whole* Earth can only be obtained if measurements for the poorly represented regions are somehow supplemented. An attempt to do this has been made by Chapman and Pollack[292], who "predicted" values for the areas in which measurements are lacking, according to their geotectonic character. From this, then, a degree-12 spherical-harmonic analysis for the whole globe was made. The result is shown in Fig. 55.

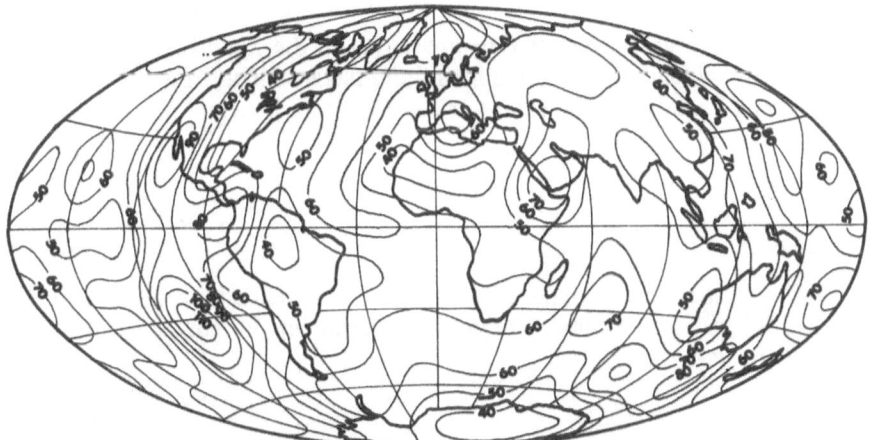

Fig. 55. Degree-12 spherical harmonic representation of global heat flow from observations supplemented by predictor. Heat flow in mW m⁻². (After Chapman and Pollack[292])

The regions with the highest (positive) heat-flow anomalies have been called "hot spots"[293]. There is some controversy as to whether these hot spots move with the tectonic plates during their drift or whether they have a deeper origin in the mantle.

293 Burke, K. C., Wilson, J. T.: Sci. Am. 235(2):46 (1976)

2.6.3 Temperature in the Earth's Interior

Attempts at estimating the temperature in the Earth's interior were originally based on linear extrapolations of the thermal gradient at the surface. Using a gradient of 25 °C/km yields a temperature of 157,500°C at the center of the Earth. It is difficult to believe that such high temperatures can exist in the Earth's interior, since they would have thermodynamic-chemical consequences which would be difficult to understand. Somewhat, therefore, the thermal gradient must be much less in the interior than near the surface of the Earth.

It was Jacobs[294,295] who gave a reasonable thermodynamic explanation of a decrease of the thermal gradient with depth. In order to estimate the temperature distribution in the Earth, one starts in principle with the density distribution from which the pressure distribution can be calculated by an application of Newton's law of gravitation (see Fig. 30, Sect. 2.1.2). Next, it is necessary to introduce a hypothesis regarding the dependence of the volume coefficient of thermal expansion α on the pressure p:

$$\frac{1}{\alpha} + \frac{1}{\alpha_0} + bp. \tag{2.6.3.-1}$$

This assumption is suggested by analogy with the compressibility-pressure relationship which is of the same form, and has been confirmed by results of the theory of solids[294]. The values of the constants $1/\alpha_0$ and b are as follows

$$\frac{1}{\alpha_0} = 2.4 \times 10^4 \, \text{K}, \tag{2.6.3-2}$$

$$b = 6.2 \times 10^{-7} \, \text{Km}^2/\text{N}. \tag{2.6.3-3}$$

It is now possible to make an estimate of the temperature distribution in the Earth[294,295]. The adiabatic temperature gradient (for constant entropy S) satisfies the equation

$$\left(\frac{\partial T}{\partial p}\right)_{S=\text{const}} = \frac{\alpha T}{\varrho C_p}. \tag{2.6.3-4}$$

Here, T is the temperature, ϱ the density, and C_p is the specific heat at constant pressure. Inserting the hypothesis formulated in Eq. (2.6.3 – 1) into Eq. (2.6.3 – 4), one obtains

$$\frac{dT}{T} = \frac{dp}{\varrho C_p \left(\dfrac{1}{\alpha_0} + bp\right)}. \tag{2.6.3-5}$$

294 Jacobs, J. A.: Can. J. Phys. 31:370 (1953)
295 Jacobs, J. A.: Nature (London) 170:838 (1952); also Adv. Geophys. 3:183 (1956)

If one further assumes that C_p is a constant, the above relationship can be integrated to yield:

$$\log \text{nat } T = \frac{1}{C_p} \int \frac{dp}{\varrho \left(\dfrac{1}{\alpha_0} + bp \right)} . \qquad (2.6.3-6)$$

This can be evaluated numerically, since the dependence of the density ϱ on depth (and hence on pressure) is approximately known.

If one does this and begins at a depth of 100 km with 1500 K, one arrives at the center of the Earth at about 6000 K. Figure 56 shows a recent result of calculations based upon the above principles (after Mayeva[296]).

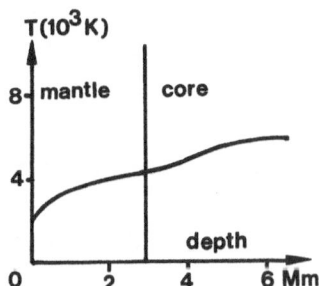

Fig. 56. Temperature distribution in the interior of the Earth. (Modified after Mayeva[296])

This curve estimates the temperature in the Earth only below a certain depth. Information regarding the upper layers must be sought from different sources. Various authors have made estimates, based on model assumptions regarding the constitutive equation of the upper layers of the Earth in various regions (continental, oceanic, etc.). Furthermore, the high observed thermal gradient at the surface must somehow be joined to the low gradient in the deeper parts of the Earth. The calculations are generally based on the heat *conductivity* equation, so that the assumed distribution of heart sources becomes important. The problem, thus, boils down to one of constructing suitable *models* of the crust and upper mantle, assuming appropriate cross-sections and appropriate distributions of heat sources and conductivities. For various continental areas, this has been done by Buntebarth[297] (Alps), Balling[298] (Fennoscandia), and by Crough and Thompson[299] (general); for oceanic areas by Schubert et al.[300] and by Davis and Lister[301]. Special areas,

296 Mayeva, S. V.: Bull. (Izv.) Acad. Sci. USSR Geophys. Ser. 1971(1):3 (1971)
297 Buntebarth, G.: Z. Geophys. 39:97 (1973)
298 Balling, N. P.: J. Geophys. 42:237 (1976)
299 Crough, S. T., Thompson, G. A.: J. Geophys. Res. 81(26):4857 (1976)
300 Schubert, G., Froidevaux, C., Yuen, D. A.: J. Geophys. Res. 81(20):3525 (1976)
301 Davis, E. E., Lister, C. R. B.: J. Geophys. Res. 82(30):4845 (1977)

such as the Ivrea geothermal zone[302] or Rift Valleys[303] were also analyzed. Finally, general correlations were proposed between crustal upper mantle structure and the thermal gradient[304,305]. From these various investigations a general qualitative agreement for the course of the geothermal gradient in type areas was observed. As an example, an average curve for continental plains areas composed by the writer from the mentioned literature is shown in Fig. 57.

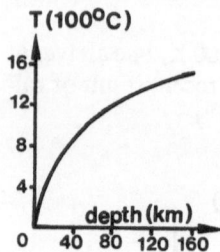

Fig. 57. Temperature distribution in the uppermost 160 km of the Earth (plains area)

2.7 Electromagnetic Effects

2.7.1 The Earth's Magnetic Field

It has been well known for a long time that a magnetic field is associated with the Earth[306]. The main part of the field can be described as that of a magnetic dipole whose axis is somewhat offset with regard to the present axis of the Earth's rotation. Short term disturbances of the magnetic field are known to be caused by stray currents in the ionosphere, pressumably induced by cosmic radiation. Other characteristic disturbances of the magnetic field are called secular variations. These appear to be regional phenomena; they also include the westward drift of non-axial characteristics[307,308].

The origin of the magnetic field of the Earth and of its characteristic disturbances are not properly understood. At present it is usually held that some magneto-hydrodynamic phenomenon in the core of the Earth is responsible for the magnetic field[309,310]. If this is true, then there is reason to believe that

302 Höhndorf, A., Haenel, R., Giesel, W.: J. Geophys. 41:179 (1975)
303 Chapman, D. S., Pollack, H. N.: Nature (London) 256(5512):28 (1975)
304 Chapman, D. S., Pollack, H. N.: Geology 5:265 (1977) also Tectonophysics 38:279 (1977)
305 Stromeyer, D.: Rev. Roum. Geol. Geophys. Geogr. Ser. Geophys. 22:83 (1978)
306 Balmer, H.: Beiträge zur Geschichte der Erkenntnis des Erdmagnetismus. Aarau: Sauerländer 1956
307 See e.g. Runcorn, S. K.: Handbuch der Physik, Bd. 47, S. 498. Berlin-Göttingen-Heidelberg: Springer 1956. For many references
308 Chapman, S., Bartels J.: Geomagnetism. Oxford: Univ. Press 1948
309 Elsasser, W. M.: Rev. Mod. Phys. 22:1 (1950), 28:135 (1956)
310 Herzenberg, A.: Philos. Trans. R. Soc. London Ser. A 250:543 (1958)

the axis of the dipole field approximating the Earth's magnetic field should always more or less coincide with the axis of rotation.

The thermohydrodynamic theories of the generation of magnetic fields in celestial bodies are of extreme complexity and only of secondary importance with regard to geodynamic problems. Recent reviews have been published by Gubbins[311] and by Levy[312].

From analyses of the geomagnetic field, in turn, inferences have been made regarding the "topography" of the Earth's core – mantle interface. In particular, statements regarding its asymmetry have been made[313,314]. Many results of studies of this type have been summarized by Jacobs in a monograph on the Earth's core[315].

2.7.2 Paleomagnetism

It has been observed that, in sediments containing iron-oxide minerals, the magnetized grains are oriented in a definite direction. It must therefore be postulated that an orienting magnetic field was present during the deposition of the sediments, and the contention is[316] that, under sufficiently quiet conditions, the Earth's magnetic field is strong enough to effect this. An even stronger effect occurs in magnetizable materials that passed through the Curie point after emplacement. This refers mainly to extrusive rocks which are hot in the beginning and then cool to surface temperature, passing thereby through the Curie point. The contention is that these rocks assume the direction of the magnetic field prevalent at the time of their passage through the Curie point, leading thereby to "thermoremanent" magnetization. The theory of this physical process is not yet completely understood; Soffel[317] has given a review of the possibilities.

If suitable rocks from various parts of the world are analyzed regarding their remnant magnetization, it may be hoped that the results can be interpreted in terms of a path of wandering of the Earth's magnetic pole. By implication this, then, would also indicate the path of the pole of rotation.

There are now a vast number of such pole determinations available, some of them summarized in catalogs[318,319]. These pole positions, naturally, refer to the present-day coordinate system on the Earth and reflect the relative

311 Gubbins, D.: Rev. Geophys. Space Phys. 12(2):137 (1974)
312 Levy, E. H.: Ann. Rev. Earth Planet. Sci. 4:159 (1976)
313 Petrova, G. N.: Bull. (Izv.) Acad. Sci. USSR Phys. Solid Earth 13(11):2 (1977)
314 Jones, Q. M.: J. Geophys. Res. 82:1703 (1977)
315 Jacobs, J. A.: The Earth's Core. London, New York: Academic Press 1975
316 Runcorn, S. K.: Handb. Phys. 47:370 (1956)
317 Soffel, H.: Z. Geophys. 36:237 (1970)
318 Cox, A., Doell, R. R.: Bull. Geol. Soc. Am. 71:645 (1960)
319 Hicken, A., Irving, E., Law, L. K., Hastie, J.: Publ. Earth Phys. Branch, Dep. Energy, Mines and Resource, Ottawa 45(1):1 (1972)

ancient positions to the present-day location of the outcrops or group of outcrops which were chosen for the determination.

It is found that the various ancient positions of the pole as "seen" from
any one region can be joined by a fairly smooth curve. This, however, is
possible only if no distinction is made between North and South Pole. It must
be assumed that either the magnetic field of the Earth can undergo a spon-
taneous reversal, or that a self-reversal of the remnant magnetization of the
rocks can occur during consolidation. Only if this is assumed can one define a
coherent path of "polar wandering". The polar path determined in this
fashion from European and North American rocks turns out be fairly
resonable and more or less in agreement with that inferred from paleoclimatic
evidence.

The paleoclimate evidence suggests not only that a shift of the poles took
place, but also large continental drifts occurred, particularly in view of the
simultaneous glaciation of several southern continents which are now widely
separated. It is interesting to observe that similar conclusions are suggested by
paleomagnetic work. Clegg et al.[320] made a very careful study of this question,
using rock samples collected in India. These authors arrived at a suggested
position of the North Pole for the Eocene at 28°N, 85°W. This is widely
different from the position of 75°N, 120°W obtained from North American
rocks for the same epoch. "Looking" from various continents, one obtains,
therefore, entirely different "polar wandering paths". The discrepancy can be
resolved if it assumed that continents have moved with regard to each other.
Thus, India would have to have drifted some 6000 km northward toward
North America in 60 million years. In other areas, similar results have been
suggested[321-324]. The apparent polar wandering paths were interpreted in
terms of drifting continents by Irving[325] (see Fig. 58) since the Paleozoic.
Extensions to earlier epochs are less certain[326].

The fact that pole positions are only relative to moving continents,
suggests the question whether there was, in addition to the drift of the continents, a polar wandering with regard to the lithosphere as a whole. This
question was particularly investigated by Jurdy and van der Voo[327] with somewhat inconclusive results. These authors attempted to separate a random
(drift) motion from a systematic rotation by the application of statistical

320 Clegg, J. A., Deutsch, E. R., Griffiths, D. H.: Philos. Mag. 1:419 (1956)
321 Deutsch, E. R.: In: Polar Wandering and Continental Drift, Publ. by Society of Economic
    Paleontol. and Mineral., AAPG, Tulsa, p. 3 (1963)
322 Deutsch, E. R.: In: Continental Drift (ed. Garland, G. D.) R. Soc. Can. Spec. Publ. 9:28
    (1966)
323 Zonenshtayn, L. P., Gorodinskiy, A. M.: Geotektonika (Moscow) 1977(2):3 (1977)
324 Zonenshtayn, L. P., Gorodinskiy, A. M.: Geotektonika (Moscow) 1977(3):3 (1977)
325 Irving, E.: Nature (London) 270(5635):304 (1977)
326 Morel, P., Irving, E.: J. Geol. 86:535 (1978)
327 Jurdy, D., Van der Voo, R.: J. Geophys. Res. 79(20):2945 (1974), Science 187, 28 March
    1975

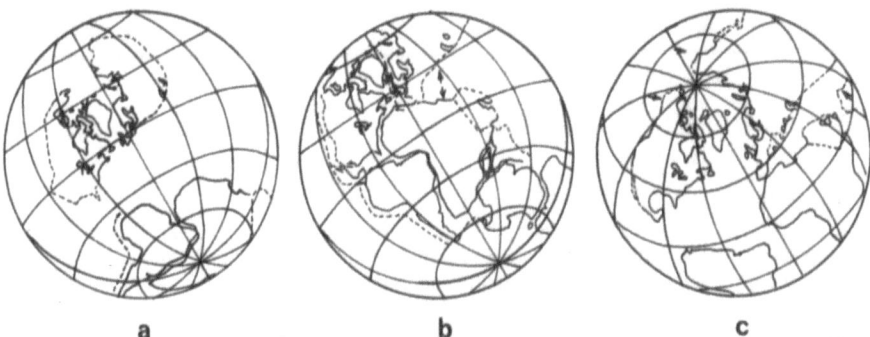

a   b   c

Fig. 58. Continental drift since the Devonian based on paleomagnetic evidence. (After Irving[325])

methods, but the errors turned out to be of the order of the suspected polar wandering. The need to postulate drifting continents from an analysis of paleomagnetic data constitutes one of the strongest arguments in favor of plate tectonics.

2.7.3 Magnetic Reversals

A further result of studies of paleomagnetism is that the Earth's magnetic field may change its polarity. In effect, about 50% of the rocks appear to have a polarity which is opposite to the "normal" (present-day) polarity. A study of the polarity-sequences has yielded the picture shown in Fig. 59. This figure shows results only from the Cretaceous to the Paleozoic. During the Tertiary and Cenozoic very frequent reversals occurred, the latest (the so-called Brunhes-Matuyama transition) about 600,000 years ago[328,329].

The changes in polarity yield rock layers which are successively magnetized in opposite directions. Attempts have been made to determine the details of magnetic polarity transitions recorded in the rocks. Indications in some cases are that the field reversals are characterized by a rotation of the dipole axis in the meridional plane so that the Earth's field remains dipolar during the transition[330], although the intensity may become very low at times. The duration of a polarity transition has been estimated as 4600 years[331]. On the other hand, Hillhouse and Cox[332] found a non-dipolar field during the most recent polarity reversal.

The polarity bands are not only found in vertical sections, but also in horizontal surface strips paralleling the mid-oceanic ridges. This leads to a corro-

328 Tarling, D. H., Mitchell, J. G.: Geology 4:133 (1976)
329 Labrecque, J. L., Kent, D. V., Cande, S. C.: Geology 5:330 (1977)
330 Steinhauser, P., Vincenz, S. A.: Earth Planet. Sci. Lett. 19:113 (1973)
331 Opdyke, N. D., Kent, D. V., Lowrie, W.: Earth Planet. Sci. Lett. 20:315 (1973)
332 Hillhouse, J., Cox, A.: Earth Planet. Sci. Lett. 29:51 (1976)

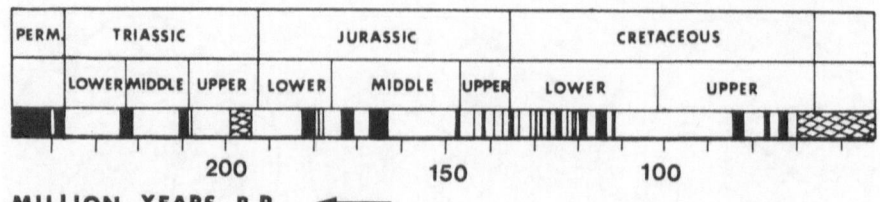

Fig. 59. Field reversals from Tertiary to Permian. *White:* normal, *black:* reversed, *cross-hatched:* uncertain (due to frequent reversals) magnetization. (Data after Irving and Couillard[333], for the Cretaceous and from Creer[334], for the earlier epochs)

boration of the idea that material wells up at the ridges and moves outward from them. As it cools through the Curie point, it obtains the magnetization corresponding to the prevailing magnetic field. A correlation (achieved by statistical methods on a computer) of the distances of the polarity strips with the polarity epochs determined on vertical sections permits one to calculate the *velocity* of horizontal motion of the ocean floor as it moves away from the ridges. In the Atlantic, this velocity is $10 - 50$ mm/year[335].

2.7.4 Electrical Effects

It has been found that electric currents in the Earth are associated with the time changes of the magnetic field. These currents are termed "telluric currents". If one measure the magnetic and electric field variations simultaneously, then it is possible to obtain an idea of the distribution of the electric conductivity at depth. The theory of such measurements was first given by Cagniard[336]; a good review has later been published by Porstendorfer[337].

Unfortunately, not too many surveys of the above type have been made. The area analyzed best is undoubtedly Japan where Rikitake has made many investigations. The latter author came to the conclusion that a hypothetical dipole is located beneath the central part of Japan whose explanation is not yet entirely certain[338]. His investigations also led Rikitake[339-341] to postulating a model of the Earth's crust in the vicinity of Japan which is based on the assumption that there is a wedge-shaped intrusion of low-conductivity material that lies ordinarily near the surface, into great depths (to 700 km).

333 Irving, E., Couillard, R. W.: Nature (London) Phys. Sci. 233:10 (1973)
334 Creer, K. M.: Nature (London) 233:545 (1973)
335 Krause, D. C., Watkins, N. D.: Geophys. J. R. Astron. Soc. 19:261 (1970)
336 Cagniard, L.: Geophysics 18:605 (1953)
337 Porstendorfer, G.: Tellurik. Freiberg. Forschungsh. C 107 (1961)
338 Rikitake, T.: Bull. Earthquake Res. Inst. 34:291 (1956), 36:1 (1958)
339 Rikitake, T.: Geophys. J. R. Astron. Soc. 2:276 (1959)
340 Rikitake, T.: Bull. Earthquake Res. Inst. 34:291 (1956)
341 Rikitake, T.: Tectonophysics 7:257 (1969)

This fits well together with the idea that there is some sort of a geosyncline where an island arc is located; the deepest deep-focus earthquakes would occur at the lower boundary of the wedge. A schematic cross-section of Rikitake's model is shown in Fig. 60.

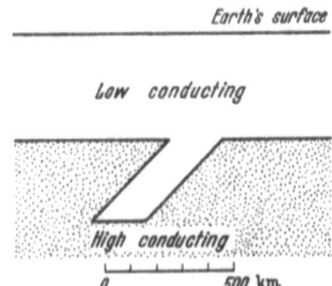

Fig. 60. Schematic cross-section of the electrical conductivity beneath Japan. (After Rikitake[338])

Magnetotelluric studies have been made in other parts of the world as well, usually in conjunction with seismic crustal studies. In this fashion, some information regarding the conductivity profile in various areas has been obtained. From such studies, the existence of water in the lower crust beneath the Adirondacks in North America has been inferred[342].

2.8 Geochemical Data

2.8.1 General Remarks

A further body of information regarding the Earth may be obtained from chemical investigations. From the spectroscopic analysis of celestial bodies it has been inferred that the chemical composition of the Universe is quite uniform. The latter consists throughout of the same elements and their relative abundances are everywhere more or less identical. Such differences as exist can usually be explained by the prevailing local conditions such as high temperatures which permit certain elements to be used up in thermonuclear reactions.

The Earth, then, has been formed by differentiation from the "cosmic soup". Fundamental monographs on cosmo-chemical questions have been written by various people[343-346]. Accordingly, the relative abundances of elements in the solar system was established authoritatively by Suess and

342 Nekut, A., Commerney, J. E. F., Kuckes, A. F.: Geophys. Res. Lett. 4(6):239 (1977)
343 Mason, B.: Principles of Geochemistry. New York: Wiley 1952
344 Rankama, K., Sahama, T. G.: Geochemistry. Chicago: Univ. Press 1950
345 Turekian, K. K.: Chemistry of the Earth. New York: Holt-Rinehart 1972
346 Manghnani, M. H., Akimoto, S. E.: High Pressure Research, Applications in Geophysics. London, New York: Academic Press 1977

Urey[347] and Rösler and Lange[348]; their results are shown in Fig. 61. Inasmuch as the abundances decrease very rapidly with atomic number, only very few elements can be of significance with regard to the composition of the Earth. In addition, all those elements are of no significance which are either very volatile themselves or which possess simple compounds that are volatile (such as H_2, H_2S, and CO_2). What is left, is a "mix" of about 35% of iron, 28% of oxygen, 17% of magnesium, 13% of silicon, and 7% of other elements as the most likely average composition of the Earth as a whole.

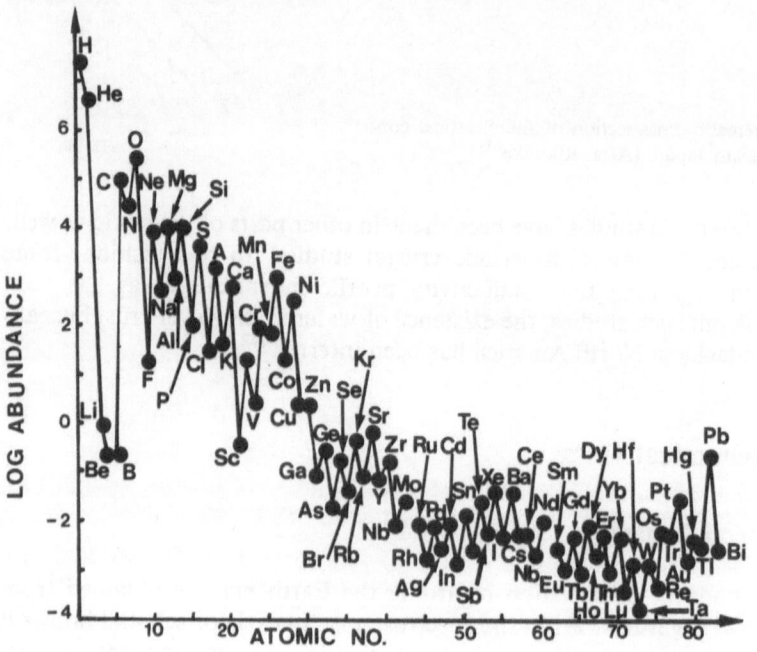

Fig. 61. Relative abundances of the elements, referred to Si = 10,000, plotted against atomic number. (After Ahrens[349])

2.8.2 Geochemistry of the Crust

The crust is the only part of our globe which is directly accessible. The materials found in it show an extreme variety; however, silica (silicium dioxide) is the most important compound of elements. This follows from an analysis of igneous rocks and lavas. Equally, if one were to melt an "average" collection of sedimentary rocks, one would also obtain a silica-containing mixture. It is therefore convenient to classify igneous rocks according to their

347 Suess, H. E., Urey, H. C.: Rev. Mod. Phys. 23:53 (1956)
348 Rösler, H. J., Lange H.: Geochemische Tabellen. Stuttgart: F. Enke 1976
349 Ahrens, L. H. (ed.): Origin and Distribution of Elements. Oxford: Pergamon Press 1968

silica content; in addition, they are also classified according to the size of crystal grains which they contain. The usual classification thus obtained is shown in Table 6.

Table 6. Classification of rocks according to silica content and size of crystal grains

Silica content	Ultrabasic <45%	Basic 45 – 55%	Transitional 55 – 65%	Acid >65%
Size of crystal grains				
fine	–	Basalt	Andesite	Rhyolite
coarse	Peridodite Pyroxenite	Gabbro	Granodiorite Syenite Gneiss	Granite

Most of the *continental* material can be classified as andesitic. When it becomes worn down, it causes various types of sedimentary rocks to be formed, finally due to subsequent metamorphism it becomes granite and granodiorite. Thus, the evolution of continental rocks can be considered as a geochemical cycle. Starting from andesitic lava, one has first crystallization, then weathering and transportation which produce sediments. The sediments become consolidated to form sedimentary rocks; thence metamorphism takes over and produces metamorphic rocks[350,351], and finally anatexis (remelting) may take place and recreate the lava. The passage of any particular chemical element through its "geochemical cycle"[352,353] can be followed and it is found that the picture created in this fashion is essentially consistent.

In contrast to the andesitic rocks of the continents, there are the basaltic ones of the *ocean bottoms*. The basalt cycle is similar to the andesite cycle. Starting out with basaltic lava, one has crystallization and sedimentation. However, upon metamorphosis one obtains gabbro and not granite. If anatexis takes place, one obtains again the lava.

As mentioned above, basalt seem to be the material of the ocean bottom. Basaltic type rocks are mostly found in oceanic islands. Volcanoes at the margins of continents seems to spew out mostly andesitic lava, whereas the lava found on mid-oceanic islands is mostly basaltic. The reason for the difference in chemical composition of the continental crust and the oceanic crust is not clear; it seems to run parallel with the marked differences in physiographic and geophysical properties. The boundary between basaltic and

350 Fyfe, W. S. et al.: Metamorphic Reactions and Metamorphic Facies. Geol. Soc. Am. Mem. No. 73 (1958)
351 Read, H. H.: The Granite Controversy. London: Murby 1957
352 Engelhardt, W. v.: Nova Acta Leopold. 21, No. 143:85 (1959)
353 McLennan, S. M., Taylor, S. R.: Nature (London) 285:621 (1980)

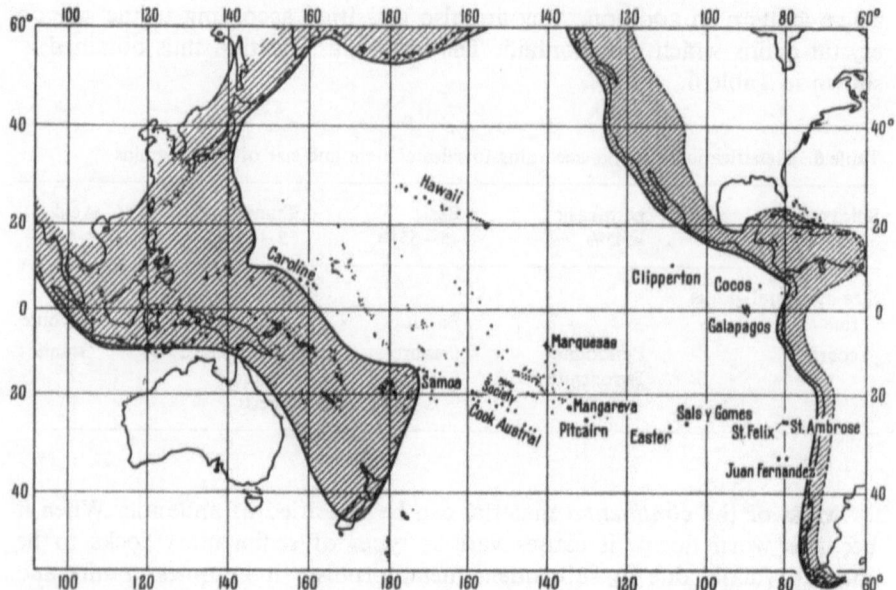

Fig. 62. The Andesite Line, separating andesitic (*shaded*) from non-andesitic areas in the Pacific. (After Chubb[354])

andesitic rocks and lavas in the Pacific ocean has been called the "*Andesite Line*"[354] (see Fig. 62).

Inasmuch as the continental crust is characterized by a more andesitic composition than the oceanic crust, it seems unlikely that parts of the continental crust can become oceanic and vice versa. Speculations regarding the origin of such oceanic areas as the Mediterranean Sea by "oceanization" of a formerly continental crust[355] run counter to the opinion of most scientists at this moment who seem to favor the idea that continental and oceanic crusts represent two different chemical entities[356]. This view also corresponds to the idea that the "lithospheric plates are permanent"; an oceanic area in a formerly continental one would be formed by rifting and drifting asunder of the pieces (cf. also Sect. 1.5.3). At most, there may be some specific differentiation[357] of continental material from a more oceanic one.

Based upon the above data, attempts at setting up geochemical models (analogous to seismic and density models) of the Earth's crust have been made. For the setting up of such models, the structural and chemical constraints can be clearly stated[358]. Based on phase-transition data it is then pos-

354 Chubb, L. J.: Geol. Mag. 71:289 (1934)
355 Van Bemmelen, R. W.: Verh. K. Ned. Geol. Mijnbowk. Genoot. 26:13 (1969)
356 E. g., Ashgurei, G. D.: Geology 1974:401 (1974), Lowman, P. D.: J. Geol. 84(1):1 (1976)
357 Rogers, J. J. W., Novitsky-Evans, J. M.: Geophys. Res. Lett. 4(8):347 (1977)
358 Smithson, S. B.: Geophys. Res. Lett. 5(9):749 (1978)

sible to construct numerically possible model cross-sections for specific features, such as geosynclines[359].

2.8.3 Geochemistry of the Mantle

The mantle reaches from the Mohorovičić discontinuity to the core. It appears that it rises very close to the Earth's surface in the mid-oceanic ridges; the volcanic eruptions in these regions produce basaltic lavas so that the current opinion is that the mantle consists of some high-pressure phase of basalt, such as eclogite[360]. Materials of this type have the right density and have elastic constants of the right order of magnitude to yield the seismic velocities observed at that depth. Admixtures of various materials, such as iron, magnesium, alkali metals, and water may produce various modifications of pure eclogite, such as "pyrolite" which is a mineral consisting of about 45% of silica, 37% of magnesium oxide, and 8% of iron oxide, the rest being other oxides[361].

Within the mantle, at a depth of about 900 km, there is perhaps a further discontinuity (Birch discontinuity) involving a change of phase, or a chemical change, or both[362]. This follows from the observation that the variation of the observed compressibility with depth cannot be accounted for by the compression of a homogeneous material between 200 km and 800 km of depth (Bullen's B-layer), but that the compression alone can account very well for the observed compressibility between 900 and 2,900 km of depth. Hence, there may be a discontinuity at a depth between 800 and 900 km. However, no seismic evidence for such a discontinuity exists. Thus, the "transition layer" (Bullen's C-layer) poses certain problems, and a number of conjectural calculations have been made[363,364].

In addition to inhomogeneities with depth, there appear also lateral inhomogeneities to exist in the mantle. These have been established through the investigation of characteristic trace elements representative of mantle material from different locations in the Atlantic and Pacific oceans[365].

Inasmuch as there appears to be a fundamental chemical difference between mantle material and continental crust, the Mohorovičić discontinuity would appear to be a major chemical discontinuity. Speculations that the Mohorovičić discontinuity is merely a phase boundary are therefore currently out of favor. In addition, there may be a difference between the oceanic and the continental Mohorovičić discontinuity[366], because the mantle material

359 Spohn, T., Neugebauer, H. J.: Tectonophysics 50:387 (1978)
360 Anderson, D. L.: Geophys. Res. Lett. 6(6):433 (1979)
361 Turekian, K. K.: Chemistry of the Earth. New York: Holt-Rinehart 1972
362 Birch, F.: Trans. Am. Geophys. Union 32:533 (1951)
363 Shimazu, Y.: J. Earth Sci. Nagoya Univ. 6(12):31 (1958)
364 Knopoff, L., Uffen, R. J.: J. Geophys. Res. 59:471 (1954)
365 Bougault, H., Joron, J. L., Treuil, M.: Philos. Trans. R. Soc. London Ser. A 297:203 (1980)
366 McKenzie, D.: Geophys. Monogr. 13:660 (1969)

upwelling in the oceanic ridges becomes altered by the admixture of water. Present indications are that the oceanic Mohorovičić discontinuity is a boundary separating partially (35%) serpentinized ultramafic rocks above from fresh tectonized ultramafic rocks below[367], whereas the continental Mohorovičić discontinuity is a much stronger chemical boundary.

2.8.4 Geochemistry of the Core

The *core* which reaches from 2,900 km to the center of the Earth, presumably consists of a mixture of iron and nickel. The latter hypothesis is supported by the abundance of iron in meteoritic material; meteorites being thought of as débris from a planet of similar constitution as the Earth which for some reason disintegrated. However, all that is *really* known about the core of the Earth is that it must have a high density and that it probably has a high electrical conductivity. The latter follows from the customary explanation of the Earth's magnetic field in terms of magneto-hydrodynamic convection currents[368]. Both these properties conceivably could also be exhibited by a metallic phase of magnesium-iron silicates (Ramsey's hypothesis[369]) or hydrogen (Kuhn-Rittmann's hypothesis[370]) which would be stable at the high pressures prevailing at the depth in question. It is not quite clear, though, whether such a hypothesis can be maintained in the light of quantum mechanical estimates of the density to be expected in such a phase. It seems, therefore, that the assumption of a core consisting of an iron-nickel alloy is still the most satisfactory one in the light of present knowledge.

A slight modification of this view would be the assumption that the core contains iron *oxide*[371] rather than pure iron which would accommodate the known high cosmic abundance of oxygen. Equally, various modifications of the pure iron-nickel hypothesis have been attempted by the admixture of other elements, such as hydrogen, sulfur, and oxygen[372, 373].

367 Clague, D. A., Straley, P. F.: Geology 5:133 (1977)
368 Brett, R.: Rev. Geophys. Space Phys. 14(3):375 (1975)
369 Ramsey, W. H.: Mon. Not. R. Astron. Soc. 108:404 (1948)
370 Kuhn, W., Rittmann, A.: Geol. Rundsch. 32:215 (1941)
371 Altshuler, L. V., Sharipdzhanov, L. S.: Bull. (Izv.) Acad. Sci. USSR Geophys. Ser. 1971(4):3 (1971)
372 Stevenson, D. J.: Nature (London) 268(5615):130 (1977)
373 Jacobs, J. A.: The Earth's Core. London, New York: Academic Press 1975

3. The Mechanics of Deformation

3.1 Finite Strain in Rheological Bodies

3.1.1 The Physics of Deformation

The basic problem in the study of the Earth's crust is to understand its deformations. Unfortunately, the general physics of deformations is not as well understood as one might desire.

Deformations can occur in two fundamentally different ways: viz. continuously or discontinuously. Under continuous deformation we understand a condition where neighboring points in a material always remain neighboring points, whereas in discontinuous displacements this is not the case.

Several branches of the theory of continuous deformation have been very intensively developed. It is well known that the theory of elasticity has been carried to a high degree of refinement; the same is true for the hydrodynamics of viscous fluids. Unfortunately, the materials of which the Earth's crust is composed are very unlikely to fit either of these theories. Of those branches of the continuous displacement theory that are more or less well developed, the theory of plasticity has the most bearing upon the displacements observed in the Earth's crust. However, the theory of plasticity has been developed for the description of the behavior of a metal during cold working, and it cannot be expected, therefore, that its application to the Earth's crust will lead to entirely satisfactory results. The material in the Earth's interior shows a very complicated behavior, possibly of the types discussed in the various theories of "rheology". However, these theories all appear to be more or less heuristic and therefore incomplete.

Very important is the discussion of discontinuous displacements. There are many instances where ruptures, fissures, fractures and such like occur in the Earth's crust. Although humans have been breaking things since the inception of civilization, it is an unfortunate fact that the whole subject of fracture is only very incompletely understood. There are quite a number of rule-of-thumb criteria of fracture or better: of when a structure is supposed to be safe so as *not* to fracture, – but the basic problem of describing the progress of a fracture surface in a given body under given external stresses has not yet been solved.

We shall, in the following sections, consider the various aspects of the theory of deformations one by one.

3.1.2 The Structure of a Finite Strain Theory [1]

In order to obtain a description of the dynamics of continuous media, there are various steps that must be observed [2,3]. In the first place, one must decide upon a description of the deformation. Once this has been achieved, one must express various physical laws: the condition of continuity, the law of motion, and boundary conditions. We shall discuss these steps one by one.

a) Measure of Displacement. Let us assume that a certain volume W (which may be infinite) of space is filled with matter. The volume W, and the way in which it is filled, changes with time t.

The points of space may be specified by giving three Cartesian coordinates x_i such that the line element is defined as follows:

$$ds^2 = dx_i dx_i. \qquad (3.1.2-1)$$

In this formula, the summation convention has been used, which stipulates that one has to sum over all indices that occur twice.

The above scheme characterizes the geometrical space ("coordinate space") occupied or potentially occupied by the continuous medium. The next task is to characterize the medium. It is well known [2] that this can be done by introducing three parameters ξ_α. The whole medium is characterized if the parameters run through all points of a volume Ψ in "parameter space", i.e., the space of the ζ's.

We shall assume that the space Ψ of the parameters is endowed with a Cartesian metric, so that a line element $d\sigma$ can be defined as follows:

$$d\sigma^2 = d\xi_\alpha d\xi_\alpha. \qquad (3.1.2-2)$$

It is always possible to make such a parameter transformation that the parameters are equal to the coordinates of all the particles at a particular time, say t_0. It is often convenient to do this.

With the above characterization of a continuous medium and the geometrical space occupied by it, one can proceed to describe motions. The complete motion is obviously given if the geometrical coordinates of each particle of the medium are known for all times:

$$x_i = x_i(\xi_\alpha, t). \qquad (3.1.2-3)$$

Thus, the specification of three functions of the three parameters plus time determines the motion. This type of description is often called the *material* form of the description of motion.

One also could have provided a description of the motion in another way, viz. by solving Eq. (3.1.2 – 3) for the ζ's:

---

1 The following is after Scheidegger, A. E.: Can. J. Phys. 34:498 (1956)
2 Truesdell, C.: J. Rat. Mech. Anal. 1:125 (1952)
3 Prager, W.: Introduction to Mechanics of Continua. Boston: Ginn & Co. 1961

$$\xi_\alpha = \xi_\alpha(x_i, t) .$$ (3.1.2-4)

Physically, this means that one states which "particle" is at a given time at any given spot. This is called the *spatial* form of the description of motion.

Although the specification of the functions in Eq. (3.1.2-3) completely describes the motion, it is often convenient to introduce various other quantities. This is so because it is cumbersome to express the equations of motion directly in terms of the quantities introduced heretofore.

An important kinematical notion is the concept of *strain*. Strain, in the finite theory, is defined as half the difference of the squared distance between neighboring points in two states, one of which is arbitrarily called "state of zero strain". The element of distance between neighboring particles is:

$$ds^2 = \frac{\partial x_i}{\partial \xi_\alpha} \frac{\partial x_i}{\partial \xi_\beta} d\xi_\alpha d\xi_\beta = \varkappa_{\alpha\beta} d\xi_\alpha d\xi_\beta$$ (3.1.2-5)

where

$$\varkappa_{\alpha\beta} = \frac{\partial x_i}{\partial \xi_\alpha} \frac{\partial x_i}{\partial \xi_\beta} .$$ (3.1.2-6)

Thus, the element of distance between two particles defines a symmetric tensor $\varkappa_{\alpha\beta}$ in parameter space. If we denote the (time-independent) tensor of the state of zero strain by $\zeta_{\alpha\beta}$, we can define the "material strains" as follows:

$$\varepsilon_{\alpha\beta} = \frac{1}{2}(\varkappa_{\alpha\beta} - \zeta_{\alpha\beta}) = \frac{1}{2}\left(\frac{\partial x_i}{\partial \xi_\alpha} \frac{\partial x_i}{\partial \xi_\beta} - \zeta_{\alpha\beta}\right).$$ (3.1.2-7)

Eq. (3.1.2-5) permits a different interpretation of strain from that given above. For, this Eq. (3.1.2-5) can be taken as fundamental metric form in a certain space $\Sigma(t)$ which has been called the "material strain space". In this instance, one should note that the $d\xi$'s are contravariant vectors and could be expressed by using superscripts instead of subscripts (employing the notation of Riemannian geometry):

$$ds^2 = \varkappa_{\alpha\beta} d\xi^\alpha d\xi^\beta .$$ (3.1.2-8)

The elements (points) of this strain space are the parameters; the line element is ds, which is the line element of the coordinates. The metric in strain space is a function of time. In the strain space $\Sigma(t)$, the tensor $\varkappa_{\alpha\beta}$ can be used to raise and lower indices, if the contravariant metric tensor is defined as follows:

$$\varkappa_{\alpha\beta} \varkappa^{\beta\gamma} = \delta_\alpha^\gamma$$ (3.1.2-9)

where δ_α^γ signifies the Kronecker symbol.

Because of Eq. (3.1.2-5) the metric in strain space must be flat. This means that the contracted Riemann-Christoffel curvature tensor in Σ must be zero. This imposes six conditions upon the $\varkappa$'s and hence upon the strains. These conditions are very well known in the theory of elasticity where they are called "compatibility conditions".

Henceforth, we shall consider $\varkappa_{\alpha\beta}$ (and therewith $\varepsilon_{\alpha\beta}$) as a tensor in parameter space, and not as a metric. The problem, therefore, will be to determine the components of a tensor "field" $\varepsilon_{\alpha\beta}$ as a function of time.

As a final kinematical notion, one can introduce the concept of density. Postulating the medium as homogeneously of density ϱ_0 in the unstrained state, we define, in accordance with the principle of mass conservation:

$$\varrho(\xi_\alpha, t) = \varrho_0 \sqrt{\det \zeta} / \sqrt{\det \varkappa} = \varrho_0 \sqrt{\det \zeta_{\alpha\beta}} / \sqrt{\det (2\varepsilon_{\alpha\beta} + \zeta_{\alpha\beta})} . \quad (3.1.2-10)$$

The above definition of strain was termed "material" because it gives the line element ds at time t in terms of the parameters. It is customary in hydrodynamics to term all description of motion in terms of the parameters (which are identifiable with the coordinates at t_0) as "material" and this terminology is being retained here.

Since parameter and coordinate space are entirely homologous, it is obviously possible to reverse the rôles played by the two[4,5]. Thus, let the line element $d\sigma$ of parameter space (i.e., the line element of the material at time $t = t_0$) be expressed in terms of the coordinates at the time t. One obtains:

$$d\sigma^2 = \frac{\partial \xi_\alpha}{\partial x_i} \frac{\partial \xi_\alpha}{\partial x_j} dx_i dx_j = k_{ij} dx_i dx_j \quad (3.1.2-11)$$

where

$$k_{ij} = \frac{\partial \xi_\alpha}{\partial x_i} \frac{\partial \xi_\alpha}{\partial x_j} . \quad (3.1.2-12)$$

This means that the distance $d\sigma$, which was taken up at t_0 by two points of the medium differing at time t by dx_i, is specified by a symmetric tensor k_{ij} in coordinate space.

It is fairly easy to calculate the connection between spatial and material (as defined above) distances. One obtains:

$$k_{ij} = \varkappa_{\alpha\beta} \frac{\partial \xi_\alpha}{\partial x_m} \frac{\partial \xi_m}{\partial x_i} \frac{\partial \xi_\beta}{\partial x_r} \frac{\partial \xi_r}{\partial x_j} . \quad (3.1.2-13)$$

In particular, the state of zero strain is characterized by the following tensor:

$$z_{ij} = \zeta_{\alpha\beta} \frac{\partial \xi_\alpha}{\partial x_m} \frac{\partial \xi_m}{\partial x_i} \frac{\partial \xi_\beta}{\partial x_r} \frac{\partial \xi_r}{\partial x_j} . \quad (3.1.2-14)$$

Thus, one can define the "spatial strain" as follows:

$$e_{ij} = \frac{1}{2}(k_{ij} - z_{ij}) = \frac{1}{2}\left(\frac{\partial \xi_\alpha}{\partial x_i} \frac{\partial \xi_\alpha}{\partial x_j} - z_{ij} \right) . \quad (3.1.2-15)$$

4 Deuker, E. A.: Dtsch. Math. 5:546 (1941)
5 Eckart, C.: Phys. Rev. 73:373 (1948)

In the above formulas, it should be noted that the tensor z_{ij} characterizing the state of zero strain is no longer time-independent. The spatial density becomes:

$$\varrho(x_i, t) = \varrho_0(\sqrt{\det z}/\sqrt{\det k}) = \varrho_0\sqrt{\det z_{ij}}/\sqrt{\det(2e_{ij} + z_{ij})} \quad (3.1.2-16)$$

which is the same as the material density. The latter statement is easy to check as the expression for z as well as k can be written as the product of the matrix ζ or $\varkappa$, respectively, with the same matrices. Upon the formation of the determinants, the determinants of these matrices can be factorized out and cancel as they are the same in numerator and denominator.

It will be noted that the tensor k_{ij}, also, if taken as a metric tensor, describes a flat metric. Hence it must satisfy the condition that the Riemann-Christoffel curvature tensor, formed with k_{ij}, must be the zero tensor. This leads to six compatibility relations for k_{ij}, and hence for e_{ij}, as was the case for the material strains.

Finally, one may make a few remarks regarding "convected coordinates". We have seen above that the tensors $\varkappa_{\alpha\beta}$ and k_{ij}, being symmetric tensors, can be thought of as metric tensors in certain spaces. This is the approach to finite displacement rheology which has been taken by Oldroyd[6]. The tensor $\varkappa_{\alpha\beta}$, in fact, can be regarded as describing the metric in a "convected" coordinate system, viz. in a system whose coordinate lines are given at $t = t_0$ as Cartesian coordinate lines, and are moving along with the medium. If the motion is expressed in terms of such convected coordinates, one has to introduce the whole formalism of Riemannian geometry, which is actually quite unnecessary in view of the fact that, after all, the medium is moving in an ordinary Euclidean space. Convected coordinates, therefore, appear as rather clumsy means of describing the dynamics of continuous matter.

b) Continuity Condition. In order to expand the theory further, it is necessary to define time derivatives of the various quantities introduced above.

Owing to the occurrence of parameters and of coordinates, time differentiation of any function can be performed either with the parameters held constant, or with the coordinates held constant. Although the complete reciprocity between parameter space and coordinate space has already been demonstrated by Deuker[7], this fact is not commonly brought to light in the customary presentation of rheological theories.

We shall denote the time derivative of a scalar function of the parameters and coordinates, for constant parameters, by D/Dt:

$$\frac{D}{Dt}f(\xi, t) = \frac{\partial f}{\partial t}\bigg|_{\text{constant }\xi} \quad (3.1.2-17)$$

6 Oldroyd, J. G.: Proc. R. Soc. London Ser. A 200:523 (1950)
7 Deuker, E. A.: Dtsch. Math. 5:546 (1941); see also Green, A. E., Zerna, W.: Philos. Mag. (7) 41:313 (1950)

and the time derivative with the coordinates held constant, by

$$\frac{\Delta}{\Delta t} f(x, t) = \frac{\partial f}{\partial t}\bigg|_{\text{constant } x} .$$
(3.1.2 – 18)

With the definition of time derivatives, one is now in a position to formulate the continuity condition. It is

$$\frac{\Delta \varrho}{\Delta t} + \frac{\partial}{\partial x_i} \left(\varrho \frac{D x_i}{D t} \right) = 0 .$$
(3.1.2 – 19)

In the presentation of the continuity equation, care has been taken to indicate the various types of time derivatives, according to our notation. It becomes then apparent that the usual form of the continuity equation is somewhat cumbersome since it contains functions which have x as well as ζ as arguments.

c) The Equations of Motion. A similar situation occurs in the equations of motion which are usually written as follows:

$$\frac{\partial \tau_{ik}}{\partial x_k} + \varrho \left(f_i - \frac{D^2 x_i}{D t^2} \right) = 0 .$$
(3.1.2 – 20)

The equations of motion contain several important concepts. The quantity τ_{ik} is called the stress tensor, f_i the specific mass force; ϱ is as usual the density of the material. It is customary to define the stress tensor as a function of x_i. The stress tensor τ_{ik} is postulated in such a fashion that, upon any imagined closed surface within a body, there exists a distribution of stress vectors (tractions) $p_{(n)i}$ whose resultant and moment are equivalent to those the actual forces of material cohesion exerted by the material outside upon that inside[8], and that these stress vectors can be written as follows

$$p_{(n)i} = \tau_{ij} n_j$$
(3.1.2 – 21)

where n_j is the normal unit vector to any surface element under consideration. It is customary to represent the normal component of the stress vector upon a given surface by σ, the tangential one by τ, and to call them tension and shear, respectively.

A much used representation of the stress tensor has been devised by Mohr[9]. In a two-dimensional stress state, one can represent the stress tensor by the locus of all the corresponding points in a $\sigma - \tau$ diagram. This locus turns out be a circle (Mohr's circle, see Fig. 63), for one has:

---

8 Truesdell, C.: J. Rat. Mech. Anal. 1:125 (1952)

9 Mohr, O.: Abhandlungen aus dem Gebiete der technischen Mechanik, 3. Aufl. Berlin: Wilh. Ernst & Sohn 1928

$$\sigma = \frac{\sigma_1 + \sigma_2}{2} + \frac{\sigma_1 - \sigma_2}{2} \cos 2\,(n, 1)\,, \qquad (3.1.2-22)$$

$$\tau = (-)\,\frac{\sigma_1 - \sigma_2}{2}\sin 2\,(n, 1) \qquad (3.1.2-23)$$

where σ_1 and σ_2 denote the principal stresses, i.e., the tractions for those orientations of the surface in the point under consideration for which the shear is zero. In the three-dimensional case, one can accordingly represent a stress tensor by three limiting circles, each corresponding to the plane defined by each pair of dimensions.

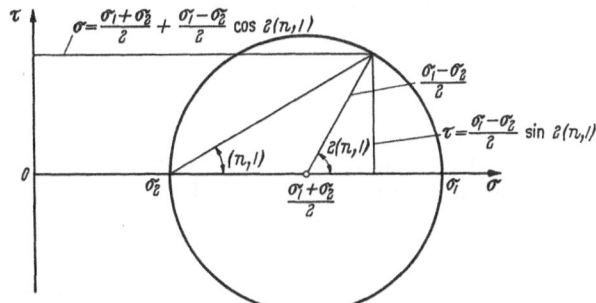

Fig. 63. Mohr circle for a two-dimensional stress state

d) Rheological Condition. The relations noted heretofore are not sufficient to determine the behavior of continuous matter. What is needed is a connection between the kinematical quantities and the dynamical quantities. Such relations are commonly termed "rheological conditions", often also called "constitutive equations".

The rheological condition, being an equation of state, must be independent of the motion of the medium as a whole. It, therefore, appears as natural to use as kinematical variables the strains $\varepsilon_{\alpha\beta}$ or the components of the tensor $\varkappa_{\alpha\beta}$ describing the state of the medium in parameter space.

Many types of rheological equations have been postulated (see Sect. 3.2), but it is to the credit of Oldroyd[10] to have made a satisfactory enumeration of possible variables and the form in which they may enter the rheological condition. Oldroyd states that, in its general form, the rheological condition can be written as a set of integrodifferential equations in parameter space, of which six are independent, relating the stresses τ_{ik}, the finite strains $\varepsilon_{\alpha\beta}$ (or else the components of the tensor $\varkappa_{\alpha\beta}$), the temperature $T(\xi, t)$, and such physical constants as may pertain to the medium.

The quantities τ_{ik}, $\varepsilon_{\alpha\beta}$ and T are functions of time. The fact that it is permissible for the rheological condition to contain integrations of time (with

10 Oldroyd, J. G.: Proc. R. Soc. London Ser. A 200:523 (1950)

fixed ξ) allows for the possibility of occurrence of effects which depend on the total previous strain history at a certain point of the medium. In simple cases, the integrodifferential equations can be reduced to differential equations by the processes of rearrangement and differentiation.

e) Boundary Conditions. With the conditions enumerated above expressed mathematically, it is, in principle, possible to solve any given problem: The conditions reduce to a set of integro-differential equations. In order to make the solution of the latter specific to a given problem, the appropriate boundary and initial conditions must be stated.

Finally, in order to solve a concrete dynamic problem, it is necessary to express all the equations in one type of independent variable. The functions sought will then be either $x_i(\xi_\alpha, t)$ or $\xi_\alpha(x_i, t)$, or the corresponding strains.

3.1.3 Inhomogeneous Media

Earth materials, particulary near the surface, are often sufficiently in-homogeneous so that they cannot be treated as a single continuous system. This is particulary the case for fluid-filled porous or granular rocks. In such cases, it is possible to consider the rock matrix and the fluid as two separate, macroscopically homogeneous media which interact with each other.

In connection with geodynamic problems, the macroscopic motion of the fluid is of relatively little importance, but the stresses (i. e., the pressure) prevailing therein can have a great influence on the displacements in the rock matrix. The description of the deformation dynamics of a fluid-filled porous medium (with the fluid under pressure) in its full generality poses quite a problem. Essentially, there are three types of stesses that are involved:

The bulk stress σ_T

the fluid stress σ_F

the skeleton stress σ_S .

Matters become simpler, if the total stress σ_T is a pressure (σ_F is a pressure, viz. the fluid pressure p_F, in any case). Then one has

p_T = overburden pressure

p_F = fluid pressure

p_S = skeleton pressure .

These tree quantities are not independent. From a simple consideration of equilibrium follows

$$p_T = p_F \Phi + p_S(1 - \Phi) \tag{3.1.3-1}$$

where Φ is the (areal as well as spacial) porosity of the porous medium. Now, a simple theory accounting for the effects of pore pressure on the strains and/or displacements in a (porous) medium has been given by Terzaghi[11]. Accordingly, one may state that the deformation of the medium is caused solely by the "effective" stress p_E

$$p_E = p_T - p_F .$$ (3.1.3 – 2)

This is called the "principle of effective stress" of Terzaghi. It can be easily deduced from the law of Archimedes for a fluid-filled column of grains of height h. In such a column, the total pressure is at the bottom

$$p_T (1 - \Phi)\varrho_m gh + \Phi\varrho_f gh$$ (3.1.3 – 3)

where ϱ_m is the density of the grains and ϱ_f the density of the fluid, g is the gravity acceleration and Φ is again the porosity. As a next step, it is assumed that *deformations* in the *grains* can only be caused by the pressure p_w corresponding to the immersed weight of the grains. The latter is, according to Archimedes' law

$$p_w = (1 - \Phi)(\varrho_m - \varrho_f)gh .$$ (3.1.3 – 4)

However, this expression for p_w is exactly equal to $p_E = p_T - p_F$, as can easily be verified:

$$\begin{aligned}
p_T - p_F &= (1 - \Phi)\varrho_m gh + \Phi\varrho_f gh - \varrho_f gh \\
&= (1 - \Phi)\varrho_m gh - (1 - \Phi)\varrho_f gh \\
&= (1 - \Phi)(\varrho_m - \varrho_f)gh = p_w .
\end{aligned}$$

Thus, the "effective" pressure p_E is equal to the pressure caused by the weight of the submersed grains against each other; the physical principle that only *this* pressure can cause deformations in the porous medium is called the "Terzaghi principle of effective pressure". This principle, then, has been extrapolated to the general case; it then states that *all* deformations in the porous medium are solely caused by the effective *stress* σ_E defined as follows

$$(\sigma_E)_{ik} = (\sigma_T)_{ik} - p_F \delta_{ik} .$$ (3.1.3 – 5)

The validity of the Terzaghi principle has been confirmed by many experimental investigations[12]. In particular, it has been checked repeatedly for geological materials. Thus, even in crystalline rocks of low porosity the law was found to be valid[13, 14] provided the rate of loading is slow enough.

11 Terzaghi, K.: Sitzungsber. Akad. Wiss. Wien Math.-Natwiss. Kl. Abt. IIa, 132:105 (1923)
12 Skempton, A. W., Proc. Conf. Pore Press and Suction in Soils, p. 4. London: Butterworths (1960)
13 Brace, W. F., Martin, R. J.: Int. J. Rock Mech. Min. Sci. 5:415 (1968)
14 Brace, W. F.: Can. Geol. Surv. Pap. 68 – 52:113 (1969)

3.1.4 Additional Stress and Strain

It is evident from the above exposition that the problem of determining the displacement from a set of boundary conditions and from the knowledge of the appropriate rheological condition is, in its full generality, a very difficult one. It is therefore often necessary to make certain simplifications.

A device which is often used is the restriction that all that one attempts to calculate is the deviation from a certain "standard" stress (and therefore also strain) state. Thus, assuming a certain stress state, one aims at determining the "additional" stress and the corresponding "additional" strains. It is in this instance often possible to assume that the additional strains are small.

A justification of the above-indicated procedure lies in the fact that the tensor of zero strain is, in reality, a badly defined quantity; − at least in the manner as it was introduced in Eq. (3.1.2 − 7). One can therefore argue that one could have just as well taken any other tensor and called it "zero strain tensor", − thus, for instance, the tensor of the state from which one wants to calculate the deviation. It should be mentioned, however, that there is really a physical way conceivable in which the zero strain state can be defined. This is as follows: Imagine that one cuts a small volume element surrounding the point in which the zero strain state is to be defined. Assume further that all the tractions across its surface be removed. The material, then, presumably, will reach (at least after a long time) an equilibrium state; the latter may be taken as the zero strain state. It is evident that the zero strain state defined in this manner may not be integrable; i. e., it is quite possible that no position of the body exists in which all the strains (and therewith the stresses after infinite time) are removed.

The above physical definition of a zero strain state introduces certain complications. First of all, with the loss of integrability, one also loses the ordinary compatibility relations for the strains; − i. e., the latter have to be replaced by some different relations expressing that the *additional* displacements starting from a *real* state must take place in Euclidean space. It is therefore logically much more satisfactory to admit as zero strain states only actually possible states of the body although the latter may not be zero strain states physically.

Under the above conditions it seems logical to *define* the chosen zero strain state also as a zero stress state; − i. e., one concerns oneself only with the stresses *additional* to those present in the zero strain state. However, if this be done, it is evident that the rheological condition becomes, in general, a function of the zero strain state. The only instance where the rheological condition does not depend on the zero strain state is when the latter is a linear relationship between stress and strain: for only then is the additional displacement the same function of the additional stress, regardless of the amount of prestrain. This is the case only in infinitesimal elasticity theory (see. Sect. 3.2.1 *infra*).

The device of using additional stress and strain is very common in geodynamics. The "standard" state is the hydrostatic state of the geoid; all stresses and strains of interest are deviations therefrom.

3.2 Theoretical Rheology

3.2.1 Infinitesimal Elasticity Theory

3.2.1.1 Basic Assumptions

To a great extent, the behavior of a continuous medium is determined by its rheological properties. It will therefore be necessary to investigate the various types of rheological conditions that will be of importance for Earth materials. We shall start with infinitesimal elasticity theory.

Infinitesimal elasticity theory is obtained from the general scheme of rheological dynamics if it is assumed that (a) the displacements are always small, (b) the rheological condition is as expressed by Hooke's law.

These two assumptions completely define a mathematical theory of deformation.

Assumption (a) permits one to express the coordinates as follows [cf. Eq. (3.1.2 – 3)]

$$x_i(\xi_\alpha, t) = x_i(\xi_\alpha, 0) + u_i(\xi_\alpha, t) \qquad (3.2.1-1)$$

where u_i is called displacement. As indicated above, the displacements are assumed as small so that their squares can be neglected. It is also customary to identify the parameters with the coordinates at time $t = 0$ so that

$$x_i(t = 0) = \xi_i. \qquad (3.2.1-2)$$

Under these assumptions, the expression for the (material) strains (3.1.2 – 7) reduces to

$$\varepsilon_{ik} = \frac{1}{2}\left(\frac{\partial u_i}{\partial x_k} + \frac{\partial u_k}{\partial x_i}\right). \qquad (3.2.1-3)$$

Consequently, the compatibility relations for the strains (i. e., the condition that the Riemann-Christoffel curvature tensor of the metric $\varepsilon_{ik} + \delta_{ik}$ is zero) turns out to be

$$\frac{\partial^2 \varepsilon_{(i)(i)}}{\partial x_j^2} + \frac{\partial^2 \varepsilon_{(j)(j)}}{\partial x_i^2} = 2\frac{\partial^2 \varepsilon_{ij}}{\partial x_i \partial x_j}, \qquad (3.2.1-4a)$$

$$\frac{\partial^2 \varepsilon_{(i)(i)}}{\partial x_i \partial x_k} = \frac{\partial}{\partial x_{(i)}}\left(-\frac{\partial \varepsilon_{ik}}{\partial x_i} + \frac{\partial \varepsilon_{ik}}{\partial x_j} + \frac{\partial \varepsilon_{ij}}{\partial x_k}\right). \qquad (3.2.1-4b)$$

In the above scheme the strain ε_{ik} as defined in Eq. (3.2.1 – 3) represents actually a tensor as in indicated by the notation. Care should be taken in referring to the literature since the definition of "strain" is not always that given here. Notably, it will often be found that the "shear" strains are taken as twice the corresponding components as defined here. This, however, effects that the strain is no longer representable as a tensor.

Assumption (b) (i.e., Hooke's law) can be stated as follows (see e.g., Jeffreys[15])

$$\tau_{ij} = c_{ijkl}\varepsilon_{kl} \qquad (3.2.1-5a)$$

where c_{ijkl} is called the fourth-order elastic modulus (stiffness tensor).

For an isotropic body the stiffness tensor has only two independent components and can be written as follows

$$c_{ijkl} = \lambda\,\delta_{ij}\delta_{kl} + \mu(\delta_{ik}\delta_{jl} + \delta_{il}\delta_{jk}) \qquad (3.2.1-5b)$$

so that Eq. (3.2.1 − 5a) becomes

$$\tau_{ij} = \lambda\,\delta_{ij}\varepsilon_{kk} + 2\mu\varepsilon_{ij} \qquad (3.2.1-5c)$$

which expresses that there is proportionality of the isotropic and deviatoric components of stress and strain tensors separately. The constants λ and μ are called "Lamé's constants". The quantity μ is also often referred to as "rigidity".

The elasticity constants introduced in Eq. (3.2.1 − 5c) are not the only ones that are possible. In principle, any two constants that are linearly independent functions of λ and μ could be used. Thus, the following constants have been employed (cf. Love[16]):

Young's modulus E

$$E = \frac{\mu(3\lambda + 2\mu)}{\lambda + \mu} \qquad (3.2.1-6a)$$

Poisson's ratio m

$$m = \frac{\lambda}{2(\lambda + \mu)} \qquad (3.2.1-6b)$$

bulk modulus k (incompressibility)

$$k = \lambda + \tfrac{2}{3}\mu. \qquad (3.2.1-6c)$$

There are many good books on elasticity to which the reader is referred for further details[16-19].

15 Jeffreys, H.: Cartesian Tensors. London: Cambridge Univ. Press 1931
16 Love, A. E. H.: A Treatise on the Mathematical Theory of Elasticity, 4th edn. London: Cambridge Univ. Press 1927
17 E. g. Timoshenko, S., Goodier, J. N.: Theory of Elasticity, 2nd edn. New York: McGraw-Hill 1951
18 Muskhelishvili, N. I.: Some Basic Problems of the Mathematical Theory of Elasticity (transl. from Russian). Groningen: Noordhoff 1953
19 Green, A. E., Zerna, W.: Theoretical Elasticity. Oxford: Clarendon Press 1954

3.2.1.2 Special Cases

In geodynamics, some special cases and conditions of elastic bodies are of particular importance. We shall now consider some of these.

1. Elastic Equilibrium in Two Dimensions. In the present study, we shall be chiefly concerned with equilibrium problems. In this connection, the introduction of stress functions (Airy functions) has been proven to be very convenient. This is particularly true for two limiting cases: the plane strain state and the plane stress state[20].

In the *plane strain state* one assumes that

$$\left.\begin{array}{l} \varepsilon_{33} = \varepsilon_{23} = \varepsilon_{31} = 0 \\ \varepsilon_{11}, \varepsilon_{22}, \varepsilon_{12} \neq 0 . \end{array}\right\} \tag{3.2.1-7}$$

It can easily be verified that the conditions of equilibrium are satisfied if the following assumption is made

$$\tau_{11} = \partial^2 \varphi / \partial x_2^2 , \tag{3.2.1-8a}$$

$$\tau_{22} = \partial^2 \varphi / \partial x_1^2 , \tag{3.2.1-8b}$$

$$\tau_{12} = -\partial^2 \varphi / \partial x_1 \partial x_2 \tag{3.2.1-8c}$$

where φ is a stress function. It must satisfy the following differential equation (the Laplacians to be taken in two dimensions)

$$\text{lap lap } \varphi = 0 . \tag{3.2.1-9}$$

Any solution of Eq. (3.2.1-9) generates a solution of the plane strain equilibrium problem.

A similar situation holds for the *plane stress state.* We suppose

$$\tau_{33} = 0 ; \quad \tau_{23} = \tau_{31} = 0 \tag{3.2.1-10}$$

whereupon it can be shown that a solution of the problem is again obtained if one sets

$$\tau_{11} = \partial^2 \varphi / \partial x_2^2 , \tag{3.2.1-11a}$$

$$\tau_{22} = \partial^2 \varphi / \partial x_1^2 , \tag{3.2.1-11b}$$

$$\tau_{12} = -\partial^2 \varphi / \partial x_1 \partial x_2 \tag{3.2.1-11c}$$

where φ is again a stress function that must satisfy the following differential equation

$$\text{lap lap } \varphi = 0 . \tag{3.2.1-12}$$

---

20 Cf. e. g., Jaeger, J. C.: Elasticity, Fracture and Flow. London: Methuen & Co. Ltd. 1956

2. Waves. An important special case of elasticity theory concerns the existence of body waves. We note that the equations of motion can be written in terms of the displacements as follows, provided external forces are absent:

$$\varrho \frac{\partial^2 u_i}{\partial t^2} = (\lambda + \mu) \frac{\partial [\partial u_j / \partial x_j]}{\partial x_i} + \mu \operatorname{lap} u_i . \qquad (3.2.1-13)$$

If the divergence is formed of both sides of Eq. (3.2.1 – 13), one obtains

$$\varrho \frac{\partial^2 \Theta}{\partial t^2} = (\lambda + 2\mu) \operatorname{lap} \Theta \qquad (3.2.1-14)$$

with

$$\Theta = \partial u_j / \partial x_j , \qquad (3.2.1-15)$$

and, similarly, if the curl of Eq. (3.2.1 – 13) is formed, one obtains

$$\varrho \frac{\partial^2}{\partial t^2} [\operatorname{curl}(u_i)] = \mu \operatorname{lap} [\operatorname{curl}(u_i)] . \qquad (3.2.1-16)$$

Eqs. (3.2.1 – 14) and (3.2.1 – 16) have the form of wave equations. They imply that a dilatational disturbance Θ may be transmitted with a phase velocity v_p of

$$v_p = \sqrt{(\lambda + 2\mu)/\varrho} \qquad (3.2.1-17)$$

whereas a rotational disturbance may be transmitted with a phase velocity v_s of

$$v_s = \sqrt{\mu/\varrho} . \qquad (3.2.1-18)$$

Both types of body waves are found in the Earth; where they have been called P and S waves, respectively (see. Sect. 2.1.2). Dispersion is usually absent so that the above phase velocities are equal to the speed with which a disturbance travels through the Earth.

The above discussion can be extended and it can then be shown that, in addition to body waves, *surface waves* may also exist. The surface waves are usually subject to dispersion so that a distinction has to be made between group velocity and phase velocity. However, we shall not concern ourselves here with the detailed discussion of elastic waves which belongs into a treatise on elasticity theory. In that connection, it is standard textbook material[21-23].

3. Stresses Around Holes. A further interesting application of the equations of elasticity theory is Inglis'[24] determination of the stresses in a plate which

21 Cf. e. g. Ewing, M., Jardetzky, W. S., Press, F.: Elastic Waves in Layered Media. New York: McGraw-Hill Book Co. 1957
22 Brekovskikh, L. M.: Waves in Layered Media (transl. from Russian). London, New York: Academic Press 1960
23 Miklowitz, J.: Appl. Mech. Rev. 13:865 (1960) gives a good list of references
24 Inglis, C. E.: Trans. Inst. Nav. Archit. 55:Part 1, 219 (1913)

has an *elliptic hole*. Inglis[24] employed curvilinear coordinates α, β which are connected with a Cartesian system as follows:

$$\left.\begin{array}{l} x = c \cosh \alpha \cos \beta \\ y = c \sinh \alpha \sin \beta . \end{array}\right\} \qquad (3.2.1-19)$$

The stress determination of Inglis using the methods of general elasticity theory is quite straightforward although it is somewhat tedious. For details, the reader is referred to the cited paper[24].

The Inglis solution is valid for an elliptical hole. The formulas become somewhat simpler in the case of a *circular* hole in a plate. In fact, the corresponding formulas had been given much earlier by Kirsch[25]. For a single normal stress S at infinity the stresses around a circular hole of radius a become

$$\left.\begin{array}{l} \sigma_r = \dfrac{S}{2}\left[1 - \dfrac{a^2}{r^2} + \left(1 - 4\dfrac{a^2}{r^2} + 3\dfrac{a^4}{r^4}\right)\cos 2\Theta\right], \\[3mm] \sigma_i = \dfrac{S}{2}\left[1 + \dfrac{a^2}{r^2} - \left(1 + 3\dfrac{a^4}{r^4}\right)\cos 2\Theta\right], \\[3mm] \tau = -\dfrac{S}{2}\left(1 + 2\dfrac{a^2}{r^2} - 3\dfrac{a^4}{r^4}\right)\sin 2\Theta, \end{array}\right\} \qquad (3.2.1-20)$$

where the meaning of the symbols is explained in Fig. 64. For a stress state in which a second normal stress is acting at right angles to S, the solutions corresponding to Eq. (3.2.1 – 34) for each of the two stress states must simply be superposed.

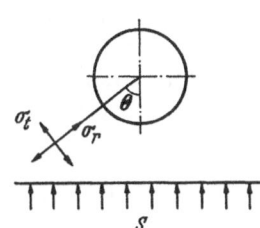

Fig. 64. Stresses around a cylindrical hole

4. *Circular Cylinders.* An analogous case to circular holes in a plate is that of circular cylinders under compression. The corresponding stress states correspond to those set up in standard laboratory tests of cylindrical specimens of concrete or rocks. Generally the stress distribution in such cylinders has been

25 Kirsch, G.: Z. Ver. Dtsch. Ing. 42:1 (1898)

assumed as linear; however, it has been pointed out by Al-Chalabi and Huang[26] that this procedure is unsatisfactory. In fact, the stress state in the specimens is triaxial because of the boundary conditions at the end faces of the cylinders which include friction. The authors cited above have obtained a solution of the elastostatic equilibrium equations which takes into account different degrees of friction at the end faces. The formulas obtained run over several printed pages and are therefore not reproduced here. The reader interested in the details is referred to the cited paper[26].

5. Buckling. Finally, we may note that the basic equations of elasticity allow solutions to exist which are unstable. This means that, in addition to the "straightforward" solution, there also exist solutions in which the deformations exceed all bounds. Such occurrences are referred to as *buckling*. A simple case where this takes place may be visualized by imagining a thin rod being compressed. The "straightforward" solution simply represents a shortening of the rod lengthwise, but this becomes unstable as soon as the compression exceeds a certain value. The rod then buckles sideways.

3.2.1.3 Dislocations

The infinitesimal theory of elasticity leads to the notion of *dislocation* by the observation that the displacement u_i corresponding to a given strain ε_{ik} in a multiply connected body may not be single-valued. In order to restore one-valuedness of the displacement, one has to introduce surfaces of discontinuity which make the body once again singly connected. Such discontinuities are called dislocations[27]. It is thus evident that dislocations require the existence of non-evanescible circuits in the body. In the limit, the multiple connection of the body can be achieved by the assumption of singular lines upon which the strains are not continuous. These singular lines must either be closed in themselves or else begin and end on the external surface of the body. They are the rims of the surfaces of discontinuity restoring the single-connectedness referred to above.

Let the strain field ε_{ik} be given and calculate the displacement u_i at a point x_i^1 proceeding from a point x_i^0. The displacement is given by the line integral[28,29]

$$u_i(x_i) = \int_{x^0}^{x^1} \frac{\partial u_i}{\partial x_i} dx_k \qquad (3.2.1-21)$$

taken along any path from x^0 to x^1. It can then be shown that the value $u_i(x_i)$ is not independent of the path.

26 Al-Chalabi, M., Huang, C. L.: Int. J. Rock Mech. Min. Sci. 11:45–56 (1974)
27 Benoit, W.: Eclogae Geol. Helv. 72 (2):571 (1979)
28 See e. g. Love, A. E. H: A Treatise on the Mathematical Theory of Elasticity, 4th edn. London: Cambridge Univ. Press 1927
29 Steketee, J. A.: Can. J. Phys. 36:192 (1958)

Owing to the definition of strain, one has in general

$$\frac{\partial u_i}{\partial x_k} = \varepsilon_{ik} - \omega_{ik} \tag{3.2.1-22}$$

with

$$2\omega_{ik} = -\frac{\partial u_i}{\partial x_k} + \frac{\partial u_k}{\partial x_i}. \tag{3.2.1-23}$$

Hence

$$u_i^1 - u_i^0 = \int_{x0}^{x1} \varepsilon_{ik} dx_k - \int_{x0}^{x1} \omega_{ik} dx_k. \tag{3.2.1-24}$$

The second integral can be written as follows (omitting the x in the limits of the integral, and writing only its superscript)

$$-\int_0^1 \omega_{ik} dx_k = -\int_0^1 \omega_{ik} d(x_k - x_k^1) = +\omega_{ik}^0(x_k^0 - x_k^1) + \int_0^1 (x_k - x_k^1) d\omega_{ik} \tag{3.2.1-25}$$

where

$$d\omega_{ik} = \frac{\partial \omega_{ik}}{\partial x_l} dx_l. \tag{3.2.1-26}$$

We now have the identity

$$\frac{\partial \omega_{ik}}{\partial x_j} = \frac{\partial \varepsilon_{jk}}{\partial x_i} - \frac{\partial \varepsilon_{ij}}{\partial x_k} \tag{3.2.1-27}$$

as one may easily verify by differentiation. Hence we have

$$u_i^1 = u_i^0 + \omega_{ik}^0(x_k^0 - x_k^1) + \int_0^1 \Lambda_{ik} dx_k \tag{3.2.1-28}$$

with

$$\Lambda_{ik} = \varepsilon_{ik} + (x_l - x_l^1)\left\{ \frac{\partial \varepsilon_{kl}}{\partial x_i} - \frac{\partial \varepsilon_{ik}}{\partial x_l} \right\}. \tag{3.2.1-29}$$

It is observed that the Λ_{ik} satisfy the following differential equations

$$\frac{\partial \Lambda_{ik}}{\partial x_l} = \frac{\partial \Lambda_{il}}{\partial x_k} \tag{3.2.1-30}$$

so that the integral in Eq. (3.2.1–28) is the same for reconcilable paths. Thus, in any singly connected body, the displacements must be single-valued, because along any circuit that returns to any point under consideration, Λ_{ik} satisfies the Eq. (3.2.1–28), and because all circuits are "evanescible", i. e., can be contracted into a point. By the same token, it is obvious that Λ_{ik} in a multiply connected body can be chosen such that the line integral in Eq. (3.2.1–28) around a non-evanescible circuit is not zero. This proves the existence of dislocations.

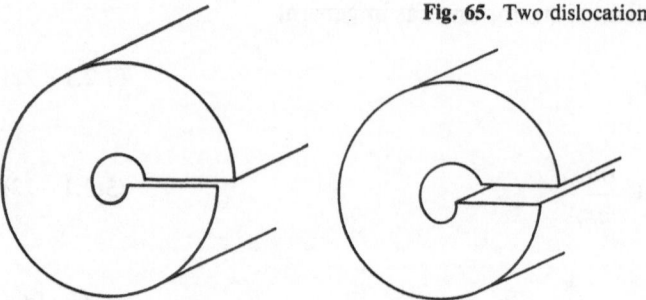

Fig. 65. Two dislocation models. (After Saito[30])

Some typical examples of dislocations are shown in Fig. 65 (after Saito[30]).

3.2.1.4 Heterogeneous Materials

A particular case arises when an attempt is made to characterize a heterogeneous elastic medium by its "average" properties. Thus, the problem may arise of finding the average elastic properties in a medium which is some sort of aggregate: e. g., if it is a polycrystalline solid or a porous medium.

The general implications of this problem have recently been reviewed by Watt et al.[31]. Accordingly, a theoretically satisfactory treatment can be based on a complete statistical analysis which requires a detailed knowledge of the correlation functions of all orders involving combinations of the material properties of the aggregate. Unfortunately, although this approach is theoretically the most satisfactory one, the determination of the required correlation functions is well-nigh impossible so that the approach is of little practical value. The same can be said of methods invoking scattering theory analogs from solid state physics.

In practise, therefore, one uses semi-empirical bounds for bracketing the elastic properties of an aggregate. Such bounds have been set up by Voigt[32], who assumed that the strain is uniform throughout the aggregate, and by Reuss[33], who assumed that the stress is uniform. The result for an isotropic mixture of isotropic phases of n components is

$$M_{\text{Reuss}} \equiv \left(\sum_{i=1}^{n} v_i/M_i \right)^{-1} \leqslant M_{\text{true}} \leqslant \sum_{i=1}^{n} v_i M_i \equiv M_{\text{Voigt}} \qquad (3.2.1-31)$$

where M is the elastic modulus (either λ or μ) in question and v_i the volume fraction of the i-th phase. The above result was deduced by Hill[34,35] who also

30 Saito, Y.: Geophys. Mag. 28(3):329 (1958)
31 Watt, J. P., Davies, G. F., O'Connell, R. J.: Rev. Geophys. Space Phys. 14 (4):541 (1976)
32 Voigt, W.: Lehrbuch der Kristallphysik. Leipzig: Teubner 1928
33 Reuss, A.: Z. Angew. Math. Mech. 9:49 (1929)
34 Hill, R.: Proc. Phys. Soc London Ser. A 65:349 (1952)
35 Hill, R.: J. Mech. Phys. Solids 11:357 (1963)

proposed to use the arithmetic mean of M_{Reuss} and M_{Voigt} as an approximation for the true modulus. This average has become known as the VRH (Voigt-Reuss-Hill) average.

In the case of porous media, recourse has generally been taken to specific models usually composed of spheres. Much of the pertinent literature has been reviewed by the author[36]. In the various models, the pores are generally taken as empty. The aim of the calculations was usually to predict the wave velocities of P and S waves in the medium. If the fluid content is taken into account, it is unavoidable to consider the communication between the pores and therewith problems of fluid flow. This leads one out of the subject of pure elasticity theory.

Finally, turning to cracks, we note that this case is even more difficult than that of pores because the former, in fact, represent singularities in the medium: The volume fraction of cracks may be zero, but their mechanical effect is pertinent. If the cracks are not randomly oriented, the medium becomes anisotropic. Again, the approach to the problem has been that of constructing theoretical models. The most obvious procedure is that of treating each crack individually as a boundary condition for a corresponding elastostatic (elastodynamic) problem, using finite-element computer techniques for the solutions. This has been successful if the cracks are relatively few, but if one has to deal with systems of cracks (joints), the computer requirements become formidable. In that case, however, one can assume the medium to consist of *elements* in which the number of cracks is small; one can, then, compute the stress concentration factors in each element (by finite-element techniques) and arrive at "effective" elastic moduli. By the nature of this approach, the results cannot be stated in general form but must be given numerically in tables[37] or graphs[38] for the various models.

3.2.2 Plasticity

3.2.2.1 Pure Plasticity

1. Foundations. Mathematical elasticity theory finds a natural extension in what is called the mathematical theory of plasticity[39]. For, is has been noted that bodies strained to a certain point (called the "elastic limit") often show a behavior which can be described with good success fairly simply without having to go into the intricacies of general finite-strain rheology.

36 Scheidegger, A. E.: Physics of Flow Through Porous Media, 3rd edn. Toronto: Univ. Toronto Press 1974
37 Singh, B.: Int. J. Rock Mech. Min. Sci. 10:311 (1973)
38 Morland, L. W.: Geophys. J. R. Astron. Soc. 37:435 (1974)
39 Hill, R.: The Mathematical Theory of Plasticity. Oxford: Clarendon Press, 1950

Thus, most metals and many other substances begin to yield in a very special way when they are strained beyond the elastic limit. The best known "yield criterion" is that of Mises[40]; it can be written as follows:

$$(\tau_{11} - \tau_{22})^2 + (\tau_{22} - \tau_{33})^2 + (\tau_{33} - \tau_{11})^2 + 6(\tau_{12}^2 + \tau_{23}^2 + \tau_{31}^2) = 6k^2.$$

$$(3.2.2-1)$$

Here, k is a parameter which depends on the amount of prestrain. Thus, if a body of the type considered here be stressed beyond its elastic limit, then the stress state is such in each point of the "plastic" region, that Eq. (3.2.2 – 1) is satified. The assumed variability of the coefficient k automatically takes into account the experimentally observed phenomenon of "strain-hardening": the coefficient is found to increase with the amount of work W_p that has been put into the *plastic* deformation:

$$6k^2 = f(W_p) .$$

$$(3.2.2-2)$$

The increment of plastic work dW_p in a plastic-elastic body may be written

$$dW_p = \tau_{ij} \left(d\varepsilon_{ij} - \frac{\partial \tau_{ij}}{2\mu} \right) = \tau'_{ij} \left(d\varepsilon_{ij} - \frac{d\tau_{ij}}{2\mu} \right)$$

$$(3.2.2-3)$$

where the dash (') indicates the deviatoric component of a tensor:

$$a'_{ij} = a_{ij} - \tfrac{1}{3} a_{nn} \delta_{ij}$$

$$(3.2.2-4)$$

and where use has been made of the (experimentally indicated) assumption that all volume changes during plastic deformation are elastic. A remark should perhaps be made concerning the strain increment $d\varepsilon$ occurring in the last formulas. In Eq.(3.2.2 – 3), the strain increment indicates the increment of the total material strain, the latter as defined by Eq.(3.1.2 – 7). However, since the plasticity formulas only refer to the strain *increment*, it is customary to think of the plastic strains in terms of an infinitesimal theory.

It is thus seen that the phenomenon of plasticity can best be described by saying that there are two "regions" of behavior of the body: an elastic one and a plastic one. In the elastic region, Hooke's law is satisfied, and in the plastic region, the stresses satisfy the yield condition (3.2.2 – 1). The yield condition itself depends on the strain history of the body.

Within the plastic region, the strain increment $d\varepsilon_{ij}$ can be split at any instant into an elastic component $d\varepsilon_{ij}^e$ and into a plastic component $d\varepsilon_{ij}^p$:

$$d\varepsilon_{ij} = d\varepsilon_{ij}^e + d\varepsilon_{ij}^p .$$

$$(3.2.2-5)$$

The rheological equation for the elastic part of the strain increment can be obtained from Hooke's law, but for the plastic part one has, so far, only the yield condition, which is not sufficient to make the displacement determined.

---

40 Mises, R. v.: Goettinger Nachr. Math.-Phys. Kl. 1913:582 (1913)

An additional assumption is therefore necessary for which Reuss[41] has proposed:

$$d\varepsilon_{ij}^p = \tau_{ij}' d\lambda \tag{3.2.2-6}$$

where $d\lambda$ is a scalar factor of proportionality which has to be determined experimentally. It expresses the amount of strain hardening that the body exhibits. Eq. (3.2.2−6) signifies that the principal axes of stress and strain always coincide.

2. Particular Cases. Exact solutions of the problem of finding the displacement pattern in a plastic-elastic material in its full generality are very difficult to achieve. Only few such solutions are available.

It is somewhat easier to obtain solutions if the elasticity of the material is disregarded, i. e., if one concerns oneself with a plastic-rigid material. Naturally, this can constitute only an approximation to reality. A further simplification is reached if one confines oneself to conditions of *plane strain*. This implies (a) that the (plastic) flow is everywhere parallel to a given plane (e. g., the x, y plane), and (b) that the motion is independent of z, the direction orthogonal to that plane.

In this case, the conditions for the determination of a problem reduce to

a) the yield condition

$$\tfrac{1}{4}(\sigma_x - \sigma_y)^2 + \tau_{xy}^2 = k^2 \tag{3.2.2-7}$$

b) the equilibrium conditions (cf. Eq. 3.1.2−20)

$$\partial\sigma_x/\partial x + \partial\tau_{xy}/\partial y = 0 , \tag{3.2.2-8a}$$

$$\partial\tau_{xy}/\partial x + \partial\sigma_y/\partial y = 0 \tag{3.2.2-8b}$$

c) the condition of zero volume change

$$\partial u_x/\partial x + \partial v_y/\partial y = 0 \tag{3.2.2-9}$$

d) the rheological (stress-strain) equation (from Eq. 3.2.2−6)

$$\frac{2\tau_{xy}}{\sigma_x - \sigma_y} = \left\{\frac{\partial u_x}{\partial y} + \frac{\partial v_y}{\partial x}\right\} \Big/ \left\{\frac{\partial u_x}{\partial x} - \frac{\partial v_y}{\partial y}\right\}. \tag{3.2.2-10}$$

Here, u_x and v_y are the velocity components along the x and y axes. The Eqs. (3.2.2−7/10) are sufficient to determine the unknowns σ_x, σ_y, τ_{xy}, u_x, v_y; since they are homogeneous in the velocities, they do, however, not really involve the time element. The "velocities", therefore, may be replaced by any monotonous functions of displacement.

It turns out that the above set of Eqs.(2.2.2−7/10) is hyperbolic. The characteristics are called "slip lines" [42]. It is possible to derive several theorems

41 Reuss, A.: Z. Angew. Math. Mech. 10:266 (1930)
42 On this subject, see the excellent review by Sobotka, Z.: Appl. Mech. Rev. 14:753 (1961)

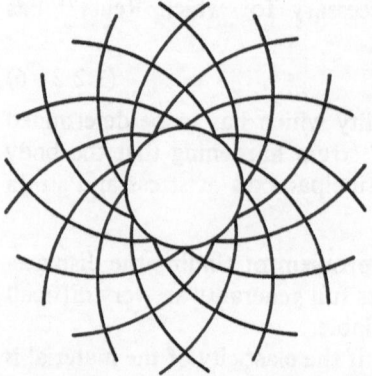

Fig. 66. Slip lines in the form of logarithmic spirals in a symmetrical stress state. (After Nadai[43])

regarding the geometry of such slip lines. This permits one to calculate the slip lines for a variety of boundary conditions. Of the many cases where slip line fields have been determined, we show here two examples. In Fig. 66 we show the slip lines resulting from a symmetrical stress state around a point as calculated by Nadai[43]. The region of plastic deformation is separated by a circle from that which remains rigid. Second, in Fig. 67 we show the plastic slip lines underneath a cylindrical stamp as determined by Hencky[44].

The slip lines are not just mathematical inventions, but do have physical reality. Characteristics allow for certain differential quotients to be discontinuous across them which permits actual physical discontinuities to exist.

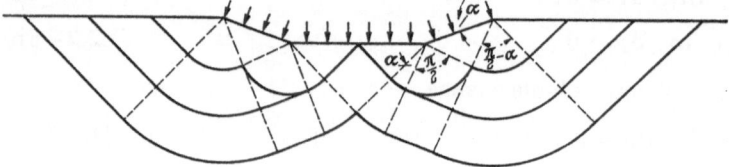

Fig. 67. Plastic slip lines underneath a cylindrical stamp. (After Hencky[44])

3. Plastic Instability. Another type of simplification of the general plasticity theory is obtained if the principles of plasticity are applied to the failure of thin steel plates under tension. This has been done by Bijlaard[45] who made the assumption that, in a thin plate, plastic flow by preference must occur in bands in whose length direction the dimensions of the material do not change during flow. This assumption immediately yields that at any point in an elastic-plastic plate there are two directions along which the above condition is fulfilled. Thus, in a homogeneous stress state, Bijlaard[45] expects two sets of

43 Nadai, A.: Z. Physik 30:106 (1924)
44 Hencky, H.: Z. Angew. Math. Mech. 3:241 (1923)
45 Bijlaard, P. P.: Rapport Assemb. Gen. Assoc. Geol. U.G.G. I. Edinburgh, 23 pp., 1936

plastic bands to develop; the angle which they enclose is of the order of 110°. Correspondingly, the angle of the bands with the principal direction of stress (minimum pressure) is 55°.

3.2.2.2 Imperfectly Plastic Materials

1. Rankine States. The theory of the behavior of imperfectly plastic materials can be described by an extension of plasticity theory. In this connection, the yield criteria given in Eq.(3.2.2−1) are generalized somewhat. In a two-dimensional stress state (σ, τ) it is taken as follows (Coulomb's equation)

$$\tau = \sigma \tan \phi + c \qquad (3.2.2-11)$$

where c is a parameter indicative of the cohesion of the material and Φ is commonly called the angle of internal friction.

The limiting shear-stress condition (3.2.2−11) is represented in Mohr's stress diagram by two straight lines (Fig. 68). So long as the stress state is such that the corresponding Mohr circle does not touch these lines, the material is in a stable condition, subject to the equations of elasticity theory. If the Mohr circle touches the limiting lines, the material is in a plastic limiting state, usually called a "Rankine"[46] state.

If the normal stress acting on an element in the 1-direction is given, then there are evidently two Rankine states possible which are represented in Fig. 68 by the Mohr circles A and B. They are called *active* and *passive* Rankine states and are attained in elements represented by the *arrows* S_A and S_B in Fig. 68; they are oriented in a direction that subtends the angles $\pm(45° \pm \phi/2)$ and $\pm(45° - \phi/2)$ toward the 1-direction, respectively.

2. Cohesionless Grains. Evidently, the plasticity theory considered in Sect. 3.2.2 is obtained if $\phi = 0$ in Eq. (3.2.2−11). The opposite limit, $\phi \neq 0$ and $c = 0$, viz.

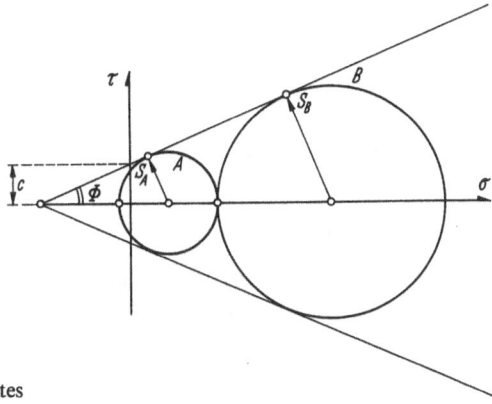

Fig. 68. Active and passive Rankine states

---

46 Rankine, N. J. M.: Philos. Trans. R. Soc. London 147:1 (1857)

$$\tau = \sigma \tan \phi \qquad\qquad\qquad\qquad (3.2.2-12)$$

corresponds to cohesionless materials, such as sands or piles of gravel. However, even sands may have a little cohesion, particularly if they are moist. If one forms a pile of a completely cohesionless material, the material will slide and not come to rest until the angle of inclination of the slope (angle of repose) becomes equal to the angle of internal friction ϕ. Materials which have cohesion as well as an angle of internal friction fall in between the two limiting cases considered above.

3. *Coulomb Slip*. In materials subject to Eq. $(3.2.2-11)$, one again encounters "slip lines" (in a plane section). The directions of the slip lines are associated with the instantaneous state of stress and make the angles ϱ

$$\varrho = \pm (45° - \phi/2) \qquad\qquad\qquad (3.2.2-12a)$$

with the σ_{min}-axis, where ϕ is, as usual, the angle of internal friction. This simply follows from an inspection of the Fig. 68 illustrating Rankine states, inasmuch as the vector S_A or S_B is reached by turning from σ_{min} through an angle of $90° - \phi$ in the corresponding Mohr diagram; hence the turning angle of the corresponding surface element is $\frac{1}{2}(90° - \phi)$. The slip lines thus obtained are called "Coulomb slip lines".

3.2.2.3 The Effect of Pore Pressure

The Eq. $(3.2.2-11)$ for the limiting shear stress is affected if there is an interstitial fluid under pressure p present in an imperfectly plastic material. In that case, Terzaghi's principle of effective stresses (see Sect. 3.1.3) applies and, according to the general rules, the stresses have to be replaced by the "effective" stresses given by Eq. $(3.1.3-5)$ in all expressions concerned. In view of the fact that this replacement affects only the normal stresses, Eq. $(3.2.2-11)$ becomes

$$\tau = (\sigma - p) \tan \phi + c \,. \qquad\qquad\qquad (3.2.2-13)$$

3.2.3 Viscous Fluids [47]

3.2.3.1 Fluid Kinematics

It seems useful to review first some of the basic facts about fluid kinematics. The theorems ensuing from that theory hold for any kind af flow.

---

47 Many good monographs exist on this subject to which the reader is referred for details; e. g., Lamb, H.: Hydrodynamics. London: Cambridge Univ. Press. 1932. Pai, S. I.: Viscous Flow Theory (2 Vols.). New York: D. van Nostrand 1957. Goldstein, J.: Modern Developments in Fluid Dynamics (2 Vols.). Oxford: Univ. Press 1938, etc.

If the *velocity vector* is drawn at every point of a moving fluid, a vector field is obtained. If this vector field is independent of time, the flow is "steady". If the curl of the velocity field is formed, a second vector field is obtained, the *vorticity field*. Finally, one can define a *vortex tube* by drawing all the vortex lines through every point of a small closed curve. The *strength* of such a vortex tube is obtained by integrating the scalar product of the vorticity vector with the unit vector normal to and of the magnitude of the element of area over any cross-section of the tube. It is a fundamental theorem of fluid kinematics that this strength is constant all along the tube. Because of this property, vortex tubes (and therefore also vortex lines) cannot begin or end in the fluid: they must be closed curves, or extend to the boundaries of the fluid.

A further important concept is that of *circulation*. Mathematically, the circulation C around a closed circuit lying entirely within the fluid is defined as follows

$$C = \oint v_s ds \tag{3.2.3-1}$$

where v_s is the component of the fluid velocity tangent to the element ds of the circuit. The integral is to be taken once completely around the circuit. In virtue of Stokes' theorem, the circulation is equal to the total strength of all the vortex tubes going through the circuit. The circulation around any given vortex tube is therefore also constant all along the tube.

The definitions of vorticity, vortex lines, vortex tubes and circulation, together with the theorems concerning the strength of a vortex tube and the connection of circulation therewith, are purely kinematic or geometrical matters, completely independent of the presence or absence of stress. They hold, therefore, for *any* kind of flow.

3.2.3.2 Dynamics of Viscous Fluids

The rheological condition for a viscous (also called "Newtonian") fluid (with η a constant of the medium called its viscosity), is for the deviatoric components

$$\tau_{ik} = 2\eta \dot{\varepsilon}_{ik} \quad (i \neq k) \tag{3.2.3-2}$$

and for the isotropic components (provided there is no bulk viscosity)

$$\partial \varrho / \partial p = \varrho \beta_f \tag{3.2.3-3}$$

where p is the pressure ($p = -\frac{1}{3}\tau_{ii}$), ϱ the density of the fluid, and β_f another constant − the compressibility. The above rheological conditions, together with the continuity conditions etc. applying in any *general* medium, completely define the dynamics of viscous (Newtonian) fluids.

The set of conditions outlined above can be combined to yield various differential equations which are applicable under various conditions. The best known of these equations is that of Navier and Stokes[48]. It is applicable to incompressible fluids. Because of its fundamental importance it is restated here:

$$v \operatorname{grad} v + \partial v / \partial t = F - (1/\varrho) \operatorname{grad} p - (\eta/\varrho) \operatorname{curl} \operatorname{curl} v \, . \qquad (3.2.3-4)$$

Here, v is the local velocity vector of a point of the fluid, t is time, and F the volume force per unit mass.

Theory and experiment show that for high flow velocities the flow pattern becomes transient although the boundary conditions remain steady: eddies are formed which proceed into the fluid at intervals. For any one system, there seems to be a "transition point" below which steady flow is stable. Above the "transition point" the steady flow becomes unstable and forms eddies. The steady flow is often termed "laminar", the unstable flow "turbulent".

Although the transition point has been calculated from the Navier-Stokes equation for certain simple systems, it is obvious that such a calculation is a very difficult undertaking. One has therefore to take recourse to experiments to determine when turbulence will set in. If some systems can be shown to be dynamically similar, then the transition point in one system will have a corresponding point in the dynamically similar system which can be calculated.

It has been shown by Reynolds that flow systems which are geometrically similar, are also dynamically similar if the following "Reynolds number" (denoted by Re) is the same in both systems

$$Re = \varrho v d / \eta \qquad (3.2.3-5)$$

where all the constants have the same meaning as before and d is a characteristic diameter of the system.

It has been observed that in straight circular tubes (which are naturally all geometrically similar), turbulence will set in if $Re = 2000$. This value, however, is tied up with the assumption that what is under consideration is a straight tube. In other systems (for instance curved tubes), the "critical" Reynolds number (at which nonlinearity sets in) may be quite different.

A further important case is that of a viscous fluid flowing in laminar, parallel flow past a sphere of radius a. The resistance R offered by the sphere has been calculated by Stokes; it is[48]

$$R = 6 \pi a \eta v \qquad (3.2.3-6)$$

where v is the flow velocity at infinity.

3.2.3.3 Flow Through Porous Media

In geodynamics, viscous fluids may be encountered that are contained in porous rock. Under such circumstances, it is no longer feasible to solve the Navier-Stokes equations for the boundary conditions represented by the walls

---

48 See Lamb, H.: Hydrodynamics. London: Cambridge Univ. Press 1932.

of the pore system, but it is advisable to use a semi-empirical approach originally suggested by Darcy[49]. Accordingly, the flow in the pore space is considered as an average macroscopic flow process wich is governed by the following law

$$q = -\frac{k}{\eta}(\text{grad } p + \varrho g) \qquad (3.2.3-7)$$

where q is the seepage velocity vector (given as the volume seeping in unit time through a unit area), k the permeability (a constant characterizing the porous medium), η the viscosity of the percolating fluid, p the pressure in the liquid in the pores, ϱ the density of the liquid and g the gravity acceleration vector[50]. Appropriately, the law represented by Eq. (3.2.3 – 7) has been called "Darcy's law".

Darcy's law alone is not sufficient to determine a flow problem; in addition, a continuity equation is needed:

$$-\phi\frac{\partial\varrho}{\partial t} = \text{div } \varrho q \qquad (3.2.3-8)$$

where ϕ is the porosity (a fractional pure number characterizing the porous medium; it is equal to the average ratio of pore volume to bulk volume) and t time. Combined, Eqs. (3.2.3 – 7 and 3.2.3 – 8) yield

$$\phi\frac{\partial\varrho}{\partial t} = \text{div}\left[\rho\frac{k}{\eta}(\text{grad } p - \varrho g)\right] \qquad (3.2.3-9)$$

which is a partial differential equation for ϱ. If the connection between ϱ and p (the rheological equation of the fluid):

$$\varrho = \varrho(p) \qquad (3.2.3-10)$$

is added, any flow problem is determined, provided the relevant initial and boundary conditions are specified.

3.2.4 General Linear Bodies

3.2.4.1 Principles

The "ideal bodies" discussed so far (excluding the plastic body) constitute the system of "classical" bodies. According to the general remarks made earlier, it is obvious that it cannot be hoped that these classical bodies will provide

49 Darcy, H.: Les fontaines publiques de la ville de Dijon. Paris: Dalmont 1856
50 Cf., e. g., Scheidegger, A. E.: The Physics of Flow through Porous Media, 3rd edn. Toronto: Univ. Toronto Press 1974

anything but a crude classification of all rheological behavior that can be
envisaged. The most general rheological condition that can be thought of is a
very complicated affair. In order to obtain a physical picture of some of the
possibilities, the system of classical bodies must be enlarged.

An extension of the classical bodies to approximate some of the more
commonly found natural bodies is usually done by, first of all, restricting
oneself to infinitesimal strains. This, of course, is a severe limitation of
generality and, in fact, usually quite an inconsistent procedure as, if any type
of true rheological effects occur at all, the displacements become large, at
least after the elapse of sufficient time. Nevertheless, a very approximate idea
of the physical possibilities may be obtained by assuming the strains as
infinitesimal. At any rate, it is customary to do this in most engineering
literature, whether it be justified or not. A convenient summary of this type of
treatment has been given by Reiner[51]. We shall present here a discussion of the
more important cases.

3.2.4.2 Maxwell Liquid

In terms of infinitesimal strain terminology, the part of the rheological
equation referring to the deviatoric components for the Hooke (elastic) solid
can be written as follows

$$\tau_{ik} = 2\mu\varepsilon_{ik} \quad (i \neq k) \tag{3.2.4-1}$$

and for the viscous (Newtonian) liquid as follows

$$\tau_{ik} = 2\eta\dot{\varepsilon}_{ik} \quad (i \neq k). \tag{3.2.4-2}$$

Combination of both gives (after Reiner[52])

$$\dot{\varepsilon}_{ik} = \dot{\tau}_{ik}/(2\mu_M) + \tau_{ik}/(2\eta_M) \quad (i \neq k). \tag{3.2.4-3}$$

A similar combination with appropriate constants can be set up for the isotropic components. The quantities μ_M and η_M are constants of the body; they
are often referred to as (Maxwell) rigidity and viscosity. However, it should be
noted that the names "rigidity" and "viscosity" are not very good ones. The
ordinary viscous fluid is obtained if, in Eq. (3.2.4-3): $\mu_M \to \infty$, in which case
η_M becomes the fluid viscosity. Similarly, for $\eta_M \to \infty$ one obtains an elastic
solid with rigidity μ_M. It is for this reason that the two constants μ_M and η_M
have been called "rigidity" and "viscosity". However, "Maxwell constants"
would be a better name: Hence the subscript "M".

The rheological Eq. (3.2.4-3) describes a liquid which shows stress
relaxation. That the latter occurs can be seen immediately if it is assumed that

51 Reiner, M.: Twelve Lectures on Theoretical Theology. Amsterdam: North-Holland Publish-
    ing Co. 1949
52 Reiner, M.: loc. cit.

the deformation is kept constant (i.e., $\dot{\varepsilon} = 0$). Then the stress diminishes exponentially with a time constant τ

$$\tau = \eta_M/\mu_M. \tag{3.2.4-4}$$

Furthermore, if a constant stress is applied, it is seen that deformation occurs at a constant *rate*. This phenomenon is called *creep*.

Stress relaxation phenomena have been studied extensively by Maxwell[53]. Hence the name "Maxwell liquid" for the material at present under consideration.

3.2.4.3 Kelvin Solid

The equations of the Hooke solid and of the viscous liquid in their infinitesimal-strain form as displayed in Eqs. (3.2.4−1/2) can also be combined in a different manner, viz.[54,55]

$$\tau_{ik} = 2\mu_K\varepsilon_{ik} + 2\eta_K\dot{\varepsilon}_{ik} \quad (i \neq k) \tag{3.2.4-5}$$

where again a similar combination with appropriate constants could be written down for the diagonal components ($i = k$). Again, the quantities μ_K and η_K are constants of the body which are again often referred to as "rigidity" and "viscosity", respectively. Again, these names are not very good. One notes that a "Kelvin solid" becomes an elastic solid if $\eta_K \rightarrow 0$, and a viscous fluid if $\mu_K \rightarrow 0$. These are the opposite limits to those required to yield the same classical bodies in the case of Maxwell liquids. The use of the terms "rigidity" and "viscosity" indiscriminately in the case of Maxwell liquids and Kelvin bodies has given rise to much confusion. Correspondingly, we should prefer to call μ_K and η_K "Kelvin constants".

The Kelvin solid is characterized by an elastic after effect: if a stress change is performed, the body will eventually reach that state which would correspond to Hooke's law, − but only exponentially. The time constant is again η_K/μ_K. A loading-unloading diagram has therefore the characteristics shown in Fig. 69.

The elastic after effect characteristic of a Kelvin body also causes free oscillations to be damped. The damping constant can be calculated simply as

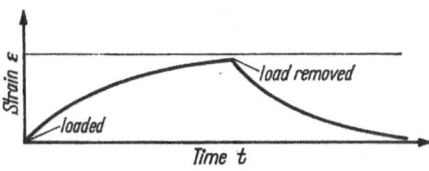

Fig. 69. Loading-unloading diagram of a Kelvin body. (After Reiner[52])

53 Maxwell, J. C.: Collect. Works 2:26 (1866)
54 Meyer, O. E.: J. Reine Angew. Math. 78:130 (1874). Ann. Phys. 1:108 (1874)
55 Kelvin, W. T.: Collect. Works 3:1 (1878)

follows. The equation of motion of a Kelvin body is

$$\mu_K x + \eta_K \dot{x} = - c\ddot{x} \qquad (3.2.4-6)$$

where c is some constant of the system. Trying for a solution

$$x = \exp\left[(\alpha + i\omega)t\right] \qquad (3.2.4-7)$$

one obtains (omitting the subscript K)

$$\mu + \eta(\alpha + i\omega) = - c(\alpha^2 - \omega^2 + 2\alpha i\omega). \qquad (3.2.4-8)$$

Equating real and imaginary parts yields

$$\mu + \eta\alpha = - c(\alpha^2 - \omega^2), \qquad (3.2.4-9)$$

$$\omega\eta = - c2\alpha\omega \qquad (3.2.4-10)$$

and hence one obtains

$$\alpha = - \eta/(2c), \qquad (3.2.4-11)$$

$$\omega = \sqrt{\frac{\mu}{c} - \frac{\eta^2}{4c^2}}. \qquad (3.2.4-12)$$

Finally, eliminating the constant c yields

$$\alpha^2 + \frac{2\mu\alpha}{\eta} + \omega^2 = 0. \qquad (3.2.4-13)$$

This allows one to estimate the ratio η/μ from measurements of ω and α. The presence of the damping in the Kelvin equation not only causes eigenoscillations to be damped, but causes also a phase shift (a lag) in *forced* oscillations. The Kelvin equation of motion in this case may be written as follows

$$C\ddot{\varepsilon} + \eta\dot{\varepsilon} + \mu\varepsilon = A \sin \omega t \qquad (3.2.4-14)$$

where η is again the Kelvin viscosity, μ the rigidity, C a constant of the system and $A \sin \omega t$ the disturbing force. The solution of the above equation is

$$\varepsilon = \frac{A/C}{\sqrt{\left(\dfrac{\mu}{C} - \omega^2\right)^2 + \dfrac{\eta^2}{C^2}\omega^2}} \sin\left[\omega(t - \delta)\right] \qquad (3.2.4-15)$$

with

$$\delta = \frac{1}{\omega}\arctan\frac{\eta\omega/C}{\mu/C - \omega^2} = \frac{1}{\omega}\arctan\frac{\eta\omega}{\mu - \omega^2 C}. \qquad (3.2.4-16)$$

Noting that

$$n = \sqrt{\mu/C} \qquad (3.2.4-17)$$

is the circular eigenfrequency of the undamped system, one can estimate the value of the Kelvin viscosity for the model in equestion. One again has to eliminate the unknown constant C

$$C = \mu/n^2 \tag{3.2.4-18}$$

and has then[56]

$$\delta = \frac{1}{\omega} \arctan \frac{\eta \omega}{\mu - \omega^2 \mu/n^2} = \frac{1}{\omega} \arctan \frac{\eta \omega}{\mu \left(1 - \dfrac{\omega^2}{n^2}\right)} \tag{3.2.4-19}$$

or

$$\tau = \frac{\eta}{\mu} = \left(\frac{1}{\omega} - \frac{\omega}{n^2}\right) \tan \delta \omega . \tag{3.2.4-20}$$

For large n (i.e., $n \gg \omega$), this becomes

$$\tau \sim \frac{1}{\omega} \tan \delta \omega . \tag{3.2.4-21}$$

The last equation enables one to get the ratio of the material constants $\tau = \eta/\mu$ from an observation of the phase shift in forced oscillations.

Finally, it may be remarked that more generalized bodies are obtained if one introduces a yield stress ϑ_{ik} into Eq. (3.2.4-1). Setting $\mu = 0$ to avoid further complications, one then has[51]

$$\tau_{ik} = \vartheta_{ik} + 2\eta \dot\varepsilon_{ik} \quad (i \neq k) \tag{3.2.4-22}$$

which is the constitutive equation of a "Bingham solid". Here, η is called the "plastic viscosity". Further complications can be achieved by including in the constitutive equations more and more terms. However, the basic shortcomings outlined in Sect. 3.2.4.1 of this type of theory are thereby not overcome, and it may be necessary to allow for finite, non-linear deformations. Indications that this is so have been presented by Knopoff and MacDonald[57] who found that the observed attenuation even of small amplitude stress waves in the Earth cannot be accounted for by a linear rheological law.

3.2.4.4 General Linear Viscoelastic Media

The linear rheological models discussed heretofore in this section are but the simplest possible cases. In a "general" linear viscoelastic substance, the stress is a linear functional of the strain. Specifically, for a rectangular Cartesian system the relationship between the stress (σ_{ij}) and strain (ε_{ij}) can be written as follows

---

56 Cf. Scheidegger, A. E.: Ann. Geofis. (Roma) 23:27 (1970)
57 Knopoff, L., MacDonald, G. J. F.: Rev. Mod. Phys. 30:1178 (1958)

$$\sigma_{ij} = \delta_{ij} \int_0^t K(t-\tau) \frac{\partial \varepsilon_{kk}}{\partial \tau} d\tau + 2 \int_0^t G(t-\tau) \frac{\partial}{\partial \tau} \left(\varepsilon_{ij} - \delta_{ij} \frac{\varepsilon_{kk}}{3} \right) d\tau \quad (3.2.4-23)$$

where $K(t)$ and $G(t)$ are, respectively, the bulk and shear relaxation moduli. With this formulation[58] the stress functional is conveniently separated into hydrostatic and deviatoric components. One further assumes that the relaxation moduli are represented by a sum of exponential functions[59].

$$\left. \begin{aligned} K(t) &= K_\infty + \sum_{n=1}^N K_n e^{-\alpha_n t}, \qquad \alpha_n > 0, \quad K_n, K_\infty > 0 \\ G(t) &= G_\infty + \sum_{n=1}^N G_n e^{-\beta_n t}, \qquad \beta_n > 0, \quad G_n, G_\infty > 0 \end{aligned} \right\} \quad (3.2.4-24)$$

where the signs of α_n, β_n, G_n, K_n and K_∞ are chosen as positive so that the material is dissipative.

A viscoelastic material "remembers" its past history; hence it is necessary to prescribe the pertinent functions *up* to some initial time t_0.

Waves in a viscoelastic material are subject to decay because of its dissipative character. A rather complete study of the propagation of plane waves in such media has been given in the cited paper of Chu[58]. Chu's method has been extended by Blake[60] to a study of the decay of spherical waves. A further extension of this to a analysis of viscoelastic porous media has been given by Boschi[61]. The details of these investigations are beyond the scope of the present treatise and the reader is referred to the cited papers for further study.

3.2.5 Non-Linear Creep

3.2.5.1 General Formulation

The rheological equations discussed heretofore are all linear. However, observational data on various materials yield the result that non-linear connections between stress and strains (or their rates or integrals) also occur.

The possible structure of an entirely general non-linear rheological equation (or set of equations) has been investigated by Strauss[62] who also gives an extensive literature survey on the subject. In his investigation, the formulation in finite strain theory has been invoked, and the possibilities of constructing stress-strain relations have been enumerated by envisaging the

58 Chu, B. T.: J. Mech. 1:439 (1962)
59 Gross, B.: Mathematical Structure of the Theories of Viscoelasticity. Paris: Hermann 1953
60 Blake, R.: Z. Angew. Math. Mech. 25:783 (1974)
61 Boschi, E.: Ann. Geofis. (Roma) 27:385 (1974)
62 Strauss, A. M.: J. Geophys. Res. 79(2):351 (1974)

corresponding non-linear integrodifferential operators. The possibilities include the description of aging in addition to that of the purely mechanical properties of the materials in question. The formalism, however, is so general that specific assumptions have to be made to describe a particular material.

3.2.5.2 Logarithmic Creep

In practice, it has been found that the rheological behavior of many Earth materials can be represented by the following equation postulated long ago by Andrade[63]

$$\varepsilon = A + BE(t) + Ct \qquad (3.2.5-1)$$

where ε is the shearing strain; A, B and C are constants, and $E(t)$ is an empirical function.

In some cases, the term $E(t)$ has been taken as of the form

$$E(t) = a(1 - e^{-kt}) \qquad (3.2.5-2)$$

which would again correspond to a Kelvin model. However, this is a rather unusual case. One often finds a function of the type

$$E(t) = a \operatorname{lognat}(1 + bt). \qquad (3.2.5-3)$$

This is called logarithmic creep. Various empirical forms of this law exist (e. g., Lomnitz[64])

$$\varepsilon = \frac{\sigma}{\mu}(1 + q \operatorname{lognat}(1 + bt)) \qquad (3.2.5-4)$$

or

$$\varepsilon = A + B \operatorname{lognat}(a + bt) + Ct. \qquad (3.2.5-5)$$

In these cases one notes the following relation for the strain rate

$$\dot{\varepsilon} = B \frac{b}{a + bt} + C. \qquad (3.2.5-6)$$

Generally, one has $a \ll bt$, so one can write for long times in any case

$$\dot{\varepsilon} \approx \frac{B}{t} + C. \qquad (3.2.5-7)$$

If a material behaves in this fashion, a plot of the strain rate $\dot{\varepsilon}$ versus $1/t$ yields a straight line. Many materials show this behavior.

Jeffreys[65] has taken Eq. (3.2.5 − 4) as a stress-strain relation containing an explicit time dependence so that he writes, for variable stresses

63 Andrade, E. N. da C.: Proc. R. Soc. London Ser. A 84:1 (1910)
64 Lomnitz, C.: J. Geol. 64:473 (1956)
65 Jeffreys, H.: Mont. Not. R. Astron. Soc. 118:14 (1958)

$$\varepsilon = \frac{1}{\mu} \int_{\tau=-\infty}^{+\infty} \{1 + q \log[1 + b(t - \tau)]\} d\sigma(\tau). \qquad (3.2.5-8)$$

However, in view of the fact that the stress-strain relation (3.2.5 − 4) is no longer linear, it does not seem entirely justified to apply the superposition principle implied in Eq. (3.2.5 − 8). Nevertheless, if Eq. (3.2.5 − 4) is taken as a stress-strain relation valid for variable stresses, Jeffreys[65a] showed that the phase lag δ induced in forced oscillations (circular frequency ω) is approximately

$$\delta = \frac{1}{2}\pi q/\omega. \qquad (3.2.5-9)$$

It would thus be possible in principle to determine the constant q from a measurement of a phase lag in an oscillatory system.

The basic form of Eq. (3.2.5 − 1) postulated by Andrade will not generally allow a steady state to exist for constant stresses. For materials in which the strain rate $\dot{\varepsilon}$ is assumed to be constant for constant stresses, the term with $E(t)$ must be assumed as zero in Eq. (3.2.5 − 1). In this case, one has after time differentiation

$$\dot{\varepsilon} = C \qquad (3.2.5-10)$$

where C may be a function of the (constant!) stresses. One generally assumes in this cases (Glen's law[66])

$$\dot{\varepsilon} = F\sigma^n \qquad (3.2.5-11)$$

with n between 2 and 9.

3.2.6 Thermohydrodynamics

3.2.6.1 Thermal Convection in Viscous Fluids

We shall turn now our attention to the problem of the free thermal convection currents in various materials between two surfaces of different temperatures, subject to a gravitational field. Heat transfer by liquids of the Newtonian type[67] has been investigated because of the importance of heat transmission in engineering problems. Some results are thus available in textbooks[68]. These results have been obtained with particular reference to water and air for which the viscosity is very small.

Dimensional analysis shows that the motion depends only on two dimensionless numbers: on the Reynolds number [see Eq. (3.2.3 − 5)] and on

65a Jeffreys, H.: Mont. Not. R. Astron. Soc. 118:14 (1958)
66 Glen, J. W.: J. Glaciol. 2:111 (1952)
67 For a summary see e. g. Scheidegger, A. E.: Bull. Geol. Soc. Am. 64:127 (1953)
68 See e. g. McAdams, W. H.: Heat Transmission, 2nd edn. New York: McGraw-Hill 1942

the product of Grashoff and Prandtl numbers (sometimes called Rayleigh number). Let c denote the specific heat per unit mass of the material, D a characteristic diameter, g the gravity acceleration, k the thermal conductivity, ΔT the temperature difference, v a characteristic velocity of the fluid in motion, β the coefficient of thermal expansion (per °C), η the viscosity and ϱ the density, then the Grashoff number G is given by the equation

$$G = D^3 \varrho^2 g \beta \Delta T / \eta^2 \qquad (3.2.6-1)$$

and the Prandtl number P by

$$P = c\eta/k. \qquad (3.2.6-2)$$

For laminar motion, the product λ of Grashoff and Prandtl numbers (Rayleigh number) governs the thermal convection. Jeffreys[69, 70] has deduced theoretically that no stable thermal convection should occur unless λ is at least 1709. In his papers, the direction of the thermal gradient is supposed to coincide with the gradient of gravity; the two surfaces are horizontal. The convective motion between the two surfaces is pictured as occurring in the form of convection cells[71]. A convection cell contains a vortex tube which is closed within the cell. The picture is thus in agreement with the kinematical properties of vorticity[72].

This theory has been tested experimentally. Schmidt and others[73, 74] have employed an optical method to study the mechanism of heat transfer by natural convection above a horizontal plate. The photographs do indeed show a cell-like pattern under certain conditions. When λ is above 2000, alternate portions of the fluid circulate upward and downward in streams of substantial width. As the characteristic product increases, the rate of fluid circulation increases, until finally turbulence ensues.

Jeffreys' results have been reviewed and extended by Low[75], Pellew and Southwell[76] extended the results of Jeffreys and Low, and made investigations under different sets of boundary conditions. Qualitatively, the previous results are confirmed, but it is shown that any *oscillatory* convective motion must of necessity decay.

The main result of the above investigations is that the conditions for which thermal convection may occur *are extremely narrow.*

---

69 Jeffreys, H.: Philos. Mag. (7) 2:833 (1926)
70 Jeffreys, H.: Proc. R. Soc. London Ser. A 118:195 (1928)
71 This has originally been postulated by Bénard, H.: Ann. Chim. Phys. 23:62 (1901)
72 Okai, B.: J. Phys. Earth 7:1 (1959)
73 Schmidt, R. J., Milverton, S. W.: Proc. R. Soc. London Ser. A 152:586 (1935)
74 Schmidt, R. J., Saunders, O. A.: Proc. R. Soc. London Ser. A 165:216 (1938)
75 Low, A. R.: Proc. R. Soc. London Ser. A 125:180 (1929)
76 Pellew, A., Southwell, R. V.: Proc. R. Soc. London Ser A 176:312 (1940)

3.2.6.2 Heat Convection in Plastic Flow

In order to obtain an idea about the thermomechanics of plastic flow, qualitative arguments have to be used. An interesting study of possible analogies between plastic and viscous flow has been made by Oldroyd[77]. The plasticity equations are solved "in the large" for cylinders of various shapes moving through each other. The plastic lines of flow are compared with the viscous lines of flow (see Fig. 70). The difference is not very great, apart from the fact that in plastic flow one finds a solid kernel in those regions where in viscous flow there is a small velocity gradient.

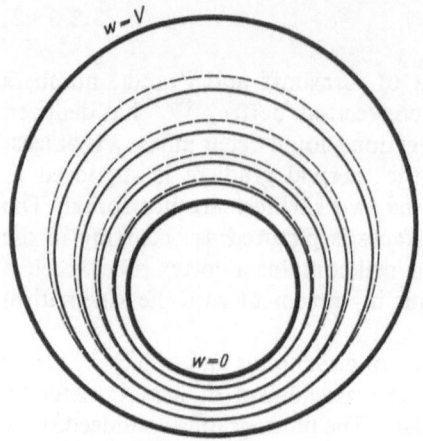

Fig. 70. Comparison of plastic and viscous flow lines in a special case. (After Oldroyd[77])

It is now our task to obtain an idea of what may happen when a plastic material is heated from below. For vanishing yield stress, the Bingham body is a Newtonian liquid. The presence of a yield stress will affect the formation of convection cells to such an extent that it will need greater temperature differences to start the motion than that corresponding to a characteristic product of Grashoff and Prandtl numbers equal to 1709. Moreover, in view of Oldroyd's results, it is very likely that the center of the vortex tube in the cell will rotate as a solid. However, it must be expected that, under very great temperature differences, vortices would be formed at the boundaries which would dissipate into the plastic medium, as this was the case with viscous liquids. The flow is then no longer steady.

77 Oldroyd, J. G.: Proc. Cambr. Philos. 53:396, 521 (1947)

3.3 The Physics of Fracture

3.3.1 Fundamentals

As outlined earlier, the analysis of fractures has a very important bearing upon the physics of the Earth's crust, since fissures, faults an related phenomena are very common occurrences. The physics of fracture is to this day a somewhat involved and imperfectly understood discipline, as evidenced, for instance, by the seven(!)-volume treatise on the subject edited by Liebowitz[78].

Although man has been very much concerned with fracture and related phenomena, most of his interest has been directed toward establishing criteria for the safety of structures rather than toward a detailed analysis of how a collapse occurs. One is therefore faced with the situation that there is a multitude of fracture criteria which in conjunction with assumed "safety factors" can be used to build safe structures, but that there are not many investigations into the detailed mechanism of fracture that could be used to explain observed conditions in the Earth's crust.

The existing investigations into the theory of fracture can be separated into various sections. Firstly, there are the heuristic descriptions, encompassing all the engineering theories and fracture criteria; secondly, there are the "microscopic theories" which try to explain the observed values of "strength" (i. e., the resistance to fracture) of materials in terms of their known molecular forces; and thirdly, there are a few attempts to give a proper analytical expression to the heuristic investigations mentioned above – in such a fashion that the stresses are indeed components of a tensor and the displacements are defined as vector fields. It is this last group of investigations which has the most direct bearing upon the physics of the Earth's crust.

However, since the last group of investigations has not been carried to a high degree of refinement, it is necessary to extract as much information as possible from the engineering theories. These are naturally based on experiments. We shall therefore discuss all the aspects of fracture theories that might prove relevant in connection with a study of the Earth's crust.

3.3.2 Phenomenological Aspects

A great number of heuristic investigations into fracture and fracture criteria have been made – as outlined above, principally in order to establish when a structure is safe.

A survey of the types of fracture that can occur in a material has been made by Orowan[79] who distinguishes between the following cases: (a) brittle

78 Liebowitz H. (ed.), Fracture and Advanced Treatise in Seven Volumes. London, New York. Academic Press 1969

79 Orowan, E.: Rep. Progr. Phys. 12:186 (1949)

fracture, (b) ductile fracture, (c) fatigue fracture, and (d) creep fracture. We shall discuss these cases in their proper order.

a) Brittle Fracture. Brittle fracture is the only type of fracture that occurs in completely brittle substances. It is that type of fracture which is theoretically best understood[80]. It is characterized by a high velocity of propagation, producing a bright, smooth fracture surface.

Under an uniaxial stress state, brittle fracture occurs in an isotropic medium if the tension reaches a (for the material) critical value (called brittle strength of the material); the fracture surface is normal to the direction of the tensile stress.

In a triaxial stress state, the condition for brittle fracture is not so simple and thus, several criteria have been proposed. Mostly used is still the time-honored hypothesis of Coulomb[81], later modified by Mohr[82], which states: "For an isotropic medium fracturing under the action of three unequal principal stresses, the surface of fracture is parallel to the direction of the intermediate principal stress and inclined at an angle $\varphi \leq 45°$ (30° is a good average) toward the maximum principal pressure". The originators of this statement arrived at it by modifying the hypothesis that the maximum shear surface would be the fracture surface, until reasonable agreement with observations was obtained (Fig. 71). Seya[83,84] gave some theoretical reasons for the observed fracture angles.

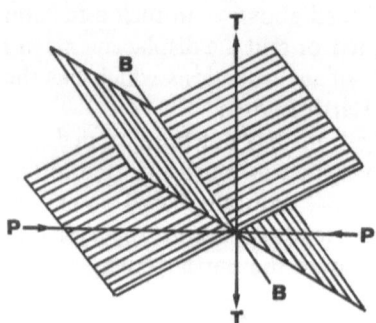

Fig. 71. Illustration of Mohr's fracture surfaces in a triaxial stress state. *P* largest, *T* smallest, and *B* intermediate principal pressure

The value of the stress at which fracture occurs, e.g., in terms of the "tensile strength" (as defined above) of the material must be inferred from

80 Paterson, M. S.: Experimental Rock Deformation: The Brittle Field. Berlin-Heidelberg-New York: Springer 1978
81 Coulomb, C. A.: Mem. Math. Phys. Acad. Sci. Paris 7:343 (1776)
82 Mohr, O.: Abhandlungen aus dem Gebiete der technischen Mechanik. 3. Aufl. Berlin: Wilh. Ernst & Sohn 1928
83 Seya, K.: J. Phys. Earth 28:191 (1980)
84 Seya, K., Suzuki, I.: Tectonophysics 64:175 (1980)

microscopic considerations and will be discussed later. However, it may be said that corresponding to the idea that the maximum shear surface would be the fracture surface, it also had been thought originally that the fracture condition would be one of critical shear stress. This was not found to correspond to observation and therefore was also modified by Mohr to yield his famous heuristic fracture criterion. Each possible stress state is represented in the $\sigma - \tau$ diagram by a family of circles (Mohr's diagram, cf. Fig. 63). Those stress states, one of whose circles touches an empirically determined envelope, are thought to produce fracture. When referring to Mohr's criterion, it should be noted that it has been applied to any type of failure, and not only to brittle fracture. Methods for the experimental determination of the strength limits have been described on many occasions. Thus, Handin[85] gave a good review of the problem up to 1969; later Kovari and Tisa[86-88] described various newer experimental techniques and the ISRM[89] proposed a set of standards for making the experiments.

If the medium being fractured is porous containing an interstitial fluid, then the limiting-stress conditions represented by envelopes in the Mohr diagram refer to the effective stresses as given by Eq. (3.1.3 – 5); for, according to the "Terzaghi principle" (cf. Sect. 3.1.3), it is only the effective stresses which can cause a deformation (and thus also fracture) in a porous medium. Thus, the modes of failure of a porous medium are exactly those encountered in non-porous substances. In addition, however, a tensile breakdown of the porous structure due to the fluid pressure from within may occur if the latter is high enough. This mode of fracture has been termed "splitting failure" by Terzaghi[90].

The high velocity of the spread of a crack in brittle fracture can be explained by noting that the only work required for the latter is that necessary to overcome the cohesion between the atoms on either side of an existing crack. This work is so small that it can be supplied by the elastic energy stored in the material just prior to its disintegration.

b) Ductile Fracture. This is characterized by a very slow propagation of existing cracks. The fracture process can be stopped at any instant simply by stopping the continuation of external deformations. It thus appears that ductile fracture is a non-elastic type of fracture in which plastic deformation and related phenomena play an essential rôle. In ductile fracture, the energy expended against the cohesive forces of the material is negligible compared with the energy of plastic deformation that has to be expended in order to

85 Handin, J. : On the Coulomb-Mohr failure criterion. J. Geophys. Res. 74:5343 – 5345 (1969)
86 Kovari, K., Tisa, A.: Mitt. Inst. Strassen- und Untertagbau. Eidg. Tech. Hochsch. Zuerich 26:1 (1974)
87 Kovari, K., Tisa, A.: Rock Mech. 7(1):17 (1975)
88 Kovari, K.: Int. J. Rock Mech. Min. Sci. 15(2):49 (1978)
89 International Society for Rock Mechanics, Comm. on Standardization: Suggested Methods for Determining Shear Strength. Document No. 1, Committee on Field Tests. 23 pp. (1974)
90 Terzaghi, K.: Proc. Am. Soc. Test. Mater. 45:777 (1945)

extend an existing crack. The crack can therefore spread only if the external forces continue to do work.

c) Fatigue. A peculiar type of fracture that may occur under cyclic stressing in a material is fatigue fracture.

The most important facts about fatigue have been stated by Orowan[91] as follows: (1) fatigue fracture may occur after a certain critical number of applications of a given stress cycle is exceeded; (2) sometimes there exists a critical amplitude of an alternating stress cycle below which the material can undergo any number of applications of the stress without breaking, but above which fatigue fracture occurs after a certain number of applications.

It appears that fatigue fracture is due to ductile fracture in weak inclusions in an otherwise elastic material, i.e., in plastic inclusions in an elastic matrix. During the repeated application of the stress cycle, the plastic inclusions get worked plastically until a ductile crack may appear therein. Once this has happened, stress concentrations at the tip of the crack may induce brittle fracture in the rest of the material.

With this model, Orowan[91] was also able to explain the curious fact, mentioned above, of the existence, in certain cases, of a "safe" stress amplitude below which no fatigue fracture occurs in any number of stress cycles. Another review of the problem has also been given by Thurston[92].

d) Creep Fracture. The term fatigue has also been applied to the strength decrease (this has also been called "stress corrosion") which is found to occur in certain materials under completely static conditions. In order to distinguish this phenomenon from that discussed earlier, it is usually termed *static* (as opposed to *ordinary* or *dynamic*) fatigue. Static fatigue is caused by some internal creep process and it, therefore, is more appropiately called *creep fracture.*

Thus, materials consisting of polycrystalline aggregates contain two different components, viz. the crystal grains and the grain boundaries. The same is true for any type of heterogeneous agglomerate. The grain and the boundaries exhibit entirely different mechanical behavior. The grain boundaries show characteristics of viscosity. If the temperature is less than a certain critical value, they are rigid; but if the critical temperature is exceeded, they behave like a viscous liquid[93]. They have no definite yield stress, so that the rate of sliding would presumably be proportional to the prevailing shear stress if the grain boundaries were smooth enough to allow uninhibited sliding. Owing to the intermittent interlocking of the grains, sliding, if the stresses are continuously increased, also occurs intermittently. Once the stresses are large enough to overcome the largest geometrical non-conformity at the grain boundaries, slow continuous sliding occurs during which cavities open up between the grains and, eventually, fracture may occur.

---

91 Orowan, E.: Rep. Prog. Phys. 12:221 (1949)
92 Thurston, R. C. A.: Trans. Can. Inst. Min. Metall. 60:390 (1957)
93 Orowan, E.: Rep. Prog. Phys. 12:228 (1949)

The creep behavior of the material need not be that corresponding to linear viscosity: A logarithmic or power law has also been proposed. Thus, Cruden[94] has shown that the strength loss with time of pillars in rock salt mines can be explained upon this basis.

A similar mechanism may occur in a somewhat related fashion in any material that contains two or more structural elements of different mechanical properties. Assume that a "surface of weakness" (e. g., an old fracture surface) representing one type of structural element exists in an otherwise homogeneous material representing the second type. The surface may be very "bumpy" and therefore geometrically interlocks. If an external strain is applied at a constant time rate, the material may give along the surface of weakness whenever the stresses have become big enough to overcome the largest geometrical non-conformity by deforming the material itself in certain spots. Once this has happened, sliding may occur along the surface (by some viscous slip mechanism) until a new interlocking occurs.

A review of slow crack propagation phenomena with a view to geophysical applications has recently been given by Anderson and Grew[95].

3.3.3 Microscopic Aspects of Fracture

3.3.3.1 The Problem

The fact that materials fracture under sufficiently high stresses must have its ultimate explanation in molecular considerations. In order to obtain such explanations, microscopic theories of fracture have been devised. In particular, the aim is to explain the experimentally observed *strength* of materials.

It has been recognized very early that the strength of a material as calculated from the physics of molecular cohesion in a homogeneous material is much higher than the actually observed fracture strength of common materials. The order of magnitude of molecular cohesion, or molecular strength, can be estimated as follows[96]. If a solid is strained uniformly, (brittle) fracture must occur if the elastic energy in the solid can provide the surface energy necessary to produce a crack. The latter is $2\,S$ for a specimen of unit thickness if S is the specific surface energy; we shall consider only the two-dimensional case. Thus, a large portion (let us assume: one-half) of the energy $2\,S$ must be present just prior to the instant of fracture in the molecules in the immediate neighborhood of the (future) surface of fracture. On the other hand, if σ_m denotes the fracture stress, the elastic energy density prior to fracture is $\sigma_m^2/2E$ (E being Young's modulus), and the strain energy e (per unit thickness), present between the two atomic planes which will be separated

94 Cruden, D. M.: Int. J. Rock Mech. Min. Sci. 11:67–73 (1974)
95 Anderson, O. L., Grew, P. C.: Rev. Geophys. Space Phys. 15(1):77 (1977)
96 Orowan, E.: Rep. Prog. Phys. 12:192 (1949)

by the ensuing fracture process, is

$$e = a\sigma_m^2/2E,$$ (3.3.3 – 1)

if the planes are originally at the atomic distance a from each other. Thus, we obtain:

$$\sigma_m = \sqrt{\frac{2SE}{a}}.$$ (3.3.3 – 2)

Using typical values for most common brittle materials, it turns out that the "molecular" fracture strength σ_m is 10 to 1000 times greater than that stress at which fracture actually occurs.

3.3.3.2 The Griffith Theory of Brittle Fracture

To overcome this discrepancy, Griffith[97–99] made the assumption that the observed values of strength were due to the presence of very small cracks inherent in every material, at which there is a stress concentration sufficient to overcome the molecular strength. Thus, during fracture, it would be actually the molecular cohesion which would be overcome; the stress causing this, however, would be not be the overall average stress in the material, but the local stress at the tips of existing cracks.

In order to calculate the magnitude of the stress concentration at the end of a very small elliptical crack, Griffith made use of the solution by Inglis (cf. Sect. 3.2.1) for the stresses at the ends of an elliptical cavity in a two-dimensional case under a given tension σ. Inglis obtained for the strain energy e_σ (per unit thickness) in this case

$$e_\sigma = -\pi c \sigma^2/E$$ (3.3.3 – 3)

where $2c$ is the length of the crack (i. e., c is the major axis of the ellipse) and E is, as before, Young's modulus. The corresponding surface energy e_s represented by the crack, according to earlier remarks, is

$$e_s = 4cS$$ (3.3.3 – 4)

where S in again the specific surface energy. The crack will spread if, for an increasing length of the crack, the decrease in strain energy is greater than the increase in surface energy[100]. Thus the equilibrium size of the crack is given by

$$\frac{\partial}{\partial c}(4cS - \pi c^2 \sigma^2/E) = 0$$ (3.3.3 – 5)

97 Griffith, A. A.: Philos. Trans. R. Soc. London Ser. A 221:163 (1920)
98 Griffith, A. A.: Proc. 1st Int. Conf. Appl. Mech. Delft A 55 (1924)
99 Orowan, E.: Rep. Prog. Phys. 12:192 (1949)
100 Petch, N. J.: Prog. Metal Phys. 5:1 (1954)

or

$$\sigma = \sqrt{\frac{2ES}{\pi c}}. \tag{3.3.3-6}$$

In view of the above remarks, it is obvious that the crack can expand only if the stress given by Eq. (3.3.3 − 6) is exceeded; hence this stress is the fracture stress in the presence of the crack.

The solution given above for the fracture stress can be recalculated for a penny-shaped crack. The resulting expression differs from that above only by a numerical factor; it is

$$\sigma = \sqrt{\frac{2ES}{\pi(1 - m^2)c}} \tag{3.3.3-7}$$

where m is Poisson's ratio.

Reviews and generalizations of the Griffith fracture theory have been published on several occasions in the literature[101−103]. An interesting application of the theory to fracture in rocks has been given by Brace[104] who took the friction of the sides into account. Steketee[105] has shown that Griffith cracks can also be regarded as dislocations. Berg[106] has calculated the deformation of Griffith-type cracks under high pressure as well as shear.

The above theory can be extended to yield a fracture criterion for brittle material. Assuming that isotropic materials contain cracks of all orientations, Griffith[107,108] used the solution by Inglis[109] for obtaining the maximum tensile stress at the tip of the crack of the most dangerous orientations, again in the two-dimensional case. Assuming that σ_1 and σ_2 are the principal tensions, with $\sigma_1 > \sigma_2$, Griffith states his criterion as follows: Fracture occurs

$$\left. \begin{array}{l} \text{1. if } 3\sigma_1 + \sigma_2 > 0 \quad \text{when} \quad \sigma_1 = K \\ \text{2. if } 3\sigma_1 + \sigma_2 < 0 \quad \text{when} \quad (\sigma_1 - \sigma_2)^2 + 8K(\sigma_1 + \sigma_2) = 0 \end{array} \right\} \tag{3.3.3-8}$$

where K is the tensile strength for uniaxial stressing.

One of the consequences of the Griffith criterion is that, in uniaxial compression, the most dangerous cracks would be those at 45° to the stress, and they should propagate in their own planes. However, Orowan[108] states that, "although fracture surfaces at about 45° are common in compressive tests, failure by cracking *parallel* to the direction of compression is almost equally

101 Sanders, J. L.: J. Appl. Mech. 27:352 (1960)
102 Sack, R. A.: Proc. Phys. Soc. London 58:729 (1946)
103 Clausing, D. P.: Q. Colo. Sch. Mines 54:No. 3, 285 (1959)
104 Brace, W. F.: J. Geophys. Res. 65:3477 (1960)
105 Steketee, J. A.: Can. J. Phys. 36:1168 (1958)
106 Berg, C. A.: J. Geophys. Res. 70:3447 (1965)
107 Griffith, A. A.: Proc. Ist Int. Conf. Appl. Mech. Delft A 55 (1924)
108 Orowan, E.: Rep. Prog. Phys. 12:192 (1949)
109 See Sect. 3.2.1

often observed with glasses and stones". On the other hand, Orowan[110] found that high pressure measurements can be correlated satisfactorily by Griffith's criterion.

Similar criteria, based on elliptical cracks, were deduced by Barron[111]; based on dislocations, by Boschi and Mulargia[112].

3.3.3.3 Dilatant Behavior of Materials During Fracture

A further extension of the basic ideas of Griffith outlined above is the supposition that the small cracks present in a medium do not only grow at their tips during the fracturing process, but also become wider. This phenomenon has been called "dilatancy"; it occurs just prior to and during the initiation of the fracturing process[113,114].

The phenomenon indicated above has been confirmed to occur in the laboratory. Thus, Tapponnier and Brace[115] determined details of the development of dilatant microcracks in granite, and Rao and Ramana[116] in ultramafic material. In these studies it was shown that not only do cracks grow which are already present in the materials, but that, in effect, new dilatant cracks are formed at around 75% of the peak stress; the density of these new cracks doubles up to the peak stress at which the material disintegrates.

3.3.3.4 The Crack Propagation Velocity

The Griffith theory gives a criterion for the stress at which a crack begins, i.e., is nucleated. However, it has been noted by Gilman[117] that the brittle fracture of a material takes place in two stages: nucleation at the atomic level, and then propagation through the material in which the crack had been nucleated. (A third stage is the spread to neighboring pieces of material if the latter is inhomogeneous.)

The Griffith theory deals only with the nucleation of a crack. In addition, it ignores the fact that at the tip of the crack where stress concentration takes place, the principal stresses are not equal so that high shear stresses are present which should cause plastic flow. This plastic flow should prevent the crack from propagating because it would relieve the stress concentration. The difficulty can be resolved by noting a remark of Mott[118] stating that it may be necessary for stresses to persist for some time in order to cause plastic flow.

110 Orowan, E.: Rep. Prog. Phys. 12:200 (1949)
111 Barron, K.: Int. J. Rock Mech. Min. Sci. 8(6):541 (1971)
112 Boschi, E., Mulargia, F.: Ann Geofis. (Roma) 30:201 (1977)
113 Brace, W. F., Paulding, B., Scholz, C. H.: J. Geophys. Res. 71:3939 (1966)
114 Kranz, A. L., Scholz, C. H.: J. Geophys. Res. 82:4893 (1977)
115 Tapponnier, P., Brace, W. F.: Int. J. Rock. Mech. Min. Sci. 13:103 (1976)
116 Rao, M. V., Ramana, Y. V.: Int. J. Rock Mech. Min. Sci. 11:193 (1974)
117 Gilman, J. J.: J. Appl. Phys. 27:1262 (1956)
118 Mott, N. F.: Proc. R. Soc. London Ser. A 220:1 (1953)

Thus, if a crack propagates fast enough, plastic flow would not have time to become established and hence the crack could indeed propagate.

The crack propagation velocity in an elastic body is determined by three factors. The driving force is the elastic energy H_E which is released during the propagation. This is balanced by the surface energy H_S ("fracture energy") necessary to separate the two sides of the crack, and by the kinetic energy H_K associated with the rapid sideways motion of the material during crack formation. The latter can be written as follows:

$$H_K = \frac{v_c^2}{v_0^2} \frac{H_E}{B^2} \qquad (3.3.3-9)$$

where B is a constant, v_s is the crack velocity, and v_0 is the velocity of sound in the material. Solving this for the crack velocity[119] yields

$$v_c = B v_0 \sqrt{1 - \frac{H_S}{H_E}} . \qquad (3.3.3-10)$$

This relation implies that the crack cannot propagate unless the Griffith criterion ($H_E > H_S$) is satisfied, and that the velocity of crack propagation can never be larger than the velocity of sound in the material. Roberts and Wells[120] calculated the constant B for an internal crack in a plate; they found B approximately equal to 0.38. Comparison of observed elastic crack velocities with theory shows that the experimental values approach the theoretical ones quite closely. This indicates that the propagation of cracks in elastic materials is quite satisfactorily understood.

For materials that are not ideally elastic, one must assume that the stress difference at the top of the crack will cause plastic flow. Thus, the driving energy H_E is not only balanced by the sum of kinetic and surface energy required in producing the crack, but in addition there is energy dissipation. This dissipation results from the plastic deformation. The energy balance equation thus reads as follows

$$H_E = H_P + H_K + H_S \qquad (3.3.3-11)$$

where H_P is the energy dissipated in plastic deformation. Gilman[119] has shown that the work of plastic deformation can be written as follows

$$H_P = W/v_c \qquad (3.3.3-12)$$

where W is a constant that depends on the stress-strain rate relation and on the size of the material. The equation for the crack velocity is thus modified to read as follows:

$$v_c = B v_0 \sqrt{1 - \frac{W}{H_E} \frac{1}{v_c} - \frac{H_S}{H_E}} . \qquad (3.3.3-13)$$

119 Gilman, J. J.: J. Appl. Phys. 27:1262 (1956)
120 Roberts, D. K., Wells, A. A.: Engineering 178:820 (1954)

This equation demonstrates that a crack must have a certain critical velocity $\tilde{v}_c$

$$\tilde{v}_c = W/H_E \tag{3.3.3-14}$$

before it can propagate spontaneously in a brittle fashion. This is the result anticipated by Mott. If the intrinsic crack velocity is less than the critical velocity, then the crack can not spread by itself at all and it can only open up in accordance with work performed by external forces.

Discussions of the kinematics of cracks, which are slightly different from that given above, have also· been published by Berry[121], Brueckner[122], McClintock and Sukatme[123] and Mulargia et al.[124].

3.3.4 Analytical Attempts

As mentioned above, not many actually analytical attempts to describe fracture are available. The geometry of a fracture can be represented by the introduction of discontinuous functions into the general pattern of finite strain theories. Describing the displacements again in terms of coordinates and parameters as in Sect. 3.1.2

$$x_i = x_i(\xi_\alpha, t) , \tag{3.3.4-1}$$

one can represent a discontinuity in the body if the x_i are discontinuous functions of the ξ_α. It appears from physical experience that discontinuities are distributed in sheets through the material. Thus, let

$$F(\xi_1', \xi_2', \xi_3', t) = 0 \tag{3.3.4-2}$$

describe a surface of discontinuity S. Equation (3.3.4-2) implies that the surface S may change its position with the passing of time. As indicated in Eq. (3.3.4-2), an arbitrary point on the surface S is denoted by ξ_α'. The rim of the surface is a line that must obviously be closed in itself or begin and end on the boundary of the body. The discontinuity at the surface S can be described by giving the vector X_i of "jump" in x_i if one passes from one side to the other. The discontinuity is thus given as a vector field, defined on the surface S:

$$X_i = X_i(\xi_\alpha', t) . \tag{3.3.4-3}$$

The vector of jump can be calculated if a circuit around the edge of the discontinuity is drawn, as follows:

$$X_i = \oint_{\substack{\xi_\alpha' \to \xi_\alpha' \\ \text{encircling rim}}} \frac{\partial x_i}{\partial \xi_\alpha} d\xi_\alpha . \tag{3.3.4-4}$$

121 Berry, J. P.: J. Mech. Phys. Solids 8:194 (1960)
122 Brueckner, H. F.: Trans. Am. Soc. Mech. Eng. 80:1225 (1958)
123 McClintock, F. A., Sukatme, S. P.: J. Mech. Phys. Solids 8:187 (1960)
124 Mulargia, F., Boschi, E., Bonafede, M.: Nuovo Cim. 1C:335 (1978)

For physical reasons, only such displacement fields are allowed where this integral does not depend on the circuit (but it will naturally depend on ξ'_a). It is seen that a proper choice of the discontinuity will physically represent a fracture sheet.

Thus it has been tried to introduce *dislocations* (as the simplest type of discontinuity) to describe fractures. One has thus a sheet of dislocations, the rim of the sheet representing the edge of the fracture, and the jump vector at every point describing the relative slip. If the fracture extends, the rim sweeps further through the body and the jump vectors change their size with time. Of particular interest is the case where one has a fracture with fixed rim, and where at a certain instant some of the jump vectors increase suddenly, i.e., some of the dislocations "snap". This can be thought of as a process sweeping at a finite velocity across the whole fracture sheet. However, such "laws" as have been postulated to describe the process are entirely heuristic. The above mechanism has been advanced as a possible description of earthquake phenomena and will be referred to again in the appropriate Section.

More elaborate models have been considered by Burridge and Willis[125] who considered an expanding elliptical crack and Madriaga[126] who studied the dynamics of an expanding circular fault. Finally, Andrews[127] has used a finite-element calculation to study theoretically the propagation of a crack in an elastic medium taking into account sliding friction at the fracture surface already created.

3.3.5 Heterogeneous Materials

We finally consider the fracture of heterogeneous materials. In connection with the Earth sciences, the most important materials of this type are rocks containing joints.

"Joints" in rock represent preexisting fracture surfaces; they may be filled with some material different from the surrounding rock or they may be empty. In either case, they generally are surfaces of weakness in the material, and the problem arises to assess the effect of the presence of the joints on the bulk fracture characteristics (e. g., the strength characteristics) of the medium as a whole. This effect is mainly due to *frictional* phenomena occurring at the joint surfaces.

The problem is not an easy one to solve. Its general implications have been enumerated, for instance, by Barton[128]. One of the difficulties is that there is an apparent lack of a connection between the fracture strength of the intact and of the jointed rock. It is even difficult to assess the true large-scale defor-

125 Burridge, R., Willis, J. R.: Proc. Cambr. Philos. Soc. 66:443 (1969)
126 Madriaga, R.: Bull. Seismol. Soc. Am. 66:639 (1976)
127 Andrews, D. J.: J. Geophys. Res. 81:5679 (1976)
128 Barton, N.: Eng. Geol. 7:287 (1973)

mation and fracture behavior of jointed rock masses inasmuch as one would have to resort to very large scale experiments with many tons of rock. The impossibility of doing this has induced investigators to make experiments with models[129-132]. These, however, only yield qualitative results. A satisfactory account in quantitative terms of the effect of joints on the strength properties of rocks was finally given a second article by Barton[133].

Accordingly, one has to distinguish between the behavior of jointed rock at low confining pressures, in which there is a strong influence of surface roughness and variable rock strength, and high confining pressures, at which first the shearing strength spectrum becomes very uniform and finally a critical state is reached in which the normal dilatant behavior in fracture is suppressed.

The laws for the behavior of jointed rock at low confining pressures are based on empirical laws of friction. For a horizontal rough joint surface this law written as follows

$$\frac{\tau}{\sigma_n} = \tan(\phi_b + 2d_n) \qquad (3.3.5-1)$$

where τ is the peak shearing strength and σ_n the effective normal stress. The term in brackets represents the effective angle of friction; it is composed of two terms. The first involves the basic angle of friction ϕ_b (of the smooth material), and the second is due to the existence of a "peak dilation angle" d_n which is equal to the instantaneous inclination of the actual shearing path relative to the mean shearing plane, i.e., d_n is the deviation angle of the local motion ("dilation") from the mean shearing motion. It further turns out that d_n depends, at low stress levels, on the (unconfined) compressive strength σ_c of the material, as follows

$$d_n = 10 \log_{10} \frac{\sigma_c}{\sigma_n}. \qquad (3.3.5-2)$$

The physical reason for the existence of such a relation is evidently that the asperities are progressively sheared off if stresses exceeding their strength are reached. Thus

$$\tau = \sigma_n \tan\left[20 \log_{10}\left(\frac{\sigma_c}{\sigma_n}\right) + \phi_b\right]. \qquad (3.3.5-3)$$

Empirically, it has been found that the coefficient "20" depends on the roughnesses of the joints and should be replaced by an empirical coefficient R that can vary from 0 to 20:

129 Müller, L., Tess, C., Fecker, E., Müller, K.: Rock Mech. Suppl. 2:71 (1973)
130 Hoagland, R. G., Hahn, G. T., Rosenfield, A. R.: Rock Mech. 5:77 (1973)
131 Morland, L. W.: Rock Mech. 8:35 (1976)
132 Schneider, H. J.: Rock Mech. 8:169 (1976)
133 Barton, N.: Int. J. Rock Mech. Min. Sci. 13:255 (1976)

$$\tau = \sigma_n \tan \left[R \log_{10} \left(\frac{\sigma_c}{\sigma_n} \right) + \phi_b \right].$$

(3.3.5 – 4)

The $\sigma_n - \tau$ relation is curved as it is also known to be in the discussion of the strength of unjointed materials. In fact, Eq. (3.3.5 – 4) gives a connection between fracture strength (σ_c) and "frictional" strength (τ/σ_n) of jointed rock. This connection has been verified by numerous experiments.

As the level of effective normal stress increases, the dilation becomes related to the fracture strength by means of the *confined* rather than by means of the unconfined compressional strength which is given by the maximum stress difference $\sigma_1 - \sigma_3$ at fracture. Thus, Eq. (3.3.5 – 2) becomes

$$d_n = 10 \log_{10} \left(\frac{\sigma_1 - \sigma_3}{\sigma_n} \right)$$

(3.3.5 – 5)

and Eq. (3.3.5 – 3) becomes

$$\tau = \sigma \tan \left(R \log_{10} \frac{\sigma_1 - \sigma_3}{\sigma_n} + \phi_b \right).$$

(3.3.5 – 6)

Thus, the spectrum of strength or frictional behavior of jointed material is greatly reduced and depends only on R and ϕ_b.

At yet higher levels of stress, the shearing stress required to fracture intact rock is equal to the shearing strength of the resulting joint. This represents the condition that there is a brittle – ductile transition in the rock which occurs at a critical state when the Mohr envelopes representing peak shearing strength of intact rock reach a horizontal tangent. At this level, the dilation angle is zero, i.e., the asperities are completely ineffective. Thus, from Eq. (3.3.5 – 5)

$$(\sigma_1 - \sigma_3)/\sigma_n = 1 .$$

(3.3.5 – 7)

However, since σ_n must be reached in the critical state at the point where the Mohr circle envelope is horizontal, one has

$$\sigma = \tfrac{1}{2}(\sigma_1 + \sigma_3)$$

and thus

$$(\sigma_1 - \sigma_3)/[\tfrac{1}{2}(\sigma_1 + \sigma_3)] = 1$$

or

$$\sigma_1 = 3 \sigma_3 .$$

(3.3.5 – 8)

The critical state represents also the maximum possible shearing strength of the rock, corresponding to the critical effective confining pressure beyond which no further increase in strength can be obtained. In this, it makes no difference whether the rock is jointed or not, because, as already noted, the shearing strength of the intact rock is now equal to the shearing strength of the resulting joint. The latter will be in the plane of maximum shear which occurs at 45° to the principal stress direction. Thus, rocks fracturing under critical conditions have a fracture angle φ (cf. Sect. 3.3.2a) equal to 45°. Thus, the theory of Barton reviewed here, in effect, provides an explanation for the fracture criterion of Mohr presented in Sect. 3.3.2. Clearly, the frictional relation (3.3.5 – 3) or (3.3.5 – 6) has the effect that in below-critical conditions fracture surfaces with $\varphi < 45°$ can be activated as implied by the Mohr criterion.

3.4 Rheology of the Earth: The Basic Problem of Geodynamics [134]

3.4.1 General Considerations

The chief problem of the science of geodynamics is to determine deformations within the Earth and upon its surface. On the surface, the present-day deformations are known and the task is to explain the latter in terms of stresses that could be considered as reasonable. In any properly defined deformation theory, it is a matter of mathematical analysis to determine the stresses from the boundary conditions. Thus, if the "deformation theory" applying to the Earth were properly defined, it would be, in principle, a straight forward (though not necessarily easy) matter to calculate the stresses from the (known) strains and then to look for causes of the latter. Unfortunately, it is a fact that the "deformation theory" applying to the Earth is not known, thus leaving the matter of finding the causes of its present-day appearance wide open to speculation.

It is thus evident that the basic problem of geodynamics is to determine the proper rheological conditions in the Earth. The explanation of the present-day physiography would then follow more or less automatically.

With regard to the geodynamically significant regions of the Earth, it may be assumed that the upper parts, say down to the low-velocity channel (cf. Sect. 2.1.4.3), which may be taken as the lower limit of the tectonic plates (Sect. 1.4.2), are of prime importance. This region has been variously referred to as "lithosphere" or "tectonosphere", the latter term being preferable because a reference is implied to tectonic activity rather than to composition (lithos = rock). We shall, below, essentially confine ourselves to the rheology of the cited tectonosphere but shall, on occasion, make reference to deeper regions.

Inasmuch as the scales of geodynamic phenomena stretch over a vast range, temporally as well as spatially (cf. Sect. 1.2.2), it was first thought [135]

---

134 This Section after Scheidegger, A. E.: Ann. Geofis. (Roma) 23:27 (1970), 23:325 (1971) and 24:311 (1971) (by permission)
135 Scheidegger, A. E.: Can. J. Phys. 35:383 (1957)

that the rheological response of the tectonosphere would be quite different in the various temporal and spatial scale ranges. However, one of the most important outcomes of recent investigations is that this is not so. As a first approximation, for small stresses, the tectonosphere reacts as an elastic body. Deviations from elasticity, below the occurrence of actual fracture, cannot be described by linear equations. The available evidence indicates that logarithmic creep is the prevailing deviation from elastic behavior of the tectonosphere, regardless of the temporal or spatial scale of the phenomena involved.

When shearing stresses are so large that a critical value is exceeded ("strength limit"), the tectonosphere reacts by a phenomenon that can heuristically be described as "fracture". The exact physical nature of this "fracture" is at present not yet clear, but for geodynamic purposes this is immaterial.

One therefore has three ranges of behavior of Earth materials, which may be referred to as "elastic range", "creep range", and "failure range", respectively. These will now be discussed individually.

3.4.2 The Elastic Range

3.4.2.1 General Remarks

For small stress changes, the upper parts of the Earth, commonly called the "tectonosphere", behave in an elastic fashion. Evidence for this comes from all time ranges. It will be reviewed below.

3.4.2.2 Laboratory Measurements

Laboratory measurements of the elastic constants can be performed by various means. The incompressibility of a rock sample can be measured directly by compression tests: more common, however, are ultrasonic measurements whereby the P and S wave velocities are found. Since the density of a rock sample can be easily measured, the elastic constants can then be deduced from Eqs. (3.2.1 – 17/18). The latter method is also suitable for use at high confining pressures. Data for the highest pressures are obtained by observations of shock waves produced by explosions.

The number of investigations along these lines is very large. Results have been collected in various summaries [136-138].

The values of the elastic constants vary of course widely for various rocks, but are, for E and μ, generally of the order of $10^{10} - 10^{11}$ Pa, the value of E being for any one rock usually about double that of μ. Poisson's ratio m is commonly around 0.2.

136 Wuerker, R. G.: Annotated tables of strength and elastic properties of rocks. AIME Pap. No. 663-G:27 (1956)
137 Birch, F.: Compressibility: Elastic constants. In: Handbook of Physical Constants. Revised edn. Geol. Soc. Am. Mem. 97:97 (1966)
138 Müller, L.: Der Felsbau. Stuttgart: F. Enke 1963. See 296 ff. therein

3.4.2.3 Indirect Measurements

Of almost greater importance for tectonophysical purposes than direct measurements of the elastic constants of rock samples are indirect inferences regarding the tectonosphere that can be drawn from seismological investigations. Here, the work of Bullen[139,140] is of greatest importance. In essence, from the seismic traveltime curves (see Sect. 2.1.2) one can obtain the wave velocities v_P and v_S as a function of depth (see Fig. 28). Using the estimates of the course of density with depth (Fig. 29), one can calculate the course of the elastic constants upon the basis of Eqs. (3.2.1 – 17/18). The result is shown in Fig. 72.

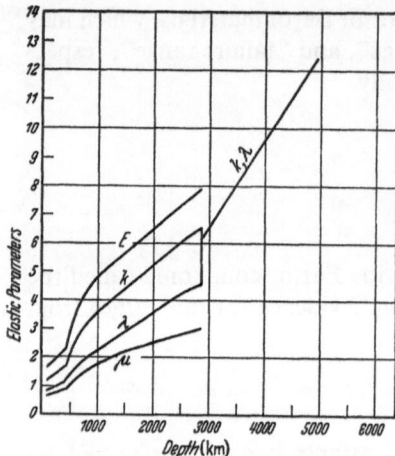

Fig. 72. Elastic parameters (scale 10^{11} Pa) in the interior of the Earth. (Based on data of Haddon and Bullen[141])

The elastic moduli mentioned above fit well with those measured on rock samples in the laboratory. Even sand grains fit the same pattern as long as they do not shift with regard to each other. Each sand grain, of course, has the elastic moduli of quartz.

The data adduced above show that the tectonosphere behaves elastically under the influence of stress cycles of very short duration (periods up to some seconds). For longer-time stress cycles, the basic elastic response of the tectonosphere to (weak) stress changes is borne out by an analysis of the Earth's tides. The elastic response in this time range (period 12 h or so) makes it possible to describe the Earth's tides in terms of Love's and Shida's numbers (cf. e.g., Melchior[142]). For yet longer time ranges, the interpretation of large-

139 Bullen, K. E.: Introduction to the Theory of Seismology, 3rd edn. London: Cambridge Univ. Press 1965
140 Haddon, R. A. W., Bullen, K. E.: Phys. Earth Planet. Int. 2:35 (1969)
141 Haddon, R. A. W., Bullen, K. E.: Phys. Earth Planet. Int. 2:35 (1969)
142 Melchior, P.: The Earth Tides. London: Pergamon Press 1966

scale gravity anomalies has also borne out elastic behavior in time ranges of 60 – 80 million years and has yielded estimates for the flexural rigidity D of the tectonosphere[143,144]. Assuming that the gravity anomalies are caused by mid-oceanic ridges yields values for D of the order of some 10^{21} to 9×10^{24} Newton meters. This is evidently quite a range; the individual values do not correlate well with crustal thickness, etc. One must therefore conclude, as admitted also by Walcott[145] that the elastic model is too simple and that relaxation effects corresponding to some other rheological behavior of the *whole* tectonosphere are of paramount importance.

3.4.3 The Attenuation Range

3.4.3.1 Laboratory Experiments

As the stresses exceed some very small limits, the response of the tectonosphere to them is no longer elastic: inelastic deviations are noted which, first of all, are seen in laboratory experiments with rocks.

In fact, the general rheological equation that was proposed long ago by Andrade (cf. Sect. 3.2.5) was based exactly on such laboratory experiments. The techniques for making such experiments range from simple calipering of rock samples under compression to such sophisticated procedures as holographic interferometry[146] and monitoring of acoustic emissions[147].

One of the essential terms in Andrade's equation [Eq. (3.2.5 – 1)] is that containing the time dependence $E(t)$. The evidence regarding this time function was collected by Morlier[148]. Accordingly, for some rocks one has

$$E(t) = A\,(1 - e^{-kt}) \qquad\qquad (3.4.3-1)$$

which corresponds to the behavior of a Kelvin material. For rocks exhibiting this type of behavior, Morlier[148] gives relaxation times of

$$\tau = \frac{1}{k} \sim 1 \text{ to } 16 \text{ days} . \qquad\qquad (3.4.3-2)$$

Using the ordinary short-term rigidity $\mu = 2 \times 10^{11}$ Pa, this yields a viscosity range of

$$\eta \cong 1.7 \text{ to } 27 \times 10^{16} \text{Pa} \cdot \text{s} . \qquad\qquad (3.4.3-3)$$

Most rocks, however, do not have an exponential time function, but a logarithmic one for a constant stress σ; a most common form is given by

143 Walcott, R. I.: J. Geophys. Res. 75:3941 (1970)
144 Walcott, R. I.: Bull. Geol. Soc. Am. 83:1845 (1972)
145 Walcott, R. I., J. Geophys. Res. 75:3941 (1970)
146 Spetzler, H., Scholz, C. H., Chi-Ping, J. L.: Pure Appl. Geophys. 112(3):571 (1974)
147 Lockner, D., Byerlee, J.: Bull. Seismol. Soc. Am. 67(2):247 (1977)
148 Morlier, P.: Ann. Inst. Tech. Bât. Trav. Publ. 19:89 (1966)

(3.2.5 – 3), so that $A = \sigma/\mu$; $B = q\sigma/\mu$; $E(t) = \ln(1 + bt)$; and $C = 0$ in Eq. (3.2.5 – 1). For practical purposes, the above equation may be modified to read

$$\varepsilon = \alpha \ln[\beta(t - t_0)] .\tag{3.4.3 – 4}$$

From experiments, Parsons and Hedley[149] obtained values for the factor α ranging from about 5.5×10^{-6} for sandstone to about 6.7×10^{-5} for potash. For potash, they required a Maxwell term as well ($C = 170 \times 10^{-6}\,\mathrm{hr}^{-1}$) but this term is very small for sandstone ($C = 4 \times 10^{-6}\,\mathrm{hr}^{-1}$). At any rate, the lower of these limits seems to be more appropriate for an extrapolation to the Earth's crust rather than the higher one. The values given above may be taken as characteristic (in order of magnitude) for the tectonosphere.

The equations discussed above do not permit the existence of a stationary creep process. However, it has been suggested that long-term geological behavior should really correspond to a steady state. As noted in Sect. 3.2.5, the term containing $E(t)$ in Eq. (3.2.5 – 1) must be zero in that case. Accordingly, many attempts have been made to fit laboratory observations to a Glen-type law as represented by Eq. (3.2.5 – 11). In particular, the dependence of the constants on absolute temperature T is of interest in this connection; the latter is generally taken into account in the following fashion:

$$\dot{\varepsilon} = A \exp(-Q_c/RT)\sigma^n \tag{3.4.3 – 5}$$

where Q_c is the creep activation energy, R the universal gas constant, and A some factor of proportionality. Many experiments to determine the constants in Eq. (3.4.3 – 5) have been reported in the literature which have been reviewed by Carter[150]. The values of n, which are of prime interest, were found as of the order of $2 – 3$ for dunite, 6.5 for dry quartzite, 8.3 for marble, and 9.1 for dolomite.

3.4.3.2 Seismic Wave Transmission

A further body of evidence regarding the inelastic behavior of the tectonosphere can also be gleaned from seismic wave transmission[151]. In this connection, it has been observed that seismic waves are *damped*. A damped harmonic wave can be described as follows.

$$A = A(x, t) \exp i(kx - \omega t) \tag{3.4.3 – 6}$$

where A is a decreasing function of x (position) and t (time), ω is the circular frequency, k the wave number vector. The function $A(x, t)$ depends on the damping mechanism, i.e., on the rheology of the material. In spherical elastic waves (distance r from source) without damping, the amplitude $A(r, t)$

149 Parsons, R. C., Hedley, D. G. F.: Int. J. Rock Mech. Min. Sci. 3:325 (1966)
150 Carter, N. L.: Rev. Geophys. Space Phys. 14(3):301 (1976)
151 Scheidegger, A. E.: Ann. Geofis. (Roma) 24:311 (1971)

decreases, because of the geometry, like $1/r$. Phenomenologically, it is often found that $A(r, t)$ can then be represented as follows.

$$A(r, t) = A_o e^{-ar}/r \qquad (3.4.3-7)$$

where a is called attenuation constant. Instead of a, one often introduces the quality factor Q, defined as follows

$$Q = \omega/2va \qquad (3.4.3-8)$$

where v is the phase velocity of the wave. In a damped standing wave, the decay of the amplitude goes with the factor $\exp(-\gamma t)$, where

$$\gamma = \omega/2Q. \qquad (3.4.3-9).$$

In a damped harmonic elastic wave, one has further

$$2\pi/Q = \Delta E/E \qquad (3.4.3-10)$$

where ΔE is the amount of energy dissipated per cycle in a given volume and E the peak elastic energy in that volume.

As with the function $A(x, t)$, the behavior of Q (particularly as a function of frequency) is characteristic for the rheology of the material in question.

The most important result of recent investigations on the damping of elastic waves in the upper mantle and crust of the Earth is that Q is essentially frequency-independent with values in the upper mantle mostly around $100-500$; the values change with depth (this may be due to partial melting[152]) and may change with horizontal distance[153] ("Q-structure of the Earth"). Although there are still some difficulties with some of the interpretations of the available data, the above fact of frequency independence of Q seems fairly well established (see Refs. [154-156] for summaries) for periods from those corresponding to seismic waves to those corresponding to the free oscillations of the Earth. A similar frequency independence of Q has also been found in laboratory experiments with rocks[154,157]. Any deductions regarding the rheology of the tectonosphere must take cognizance of this fact of Q-constancy.

The most commonly assumed imperfections of elasticity are modifications of the elasticity equations by adding a viscosity term. This can, in fact, be done in two ways so as to yield either a "Kelvin solid" or a "Maxwell liquid" (see Sect. 3.2.4).

The damping of an elastic wave in a Kelvin type of material can be calculated[158]. One finds, in approximation, that the absorption coefficient γ is proportional to the square of the frequency:

152 Gliko, A. O., Zharkov, V. N.: Bull. (Izv.) Akad. Nauk SSSR Earth Phys. 1977(5):86 (1977)
153 Brune, J. N.: Geophys. Res. Lett. 4(5):179 (1977)
154 Knopoff, L.: Rev. Geophys. 2:625 (1964)
155 Jackson, D. D., Anderson, O. L.: Rev. Geophys. 8:1 (1970)
156 Anderson, D. L., Hart, R. S.: J. Geophys. Res. 83:5869 (1978)
157 Pandit, B. I., Savage, J. C.: J. Geophys. Res. 78(26):6097 (1973)
158 Knopoff, L.: Rev. Geophys. 2:625 (1964)

$$\gamma = \text{const } \omega^2 . \tag{3.4.3-11}$$

With Eq. (3.4.3 – 9) this yields for Q

$$Q = \omega/2\gamma = \text{const}/\omega . \tag{3.4.3-12}$$

In a Maxwell material, one finds [159]

$$Q = \text{const } \omega . \tag{3.4.3-13}$$

The assumption of a Maxwell model enables one to interpret the Q-structure of the Earth in terms of a viscosity structure[160,161]. In general one finds a low-viscosity zone corresponding to the low-velocity seismic zone (cf. Sect. 2.1.4).

As is evident, none of the above two linear models yields the observed frequency independence of Q. Knopoff[159] has shown that no combination of Maxwell and Kelvin properties can give a frequency independent Q. It is possible to achieve this, by means of linear models, only if higher time derivatives are used in the stress-strain relation and if the material is no longer homogeneous. A series of such possibilities has been discussed by Caputo[162,163].

From simple macro-rheological models, a constant Q can therefore only be obtained by some sort of non-linear rheological equation. Such a model is the non-linear (logarithmic) creep model, for which the (approximate) frequency independence of Q was deduced for the first time by Becker[164] in 1925. In connection with geology, the non-linear logarithmic creep law is usually given in the form of (3.2.5 – 4).

Using this equation as a stress-strain relation containing an explicit time dependence, and employing a principle of superposition, one can show that Q becomes practically frequency independent and that

$$1/Q \cong q\pi/2 . \tag{3.4.3-14}$$

Although the logarithmic law is, in effect, the integral of a non-linear stress-strain relation[165] so that the validity of the superposition principle is somewhat doubtful, the above results can nevertheless be taken as an indication of the prevailing conditions. With an average value of $Q = 200$ for the tectonosphere (see above), one obtains

$$q = 0.0032 . \tag{3.4.3-15}$$

Many speculations have been made regarding the microscopic mechanisms causing the constancy of Q (and causing logarithmic creep). Orowan[166] has

159 Knopoff, L.: Rev. Geophys. 2:625 (1964)
160 Berckhemer, M., Auer, F., Drisler, J.: Phys. Earth Planet. Int. 20:48 (1979)
161 Meissner, R. O., Vetter, U. R.: J. Geophys. 45:147 (1979)
162 Caputo, M.: Ann. Geofis. (Roma) 9:383 – 393 (1966)
163 Caputo, M.: Geophys. J. R. Astron. Soc. 13:529 (1967)
164 Becker, R.: Z. Physik 33:185 (1925)
165 Scheidegger, A. E.: Rock Mech. 2:138 (1970)
166 Orowan, E.: Geophys. J. R. Astron. Soc. 14:191 (1967)

given an able summary of the ideas on the subject. However, for a discussion of the *macroscopic rheology* of the tectonosphere, this is of little direct significance.

3.4.3.3 Aftershock Sequences

Important information for the "intermediate" time range to be gleaned from seismology is from the source mechanisms, particularly from the time evolution of aftershock series. Commonly, one has used the *Benioff-strain rebound model* to explain the mechanism of aftershocks: This model assumes that the main shock relieves the built-up stresses and that the subsequent aftershocks are the expression of the adjustment of a Kelvin material to the stress redistribution. Scheidegger[167] has calculated the relaxation time τ required for this model from an analysis of aftershocks and obtained

$$\tau \sim 2\,\text{days}, \tag{3.4.3-16}$$

leading to an estimate of the Kelvin viscosity, using the ordinary short-term rigidity $\mu = 2 \times 10^{11}\text{Pa}$, of $\eta = 3 \times 10^{16}\ \text{Pa}\cdot\text{s}$.

However, the new information adduced over the last years showed that the Benioff model, and therefore the Kelvin model for the rheological behavior of the material in which the aftershocks take place, can no longer be maintained[168]. This contention is based on the fact that there are a number of "laws" regarding earthquake aftershock sequences with have been presented in Sect. 2.2.6.3. The three laws mentioned in Sect. 2.2.6.3 permit the setting up of a model of the response of the tectonosphere to the stress drop caused by the main shock. Because of the constancy of the magnitudes, the strain release corresponds simply to the number of earthquakes per unit time.

Thus, we have [cf. (2.2.6 – 6)]

$$\dot{\varepsilon}(t) = \text{const} \cdot n(t) = \text{const}\, at^{-1} \tag{3.4.3-17}$$

and the strain rate curve (writing the constant as k) is given by the integral of the expression above, leading to

$$\varepsilon(t) = ka \ln \varrho t \tag{3.4.3-18}$$

which represents a form of logarithmic creep; ϱ is a constant of integration. The appropriate model describing the release of aftershocks is thus one based on logarithmic creep[169]. The use of a power law may also be possible[170].

In conclusion, it should perhaps be mentioned that Pshennikov[171] tried to fit a Maxwell material to seismic aftershock sequences. However, this model

167 Scheidegger, A. E.: Can. J. Phys. 35:383 (1957)
168 Ranalli, G., Scheidegger, A. E.: Ann. Geofis. (Roma) 22:293 (1969)
169 Scheidegger, A. E.: Ann. Geofis. (Roma) 23:27 (1970)
170 Melosh, H. G.: J. Geophys. Res. 81:5621 (1976)
171 Pshennikov, K. V.: Aftershock Mechanism and the Inelastic Properties of the Earth's Crust. Moscow: Nauka 1965

fits the phenomenology of the aftershock process only very poorly and must be rejected.

3.4.3.4 Tides and Rheology

The gravitational attraction of the Moon and the Sun upon the Earth produces deformations in the latter; these are commonly called "Earth tides".

In a first approximation, the tidal deformations of the solid Earth are *elastic*, implying an instantaneous proportionality between the surface displacement (strain) and the tide potential (cf. Sect. 4.4.2). However, it turned out that there is, in fact, no instantaneous proportionality between potential and displacement but generally a significant phase lag of about 2.16° on the average (see Sect. 4.4.3).

The physical explanation of the observed phase lag has been generally sought in a linear type of anelasticity[172]. Indeed, in an oscillating system in which the damping is proportional to the velocity (strain rate), forced oscillations show a phase lag with regard to the imposed oscillation force (cf. Sect. 3.2.4). From this phase lag one can estimate the Kelvin viscosity of the tectonosphere. For the Earth, we have approximately

$$\delta\omega = 2.16°; \tan 2\varepsilon \sim 0.0378 \ .$$

Thus the "relaxation time" $\tau = \eta/\mu$ becomes [cf. Eq. (3.2.4 – 16)]

$$\tau = \left(\frac{6}{2\pi} - \frac{2\pi}{6} \times \frac{1}{4\pi^2}\right) \cdot 0.0378 = 0.0351 \text{ hrs} \simeq 126 \text{ s} \ . \qquad (3.4.3-19)$$

This is a value that is very small. Using the ordinary short-term rigidity of

$$\mu = 2 \times 10^{11} \text{ Pa} \qquad\qquad\qquad\qquad\qquad\qquad (3.4.3-20)$$

yields for the Kelvin viscosity η

$$\eta = 2.5 \times 10^{13} \text{ Pa} \cdot \text{s} \ . \qquad\qquad\qquad\qquad\qquad (3.4.3-21)$$

This is again a value which is very small. It does not agree with the values obtained from creep experiments of Kelvin-type rocks (see Sect. 3.4.3.1) nor does it agree with the value obtained from seismic aftershock sequences (see Sect. 3.4.3.3) if the latter are "forced" into a Kelvin model (Benioff-strain rebound theory).

Since, in seismology, the Kelvin model has been shown to be inadequate in the "*intermediate*" time range (as evidenced by the decay of aftershock series) in any case, one will have to search for an explanation of tidal friction by models other than that of a Kelvin body.

---

172 Melchior, P.: The Earth Tides. London: Pergamon Press 1966

A review of some such possibilities has been given by Lagus and Anderson[173]. Accordingly, it may be expected that about one-half to two-thirds of the tidal energy dissipation may occur within the shallow seas. MacDonald[174] suggests that some of the dissipation of tidal energy in the solid parts of the Earth may be due to the grating of crustal blocks against each other. This idea comes rather close to an assumption of *solid* friction as a dissipative mechanism which, if it were assumed to occur in microscopic volumes rather than macroscopic blocks, would phenomenologically be that of logarithmic creep.

Indeed, a discussion of Jeffreys[175] of the origin of the phase lag in a system with forced oscillations based on Lomnitz'[176] law [Eq. (3.2.5 – 4)], can be applied to the tidal dissipation problem. Equation (3.2.5 – 9) leads to

$$q = \frac{\delta\omega}{\frac{1}{2}\pi} = \frac{2.16°}{90°} = 0.024 . \tag{3.4.3 – 22}$$

In addition to a phase lag in the response of the solid Earth to the tidal forces, the anelasticity also causes a discrepancy between thy dynamic shear modulus for body waves and for the eigenoscillations. This difference can be correctly calculated[177] if a frequency-independent Q-factor is assumed, which again shows that the anelasticity is of the logarithmic creep type.

3.4.3.5 Rheology and the Chandler Wobble

A further effect concerning the Earth which we shall investigate regarding its implications upon the rheology of the tectonosphere is the Chandler wobble. This wobble[178] manifests itself by a periodic variation of the astronomically determined latitude of an observatory; the period is 430 days. It seems the variations become excited at irregular intervals. At the moment, there is a certain controversy as to the actual mechanism of excitation, some of the authors believing the latter to be due to the action of large earthquakes[179 – 183], others to meteorological agents[183,184]. Be this as it may, after excitation, the oscillations decay with a relaxation time (in amplitude) of some 10 (5 to 30) periods. A spectral analysis yields a corresponding value of the quality factor

173 Lagus, P. L., Anderson D. L.: Phys. Earth. Planet. Int. 1:505 (1968)
174 MacDonald G. J. F.: Rev. Geophys. 2:467 (1964)
175 Jeffreys, H.: Mon. Not. R. Astron. Soc. 118:14 (1958)
176 Lomnitz C.: J. Geol. 64:473 (1956)
177 Zharkov, V. N., Molodenskiy, S. M.: Bull. (Izv.) Acad. Sci. USSR Earth Phys. 1977(5):17 (1977)
178 Jeffreys, H.: Mont. Not. R. Astron. Soc. 117:506 (1957)
179 Smylie, D. E., Mansinha, L.: J. Geophys. Res. 73:7661 (1968)
180 O'Connell, R. J., Dziewonski, A.: Trans. Am. Geophys. Union 57(4):291 (1976)
181 O'Connell, R. J., Dziewonski, A.: Nature (London) 262:259 (1976)
182 O'Connell, R. J., Dziewonski, A.: Trans. Am. Geophys. Union 57:955 (1976)
183 Mansinha, L., Smylie, D. E., Chapman, C. H.: Geophys. J. R. Astron. Soc. 59:1 (1979)
184 Wilson, C., Haubrich, R.: Geophys. J. R. Astron. Soc. 46:707 (1976); 46:745 (1976)

Q in the range of[185] 30 to 60. In comparison with values obtained from seismic waves, this is abnormally low. It stands to reason that some inferences can be made from the pattern of decay upon the rheology of the tectonosphere causing this decay.

In this vein, Scheidegger[186] has calculated the implications of the observed damping in terms of the Kelvin model. In this case, Eq. (3.2.4 – 13) can immediately be used to make an estimate of the factor η/μ from a measurement of ω and a. One obtains, with $T = 2\pi/\omega = 420$ days and $1/a = 4200$ days,

$$\tau = \frac{\eta}{\mu} \cong 2 \,\text{days} \sim 1.7 \times 10^5 \,\text{s} . \qquad (3.4.3-23)$$

This is a value for the relaxation time τ which is not too different from that obtained by applying a *Kelvin* model to seismic aftershock series.

However, the above is not the only interpretation possible for the origin of the decay of the Chandler wobble. Thus, it has been tried to ascribe this decay to a Maxwell effect[187]. Upon this basis, using a particular model of the structure of the Earth, Gerstenkorn[187] found that the observed damping can be explained by assuming for the Maxwell material in question

$$\tau = \frac{\eta}{\mu} \cong 10^8 \,\text{s} . \qquad (3.4.3-24)$$

Using the ordinary short-term rigidity $\mu \sim 2 \times 10^{11}$ Pa, this yields a Maxwell viscosity of the Earth of the order of 2×10^{19} Pa $\cdot$ s. This is not too far from the order of magnitude of the very long-time viscosity deduced from such observations as isostatic adjustment (usually quoted as $\eta \sim 10^{21}$ to 10^{22} Pa $\cdot$ s).

Finally, some attempts have also been made to explain the damping of the Chandler wobble by the assumption of some logarithmic creep law in the tectonosphere. Thus Jeffreys[188] used the form (3.2.5 – 4) of Lomnitz' law and applied it to the damping in question. He obtained that this model would lead to a lag of 1/40; using formula (3.2.5 – 9) this yields for q

$$q = \frac{1}{40 \cdot \frac{1}{2}\pi} = \frac{1}{62.8} = 0.016 \qquad (3.4.3-25)$$

which is substantially less than the corresponding value obtained from an analysis of the Earth's tides. It does not agree with the damping values calculated from deep penetrating S waves, either. A suggested explanation has been that the damping of the Chandler wobble is only in part caused by the

185 Pedersen, G. P. H., Rochester, M. G.: In: eds. P. Melchior, S. Yumi, Rotation of the Earth, 33 – 38 (1972)
186 Scheidegger, A. E.: Principles of Geodynamics, 2nd edn. Berlin-Heidelberg-New York: Springer 1963
187 Gerstenkorn, H.: Icarus 6:292 (1967)
188 Jeffreys, H.: Geophys. J. R. Astron. Soc. 14:1 (1967)

anelasticity of the tectonosphere and that a large part of the energy is dissipated in the oceans[189].

We therefore hold that it is probable that a solution for variable stresses of some non-linear creep equation, corresponding to logarithmic creep under constant stress, will produce the required damping.

3.4.3.6 Rheology and Isostasy

The principle of isostasy requires the surface of the Earth to be in isostatic equilibrium (cf. Sect. 2.3.3). The fact that such equilibrium is not always present implies that forces are acting on the tectonosphere to which the latter responds by creep.

Thus, the rise of land areas since the end of the last ice age (cf. Sect. 1.7.6) has been interpreted as response to unloading. The rate of rise can be reconstructed backward in time by paleogeographic techniques. If this is done, the rise curves shown in Fig. 26 are obtained.

Much of the recent evidence on uplift curves has been summarized by Andrews[190,191]. Of particular interest is the form of these uplift curves. Originally, an exponential shape had been assumed, but a very bad fit with the data was obtained. An exponential uplift curve implies a viscous flow law in the substratum. If such a viscous substratum is used as a model, viscosity values of around $10^{21} - 10^{22}$ Pa $\cdot$ s are required (cf. Sect. 8.6.2).

Because of the poor fit of the uplift data with exponential curves, Andrews[191] abandoned the hypothesis of an exponential decay altogether and suggested a logarithmic equation instead:

$$U = a + r \ln t. \tag{3.4.3-26}$$

In terms of rates, this yields upon differentiation

$$\frac{dU}{dt} = \dot{U} = \frac{r}{t}. \tag{3.4.3-27}$$

Andrews[190] found a much better agreement of Eq. (3.4.3 – 26) with natural data than of an exponential equation. He[191] made later a specific comparison of the two equations for 58 uplift curves and found that the exponential dependence gives consistently present-day uplift rates that are too low in comparison with observations, whereas the logarithmic dependence gives correct values.

Equation (3.4.3 – 26) being logarithmic, is evidently not valid for $t = 0$. Thus, a cutoff must be applied near $t = 0$. This can easily be done by modifying it to read

$$U = a + r \ln (t + 1). \tag{3.4.3-28}$$

189 Jeffreys loc. cit. last page
190 Andrews, J. T.: Can. J. Earth Sci. 5:39 (1968)
191 Andrews, J. T.: Can. J. Earth Sci. 7:703 (1970)

Regarding the constants r and a in Eq. (3.4.3 – 33), Andrews[191a] found some interesting relationships. From a comparison of 21 uplift rates he deduced by a least squares procedure

$$- a = 5.6\, r + 10\ \text{meters}$$

(U in meters, t in years), and thus, using Eq. (3.4.3 – 26), and setting $t = 1000$, one finds

$$r = \frac{A + 10}{1.3}$$

where A is the uplift in meters that occurred during the first 1000 years of deglaciation. A direct correlation analysis of observational data gives a similar result, viz.

$$r = \frac{A - 9.8}{1.02}.$$

In approximation, r is thus simply equal to the uplift A that occurred during the first 1000 years.

Thus, there is a strong indication that the form (3.4.3 – 26) is the best-fitting representation of the observational evidence, and that the exponential form is actually contradicted by the data.

In spite of the above remarks, many attempts have been made to model isostatic phenomena based on various types of viscous, mostly layered[192,193], models, with or whithout an elastic layer on top. Generally, in order to obtain a fit, the course of the viscosity with depth must show a minimum at about 200 km depth[193] (coinciding with the seismic low-velocity zone). Below the minimum, the viscosity increases with depth[194,195].

Another variation of one-layer models is represented by the assumption of a viscoelastic medium[196,197].

Attempts have also been made to justify the assumption of a viscous rheological behavior of the tectonosphere not only from a phenomenological-observational standpoint, but also from considerations of solid-state theory. Accordingly, it is usually assumed that the viscosity is due to diffusion creep which appears macroscopically as viscous behavior. The viscosity η is then estimated from the Nabarro[198]-Herring[199] equation

191a Andrews, J. T.: Can. J. Earth Sci. 5:39 (1968)

192 Danes, Z.: Icarus 9:1 (1968)

193 McConnell, R. K.: J. Geophys. Res. 70:5171 (1965)

194 Brotchie, J. F., Silvester, R.: J. Geophys. Res. 74:5240 (1969)

195 Forsyth, D. W.: Rev. Geophys. Space Phys. 17(6):1109 (1979)

196 Dunbar, W. S., Garland, G. D.: Can. J. Earth Sci. 12(5):711 (1975)

197 Peltier, W. R., Andrews, J. T.: Geophys. J. R. Astron. Soc. 46(3):605 (1976)

198 Nabarro, F. R. N.: In: Report of a Conference on Strength of Solids. London: Physical Soc. p. 7, 1918

199 Herring, C.: J. Appl. Phys. 21:437 (1950)

$$\eta = A \frac{kTl^2}{a^3 D}; \quad \text{with } D = D_o e^{-E/RT}$$

where A is a constant (about $1/30$), k is Boltzmann's constant, T the absolute temperature, a an atomic lattice dimension, l the grain size, D the coefficient of self-diffusion, E the activation energy for self-diffusion, and R the gas constant. Upon the above theory, estimates of the viscosity in the mantle using reasonable values for the parameters have been made[200]; the main feature of all calculations is an increase of viscosity with depth below a low-viscosity zone at about 200 km depth confirming qualitatively the findings from uplift data[201,202].

In conclusion, it might be added that discussions similar to that referring to the isostatic rise of formerly glaciated land can also be made with regard to the observed rise of land around Pleistocene Lake Bonneville[203,204] in Utah and with regard to the submergence of guyots[205]. Generally, only viscous models have been considered which yield viscosities of the substratum of a similar order of magnitude as that found in the viscous model of the rise of formerly glaciated land.

3.4.3.7 Evaluation

When we survey the discussion of the previous paragraphs, we note that basically three different rheological models have been advanced to describe the stress-strain behavior of the tectonosphere in the intermediate time range. These are: (a) Maxwell model, (b) Kelvin model and (c) the non-linear creep model. The possible applicability of these models to the various phenomena under discussion here is set out in Table 7.

An inspection of this Table 7 shows that the only interpretation of the rheological behavior of the tectonosphere in the creep range that does not contradict *some* phenomenon, is that based on the logarithmic creep model.

To discuss the "score" for the individual models in detail, we may make the following remarks:

a) The Maxwell Model. It fits some of the direct laboratory measurements on rocks, but cannot be used for the explanation of seismic aftershock sequences. It does not seem to have been used in connection with the tides of the Earth, inasmuch as the response of the Earth to changing graviational potentials is,

200 Gordon, R. B.: J. Geophys. Res. 70:2413 (1965)
201 Ranalli, G.: Ann. Geofis. (Roma) 30:435 (1977)
202 Vetter, U. R.: J. Geophys. 44:231 (1978)
203 Crittenden, M. D.: J. Geophys. Res. 68:5517 (1962)
204 Crittenden, M. D.: New data on the isostatic deformation of Lake Bonneville. U. S. Geol. Surv. Prof. Pap. 454-E (1963)
205 Saito, Y.: J. Oceanogr. Soc. Jpn. 20th Anniv. Vol., p. 25 (1962)

Table 7

Model	Phenomenon	Direct measurements	Seismic aftershocks	Tidal friction	Chandler wobble	Isostasy	Seismic waves
Maxwell	constants	$C = 4$ to 170×10^{-6} hr^{-1} for some given stress	Estimate not possible	Not attempted	$\tau \cong 10^8$ s	$\eta = 10^{20} - 10^{22}$ Pa · s	
	fit	"Ultimate creep" seems to be of this type	Poor		Possible	Poor to possible	Poor
Kelvin	constants	$\tau = 1$ to 16 days ($\eta \sim 1.7$ to 27×10^{16} Pa · s)	$\tau \cong 2$ days ($\eta \sim 3 \times 10^{16}$ Pa · s)	$\tau \sim 126$ s ($\eta \sim 2.5 \times 10^{13}$ Pa · s)	$\tau \cong 2$ days ($\eta \sim 3 \times 10^{16}$ Pa · s)	$\tau \approx 5000$ yr	
	fit	Fits some rocks but usually poor	Poor	Possible	Possible	Poor	Poor
Logarith-mic creep	constants	$\alpha \sim 5.5 \times 10^{-6}$ to 6.7×10^{-5}	Estimate not possible	$q = 0.024$	$q = 0.016$	Estimate not possible	$q = 0.0032$
	fit	Good	Good	Possible	Possible	Good	Good

in the first approximation, obviously elastic and not viscous. A Maxwell material, however, can be used for the treatment of the Chandler wobble.

b) The Kelvin Model. It fits the direct laboratory measurements on some rocks, but for most it is a poor model indeed. It cannot be used any longer for the explanation of the behavior of seismic aftershock sequences in the light of new evidence, although it used to be very popular in this context ("strain rebound theory"). It can be used in connection with the Chandler wobble and tidal friction, but the difference in the required material constants (e. g., τ) in those two cases constitutes a severe difficulty. Thus, the Kelvin model must be rejected as a model of the rheological behavior of the tectonosphere in the creep range.

c) The Non-linear Creep Model. It fits all the phenomena discussed here, although in many instances not enough data are available as yet to make a meaningful comparison with regard to the values of the material constants that are necessary to obtain a good fit between model and observations. Nevertheless, the logarithmic creep model seems to date to be the only one which must not be a priori rejected for the explanation of the rheological behavior of the tectonosphere in the creep range.

3.4.4 The Failure Range

3.4.4.1 Introduction

When the stresses exceed a certain limit, most materials undergo a fracture-like process. In this, the material of the tectonosphere is no exception, as is evidenced by the existence of such phenomena as faults and earthquakes.

The actual mechanism of faulting, earthquake generation, etc. will be dealt with in separate chapters; in the present context we are solely concerned with the strength limits of Earth materials. The evidence for the deduction of such limits can be gleaned from three sources: Laboratory experiments, in-situ tests, and geostatic considerations.

In conformity with the general theory, it must be assumed that strength limits are determined by the Mohr fracture criterion (cf. Sect. 3.3.2), notably as expressed by Mohr envelopes or the maximum shearing stress (i. e., maximum stress difference). In this, account must always be taken of the principle of effective stress in case a fluid-filled porous medium is involved.

3.4.4.2 Laboratory Experiments

First of all, one can try to assess the strength of Earth materials by making tests[206] of samples in the laboratory. In fact, many such laboratory tests have been reported in the literature, of which only very few can be mentioned here.

---

206 Brook, N.: Int. J. Rock Mech. Min. Sci. 14:193 (1977)

Thus, for an Australian sandstone Jaeger[207] found (in MPa)

$$\sigma_1 = 61.2 + 4.6\,\sigma_3 \qquad\qquad (3.4.4-1)$$

where σ_1 is the largest, and σ_3 the smallest compressive stress. For a South African quartzite, Cook and Hodgson[208] found similarly

$$\sigma_1 = 250 + 6\,\sigma_3. \qquad\qquad (3.4.4-2)$$

As a general remark, it can be said that the more compressible a rock, the lower is its fracture strength[209].

Of great interest is the shearing strength of rocks at high confining pressures. Riecker and Seifert[210] have made a particularly noteworthy set of determinations of the shearing strengths of upper-mantle mineral analogs; some of their results are shown here in Table 8. They were able to reach about 5×10^9 Pa confining pressure which, depending on the density law chosen, corresponds to some 170 km of depth.

Table 8. Shearing strengths of some minerals at high confining pressures. After Riecker and Seifert[210]. All stress values in 10^8 Pa

Av. press.	Olivine	Enstatite	Diopside	Labradorite
5.5	1.77		2.21	3.03
9.9	3.03	4.58	3.29	
15.2	5.43			
19.3	7.58	7.74	6.32	6.32
24.8	7.60			8.21
30.3	12.60	10.11	9.23	10.39
39.9	14.15	11.38	12.38	12.14
49.7	15.17	14.90	14.03	13.15
55.1		16.10		

3.4.4.3 In-situ Tests

In-situ tests for the determination of the strength of Earth materials have particularly been developed by foundation engineers. These tests involve experimentation either with test driving of piles[211] or by static penetration tests[212] of loaded piles into the ground. Mostly, only the "bearing capacity" of the ground is of engineering interest.

207 Jaeger, J. C.: Engineering 189:283 (1960)
208 Cook, N. G. W., Hodgson, K.: J. Geophys. Res. 70:2883 (1965)
209 Balakrishna, S.: Geophys. J. R. Astron. Soc. 14:119 (1967)
210 Riecker, R. E., Seifert, K. E.: J. Geophys. Res. 69:3901 (1964)
211 Rodin, S., Corbett, B. O., Sherwood. D. E., Thorburn, S.: Proc. Eur. Symp. Penetration Test. 1:139 Stockholm (1976)
212 E. g., Rollberg, D.: Bestimmung der Tragfähigkeit und des Rammwiderstands von Pfählen und Sondierungen. Veroeff. Inst. Grundbau der RWTH Aachen 3:43 (1977)

Inasmuch as such in-situ tests refer mainly to rather loose soils, the material generally behaves like an imperfectly plastic substance (Sect. 3.2.2) subject to the Coulomb equation (3.2.2 – 11). Values for the constants occurring therein[213] range from 0 (sand) to 10^7 Pa (air-dried clay) for c, and from 0° (clays) to 40° (screes of granite) for ϕ.

3.4.4.4 Geostatic Considerations

The fact that mountains, a non-hydrostatic bulge and gravity anomalies exist indicates that there is a strength threshold in the tectonosphere. Of course, it may be that some or all of these features are the expression of a dynamic steady state (e. g., caused by subduction of the lithosphere[214]) rather than of a static equilibrium; in that case the strength of the Earth might in fact be very low. Nevertheless, values for the long-term strength of the tectonosphere have been deduced from all of these features.

Turning first to the information that can be deduced from mountain ranges, we note that from simple stability considerations one obtains for the shearing strength

$$\theta = 4 \times 10^7 \, \text{Pa} \,.$$

This comes out of the well-known stability formula of Terzaghi[215]

$$H = \frac{\theta}{\varrho g} N \qquad\qquad (3.4.4-3)$$

for a mount of height H consisting of a material of density ϱ. N is a stability factor depending on the slope angle. For a slope angle of about 45°, one has $N = 6$. Choosing for the other variables $h = 8$ km (roughly the height of Mt. Everest), $\varrho = 3000$ kg/m^3, $g = 9.8$ m/s^2, yields the above quoted value for θ. This deduction assumes, of course, that mountains are static and not dynamic steady-state phenomena.

Similar values for θ are obtained if the non-hydrostatic bulge of the Earth is interpreted as a static (and not as a lag-exhibiting transient creep) phenomenon. Munk and MacDonald[216] obtain

$$\theta = 10^7 \, \text{Pa} \,.$$

The same order of magnitude, viz.

$$\theta = 3 \times 10^7 \, \text{Pa}$$

213 E. g., Scheffer, F., Schachtschabel, P.: Lehrbuch der Bodenkunde, 7th edn. Stuttgart: Enke 1970
214 Melosh, H. J.: Geophys. Res. Lett. 5(5):321 (1978)
215 Terzaghi, K.: Theoretical Soil Mechanics. London: Chapman and Hall 1943
216 Munk, W. H., MacDonald, G. J. F.: The Rotation of the Earth. London: Cambridge Univ. Press 1960

was obtained by Jeffreys[217] for the tectonosphere below 50 km depth (up to $\theta \sim 1.5 \times 10^8$ Pa for the region above 50 km depth) from an analysis of the effect of gravity anomalies of large horizontal extent.

Again a similar value ($\theta < 2 \times 10^7$ Pa) has an upper limit of the shear strength and was obtained by Brune et al.[218] by noting the absence of a heat flow anomaly greater than about 12 mW/m^2 on the San Andreas Fault.

Perusing data on the gravity field of the Earth that have been obtained from satellite measurements, Caputo[219] arrived at an improved strength value on this basis of

$$\theta \sim 3 \times 10^6 \, \text{Pa}\,.$$

The consideration of rotational features of the Earth, other than the non-hydrostatic bulge also leads to strength values[216]. Thus, an Earth without strength would be completely unstable with regard to polar wandering. The fact that the present north pole does not move toward the pole of the continent-ocean system, if assumed due to strength properties, leads to a lower limit, of θ, viz.

$$\theta > 10^6 \, \text{Pa}\,.$$

Similar strength values were obtained by Chinnery[220] when estimating the stresses that are released in an earthquake, assuming that the latter corresponds roughly to the strength threshold of the material. Generally, he finds

$$\theta \sim 10^6 \, \text{Pa}\,.$$

The lowest strength values are indicated from isostatic rebound data. The mere fact that rebound *does* seems to occur, puts an upper limit on the strength of the tectonosphere. Thus, Crittenden[221] finds maximally

$$\theta \sim 10^5 \, \text{Pa}\,.$$

This low value comes from the observation that the Earth yields to ice surface loads that can be calculated.

In summary, one finds that values of the shearing strength θ for the tectonosphere have been estimated as between 10^5 to 10^7 Pa. A reconciliation between the various values can only be achieved if some of the theoretical models, upon which the calculations were based, are rejected. Thus it is quite conceivable that neither mountains nor the bulge of the Earth are static equilibrium features, but rather dynamic steady-state or even transient features that are dynamically supported.

217 Jeffreys, H. The Earth. 4th edn. Cambridge 1959. See p. 209 ff therein
218 Brune, J. N., Henyey, T. L., Roy, R. F.: J. Geophys. Res. 74:3821 (1969)
219 Caputo, M.: J. Geophys. Res. 70:955 (1965)
220 Chinnery, M. A.: J. Geophys. Res. 96:2085 (1964)
221 Crittenden, M. D.: Geophys. J. R. Astron. Soc. 14:261 (1967)

4. Geodynamic Effects of the Rotation of the Earth

4.1 Introduction

The Earth is a rotating celestial body and is, as such, subject to gravitation. If all other forces were absent, the Earth would exhibit the form of a sphere and its surface would be perfectly smooth. We have already discussed some of the surface irregularities, but the principal departure of the Earth from a spherical shape is due to its rotation around its axis. The first problem, thus, is that of explaining the observed figure of the Earth in terms of rotational dynamics. This is, in fact, a very difficult task and as such the subject of study of a discipline ("higher geodesy") in its own right. For our purposes, a very approximate theory will be sufficient.

The rate of rotation of the Earth is by no means constant. Even in short time intervals, it varies. It has also varied during geological history. The presumed causes and geodynamic implications of such variations are, thus, the next problem to be discussed. The rotation of the Earth exposes different parts of the Earth at different times to the gravitational forces caused by the Sun and Moon. This is felt on the Earth as changing *tides*, which can have a geodynamic effect.

Finally, questions of the *stability* of the Earth's axis of rotation with regard to a mobile crust and of the *inertial forces* experienced by the latter will be discussed.

4.2 The Figure of the Earth

4.2.1 Present-Day Parameters

The rotation of the Earth, owing to the centrifugal force it creates upon any body connected with the Earth, effects that the equilibrium figure is not a sphere. The problem of determining this "equilibrium" (geoid) figure is not entirely simple because of the surface irregularities. The equilibrium figure is represented in oceanic areas by the ("mean") sea level; in land regions one would have to regard as "geoid" surface the hypothetical continuation of the sea level into the land. The sea level represents an equipotential surface of the gravity field; its continuation would therefore be the continuation of the same equipotential surface on land. Gravity, however, is essentially measured only on the surface of (or above) the ground; if an equipotential surface is to be determined, it is therefore first of all necessary to continue the gravity field downward to the geoid surface. Inasmuch as the gravity field depends on the masses present, and the distribution of the latter is somewhat hypothetical, it is seen that the shape of the geoid depends on the model of the subsurface which one envisages, just as is also the case in the determination of the "isostatic" reduction of gravity anomalies. The problem and recent results have been presented by Gaposchkin[1].

For geodynamic purposes it is usually sufficient to replace the geoid (the equilibrium figure of the Earth's gravitational field) by an ellipsoid. The difference between an ellipsoid, properly chosen, and the true equilibrium spheroid is so small that it can usually be neglected. The data for the approximating ellipsoid are determined from geodesy. The values obtained and accepted for further calculations are listed in Table 9 (see also Fig. 73).

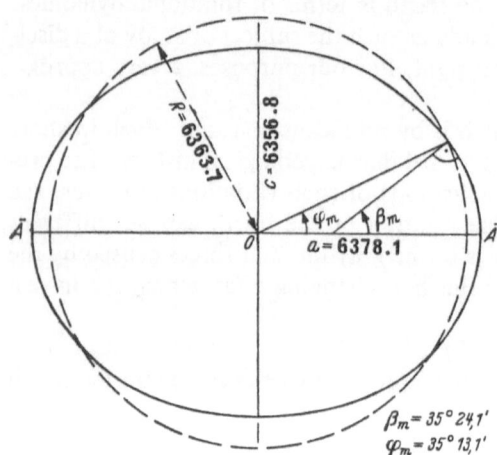

Fig. 73. Ellipsoid and sphere approximating the Earth; φ geocentric latitude, β geographic latitude; φ_m, β_m corresponding "average" latitudes (distances in km)

A further consequence of the ellipticity of the Earth is that the geographic latitude ϕ and the geocentric latitude ϕ' are not identical. The two are connected by the formula[2]

$$\tan(\phi - \phi') = \nu \sin 2\phi' + O(\nu^2) \tag{4.2.1-1}$$

where ν is the ellipticity.

The shape of the geoid, as defined above, is usually also referred to a specific reference ellipsoid. A recent example of a geoid determination is

1 Gaposchkin, E. M. (ed.): 1973 Smithsonian Standard Earth (III). Cambridge, Mass.: Smithsonian Astrophysical Observatory Special Rept. No. 353 (1973)
2 Jeffreys, H.: The Earth, 3rd edn. London: Cambridge Univ. Press, p. 130, 1952

Table 9. Geodetic parameters (IUGG Grenoble, 1975). (Errors refer to last cipher given)

Velocity of light c	299792458 ± 1.2 m/s
Newton's constant G	$6.672 \, (\pm 4.1) \times 10^{-11} \; \mathrm{m^3 s^{-2} \, kg^{-1}}$
Angular velocity ω	7.292115×10^{-5} rad/s
Newton's constant times mass of Earth GM	$3.986005 \, (\pm 2) \times 10^{14} \; \mathrm{m^3/s^2}$
Large axis (equatorial radius) a	$6378140 \, (\pm 5)$ m
Flattening $1/f$	$298257 \, (\pm 1.5) \times 10^{-3}$
Small axis (polar radius) c	$6356755 \, (\pm 5)$ m
Equatorial gravity g_e	$978.0318 \, (\pm 10)$ gal
Potential of geoid w_0	$6263683 \, (\pm 5)$ kgal/m
"Average" radius $R_0 (= GM/w_0)$	$6363676 \, (\pm 5)$ m

shown in Fig. 74. An inspection of this figure shows that there is a slight "pear"-shape present in the geoid. This is probably due to the presence of the ice caps at the poles.

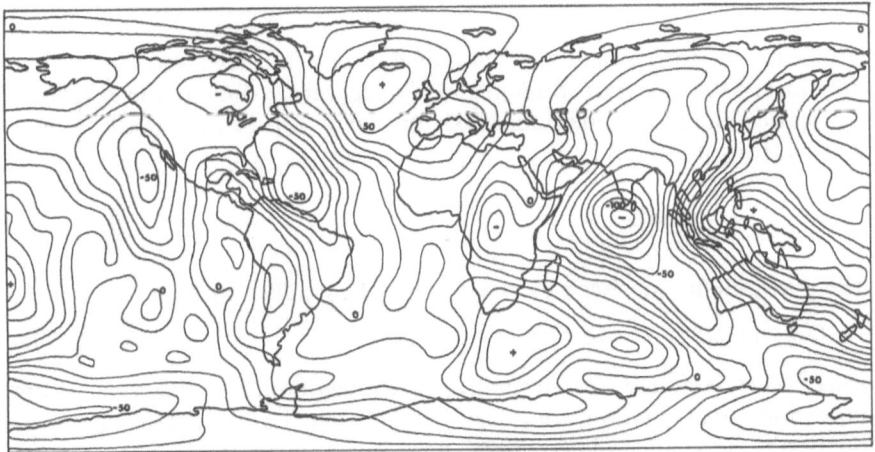

Fig. 74. Smithsonian Astrophysical Observatory Geoid of 1973. Heights in metres above an ellipsoid of flattening 1/298.256. (After Gaposchkin[1])

4.2.2 Simplified Equilibrium Theory of the Figure of the Earth

The fact that the equilibrium figure of the Earth is not a sphere, permits one to draw certain conclusions regarding the density distribution in its interior. However, such considerations are beyond the scope of the present study. The calculation of the equilibrium figure of the Earth, however, will be sketched below.

The general theory of the equilibrium figure of celestial bodies has been ably summarized by Jardetzky[3, 4]. Since it is quite involved, we present here only a simplified version given by Milankovitch[5].

Thus, if it be assumed that a liquid celestical body is rotating, then it is possible to determine its shape from the fundamental hydrodynamical equation

$$dv/dt = f - (1/\varrho)\,\mathrm{grad}\,p\,, \tag{4.2.2-1}$$

where v is the velocity-vector, f the specific mass-force due to gravity, p the pressure, and t time. In case that the liquid is assumed as highly viscous (as must be the case with the Earth), it is to be expected that, through the interaction of neighboring particles, an equilibrium figure will develop which rotates "en bloc" like a solid body at a uniform angular velocity ω. It is then possible to describe the body by using a system of axes represented by the principal axes of inertia (X, Y, Z; moments of inertia A, B, C). For the dynamical calculation, the body may be considered as being at rest if the centrifugal force F per unit mass

$$F = \omega^2 R \tag{4.2.2-2}$$

is added to the corresponding gravitational force f, R being a vector in a direction normal to the axis of rotation and of a magnitude equal to the distance of the particle under consideration from that axis. Eq. (4.1.2–1) can thus be written (for equilibrium)

$$f + F = \frac{1}{\varrho}\mathrm{grad}\,p\,. \tag{4.2.2-3}$$

The gravitational force can be expressed as the gradient of a potential U

$$f = \mathrm{grad}\,U \tag{4.2.2-4}$$

where U can be found by a straightforward integration over the body; assuming symmetry of rotation, i. e., $A = B$, which may be expected to be true for the Earth, yields (cf. Milankovitch[5], p. 115):

$$U = \varkappa\frac{M}{r} + \frac{1}{2}\varkappa\frac{X^2 + Y^2 - 2Z^2}{r^5}(C - A) \tag{4.2.2-5}$$

where $\varkappa$ is the gravitational constant, M the total mass of the body, r the distance of the point under consideration from the center of the body and X, Y, Z the coordinates of that point in the system of principal axes of inertia (see above). Similarly, the centrifugal force can be expressed as the gradient of a

3 Jardetzky, W. S.: Theories of Figures of Celestial Bodies. New York: Interscience 1958

4 See also: Ledersteger, K.: Z. Vermessungswiss. 84, No. 3:73 (1959)

5 Milankovitch, M.: Kanon der Erdbestrahlung und seine Anwendung auf das Eiszeitenproblem. Belgrade: Ed. Spec. Acad. R. Serbe Tome 133. Sec. Sci. Math. Nat. Tome 33, 1941

potential U'

$$F = \text{grad } U' = \text{grad } (\tfrac{1}{2}\omega^2 R^2) \qquad (4.2.2-6)$$

(R being the magnitude of R). Finally, setting

$$W = U + U' \qquad (4.2.2-7)$$

permits (4.2.2–1) to be written as follows:

$$\text{grad } p = \varrho \text{ grad } W. \qquad (4.2.2-8)$$

According to earlier remarks, the equilibrium figure of the Earth must be a surface of constant W

$$W = W_0. \qquad (4.2.2-9)$$

Introducing polar coordinates (ψ = longitude, φ = geocentric latitude) yields for W

$$W = \varkappa \frac{M}{r} + \frac{\varkappa}{2r^3}(C - A)(1 - 3\sin^2 \varphi) + \frac{\omega^2 r^2}{2}\cos^2 \varphi. \qquad (4.2.2-10)$$

At the surface of the Earth, one can replace r by the equatorial radius a and one obtains (to the first order of approximation)

$$r = \varkappa \frac{M}{W_0}\left[1 + \frac{C - A}{2a^2 M}(1 - 3\sin^2 \varphi) + \frac{\omega^2 a^3}{2\varkappa M}\cos^2 \varphi\right]. \qquad (4.2.2-11)$$

For $\varphi = 0$, one must obtain $r = a$, which yields

$$a = \frac{\varkappa M}{W_0}\left(1 + \frac{C - A}{2a^2 M} + \frac{\omega^2 a^3}{2\varkappa M}\right) \qquad (4.2.2-12)$$

and for $\varphi = \pi/2$, $r = c$ if c denotes the polar radius. Thus

$$c = a\left[1 - \left(\frac{\omega^2 a^3}{2\varkappa M} + \frac{3}{2}\frac{C - A}{a^2 M}\right)\right]. \qquad (4.2.2-13)$$

If we set

$$v = \frac{\omega^2 a^3}{2\varkappa M} + \frac{3}{2}\frac{C - A}{a^2 M} \qquad (4.2.2-14)$$

we obtain

$$v = \frac{a - c}{a} \qquad (4.2.2-15)$$

and

$$r = a(1 - v\sin^2 \varphi) \qquad (4.2.2-16)$$

which shows that the equilibrium figure of the Earth is (to the first order of approximation) indeed an ellipsoid with ellipticity v as given by Eq. (4.2.2 − 14).

Finally, it may be remarked that the derivative of the potential function W at any one point gives the value of gravity at that point.

4.3 The Rotation of the Earth

4.3.1 Present-Day Phenomena

The Earth is usually regarded as the most reliable clock in existence. The average angular velocity is indeed as given in Table 9, wherein one has to consider, of course, that the length of the day is commonly taken as that of the solar day. Because of the progress of the Earth along its orbit around the Sun, this is not the same as the length of the "sidereal" day (23 h 56' 4.09'') which is the time elapsed to bring a fixed star to the same longitude. Problems connected with the Earth's rotation have been discussed in well-known books[6, 7]. The rate of rotation of the Earth is, unfortunately, not as constant as is commonly believed. Changes in this rate occur on a short-time as well as on a secular scale.

The short-time variations have recently been discussed by Zagar[8]. Accordingly, internal as well as external phenomena affect the Earth's rotation. Of the internal phenomena, we have (1) different equatorial principal moments of inertia, (2) the Earth's elasticity, (3) cooling and shrinking of the Earth, (4) volcanic eruptions, (5) seismicity and internal mass transfer, and (6) growth ans shrinkage of polar ice caps. Of external phenomena, we have (1) ocean tides, (2) tides of the solid Earth, (3) atmospheric tides, (4) impact of meteorites, (5) solar eruptions and terrestrial magnetism, (6) temperature changes, and (7) mass absorption.

The short-term variations fall under two headings. First of all are the small seasonal variations and, secondly, small, somewhat irregular variations in the Earth's rate of rotation. The seasonal change of the length of the day ΔT in seconds is given as follows

$$\Delta T(\text{s}) = +0.022 \sin(2\pi t) - 0.012 \cos(2\pi t)$$
$$- 0.006 \sin(4\pi t) + 0.007 \cos(4\pi t) \qquad (4.3.1-1)$$

6 Munk, W. H., MacDonald, G. J. F.: The Rotation of the Earth. London: Cambridge Univ. Press 1960

7 Lambeck, K.: The Earth's Variable Rotation: Geophysical Causes and Consequences. London: Cambridge Univ. Press 1980

8 Zagar, F.: Il problema della rotazione terrestre. Atti Congr. Int. "Rotazione della Terra e Osservazioni di Satteliti Artificiali, Cagliari 16 − 18 April 1973, pp. 1 − 18, 1975

where t is in fractions of a year. The seasonal variations are probably caused by the temperature effects, and belong thus to those caused by "external" effects. The temperature variations cause a redistribution of atmospheric masses and thus affect the rotation rate. They may, however, also act in an indirect manner by affecting the lithospheric plates in geodynamic plate theory[9].

Small, regular variations of the Earth's rotation rate are also caused by the tides of the solid Earth. Djurovic[10] has made an attempt to correlate such variations directly with the tidal parameters.

The small, somewhat irregular variations are caused by internal phenomena. Most important of these is the inequality of the equatorial moments of inertia (denoted by A, B) which is about

$$\frac{B - A}{B} \sim \frac{1}{90,000}. \qquad (4.3.1 - 2)$$

This results in a wobble. The period of this wobble would be about 305 days; however, the effects of elasticity and energy dissipation increase the period. In fact, a wobble with a period of about 430 days has been obeserved ("Chandler wobble"). The theory of this wobble has recently been perfected by Peale[11], based upon a lengthy application and analysis of Hamiltonian dynamics. Of particular importance is the amount of energy dissipation in this wobble, which lets conclusions to be drawn with regard to the rheological state of the tectonosphere (see Sect. 3.4.3). The existence of this dissipation also requires that the wobble is relatively frequently reexcited to maintain the amplitude over the 80-year history of its observation. Usually it is assumed that this excitation is effected by major earthquakes. Other effects, like complying of core motions with the mantle may also have a triggering effect.

4.3.2 The Precession

In addition to rotating about its axis, the Earth is also subject to precession. This precession has been known for a long time as the precession of the equinoxes which effects, for instance, that the Sun, which wanders from constellation to constellation during the year, also slowly shifts its position at *given* times of the year so that the astrological zodiacal signs, fixed astronomically in Babylonian times, do not at all fit the Sun's positions at the present time. The whole zodiac returns to its original position roughly every 26,000 years. The annual precession, thus, is about $2\pi/26,000$ or $50''.37$.

The precession is due to the attraction of the Sun and Moon on the nonspherical Earth. Because of the deviations from sphericity, this attraction is

9 Proverbio, E., Poma, A.: In: Growth Rhythms and the History of the Earth's Rotation. (eds. G. D. Rosenberg, S. K. Runcorn) London: Wiley, p. 385, 1975
10 Djurovic, D.: Astron. Astrophys. J. 47:325 (1976)
11 Peale, S. J.: Rev. Geophys. Space Phys. 11(4):767 (1973)

non-uniform and results in a moment which causes the earth to precede like a spinning gyroscope.

The theory follows in a straightforward fashion from the mechanics of gyroscopes and has been presented, for instance, by Milankovitch[12]. The contribution of a gravitating mass m to the gyroscopic motion of a spheroid of mass M with moments of inertia A (equatorial) and C (polar) is (ψ = precession angle)

$$\frac{d\psi}{dt} = -\frac{3}{2}\frac{m}{M+m}\frac{v^2}{\omega}\frac{C-A}{C}\cos\varepsilon \tag{4.3.2-1}$$

where ε is the tilt of the C-axis toward the normal to the plane of motion of the bodies and v the angular velocity of the spheroid on its orbit, and ω the angular velocity around its axis. If m refers to the Sun and M to the Earth, the factor $m/(M+m)$ can be taken as unity. If m refers to the Moon and M to the Earth, the factor must be taken into account. The final result for the precession p per year of the earth (angle per year) is

$$p = \frac{d\psi}{dt}T = 3\pi\tau\frac{C-A}{C}\left(\frac{1}{T} + \frac{m_1}{M+m_1}\frac{T}{T_1^2}\right)\cos\varepsilon. \tag{4.3.2-2}$$

Here, the first term in the bracket is the contribution of the Sun, the second that of the Moon. As before, M is the mass of the Earth, m_1 that of the Moon, T is the revolution time of the Earth around the Sun, T_1 that of the Moon around the Earth, and ε the angle of the ecliptic. Furthermore, τ represents the time unit of one sidereal day; thus we have

$T = 366.25\,\tau$

$T_1 = 27.397\,\tau$.

With the observational values

$m_1 = 0.0123\,M$

$p = 50''.36$

one obtains

$$\frac{C-A}{C} = 0.003261. \tag{4.3.2-3}$$

It is possible to obtain an estimate for the difference between equatorial and polar moments of inertia of the Earth from its geometry alone. Thus, the Earth is considered as a sphere with polar radius c with the excess mass between pole and equator concentrated in a shell. The moment of inertia of this shell for the polar as well as for an equatorial axis can be calculated, assuming a reasonable mean density for the top 20 km or so of the Earth.

---

12 Milankovitch, M.: Kanon der Erdbestrahlung. Belgrad: Koenigl. Serb. Akad. 1941

Hence, a numerical estimate can be obtained for the quantity $C - A$ in Eq. (4.3.2 – 3) and hence an absolute value for C. The latter is best expressed as follows[13]

$$C = 0.33 \, Ma^2 \tag{4.3.2 – 4}$$

where M is the mass and a the mean radius of the Earth. It may be remarked that the moment of inertia of a homogeneous sphere would be $C = 0.4 \, Ma^2$. This indicates that a large amount of mass is concentrated near the center of the Earth.

4.3.3 Rotation in the Past

In addition to the short-term variations in the Earth's rotation, there have also been secular variations.

Direct evidence for secular changes has come from observations of "daily" bands of coral growth in "annual" bands. Assuming the length of the year to be constant, one deduces a change in the length of the day. The results of such studies are shown in Fig. 75 (after Runcorn[14, 15]).

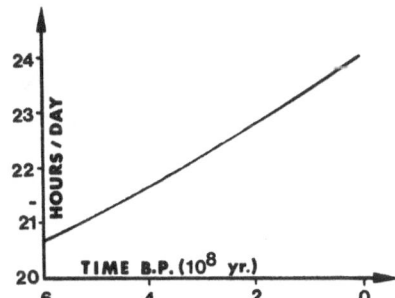

Fig. 75. Change of length of day during geological history. (Drawing based on data collected by Runcorn[14])

Accordingly, the rate of rotation of the Earth has decreased almost linearly since Paleozoic times. This can also be expressed by saying that the length of the day (in present-day hours) has increased linearly from about 20 1/2 h at the beginning of the Cambrian to 24 h today, which represents an increase in the length of the day of about 2 ms/century. In terms of angular velocity, this represents a secular change of

$$\frac{d\omega}{dt} = -5.34 \times 10^{-22} \, \text{rad/s}^2. \tag{4.3.3 – 1}$$

13 E. g., Creer, K. M.: In: Growth Rhythms and the History of the Earth's Rotation. (eds. G. D. Rosenberg, S. K. Runcorn) London: Wiley, p. 293, 1975
14 Runcorn, S. K.: Sci. Am. 215(4):26 (1966)
15 Rosenberg, G. D., Runcorn, S. K. (eds.): Growth Rhythms and the History of the Earth's Rotation. London: Wiley, 1975

In fact, astonomical observations also support the same conclusion. An angular acceleration in the longitudes of the Moon and Sun, first discovered from early observations of eclipses, has also been interpreted as a secular lengthening of the terrestrial day of about 2 ms/century. Changes in the (average) length of the day of this order of magnitude (1.8 ms/century) have also been discovered by direct comparison of the sidereal day with crystal clocks (Munk and MacDonald, loc. cit.).

The cause of the lengthening of the day (i. e., of the slowdown of the Earth's rotation) must be sought primarily in tidal friction (cf. Sect. 4.4.3). A classic discussion of this phenomenon has been given by Jeffreys[16]. The problem cannot be simply solved by assuming the rotational energy "lost" by the slowing down of the Earth as equal to that dissipated in tidal friction, because, at the same time, the Moon's rotation is affected which results in a change in the energy of the Earth-Moon system. Thus, the entire Earth-Moon system including the changing rotation of these two bodies, their relative position, and the energy dissipation has to be considered simultaneously. Thus, an accurate discussion of the problem is rather complicated. However, it is possible to make some approximate estimates.

Thus, let us assume that we have a rotating Earth and a Moon on a circular orbit around it. The angular momentum lost by the slowing down of the Earth must be taken up by an increase in angular momentum of the Moon, since the total angular momentum of the system as a whole must be conserved. Hence, according to Kepler's laws, the Moon must recede from the Earth. The basic equations of the simplified system can, then, be written down as follows, with ω_E = angular velocity of Earth, ω_M = angular velocity of Moon, r = distance Earth-Moon, dE/dt = energy loss (by dissipation) per unit time, and C = polar angular momentum of the Earth:

energy balance

$$\frac{d}{dt}\left(\frac{1}{2}C\omega_E^2 - \varkappa\frac{mM}{r} + \frac{1}{2}mr^2\omega_M^2\right) = -\frac{dE}{dt} \qquad (4.3.3-2)$$

conservation of angular momentum

$$C\omega_E + mr^2\omega_M = \text{const} \qquad (4.3.3-3)$$

Kepler's orbital law

$$r^3\omega_M^2 = \text{const}. \qquad (4.3.3-4)$$

Differentiated, this yields:

$$C\omega_E\frac{d\omega_E}{dt} + \varkappa\frac{mM}{r^2}\frac{dr}{dt} + mr\omega_M^2\frac{dr}{dt} + mr^2\frac{d\omega_M}{dt} = -\frac{dE}{dt} \qquad (4.3.3-5)$$

16 Jeffreys, H.: The Earth. 4th edn. Cambridge: Univ. Press, p. 230 ff, 1959

$$C\frac{d\omega_E}{dt} + m2r\omega_M\frac{dr}{dt} + mr^2\frac{d\omega_M}{dt} = 0 \qquad (4.3.3-6)$$

$$3r^2\omega_M^2\frac{dr}{dt} + 2r^3\omega_M\frac{d\omega_M}{dt} = 0. \qquad (4.3.3-7)$$

In these equations, the only quantities that are unknown are dr/dt, $d\omega_M/dt$, and dE/dt. Hence, they can be solved numerically and the energy loss required for the slowing-down process can be calculated. The elimination of the unknowns yields

$$\frac{dE}{dt} = \frac{d\omega_E}{dt}C\left[\omega_E - 2\omega_M + \frac{3\,\omega_M}{r} - \frac{4\varkappa M}{\omega_M r^4}\right] \qquad (4.3.3-8)$$

where the last two terms in the square bracket can evidently be neglected (as proportional to $1/r$ and $1/r^4$) with regard to the others. Thus, with $M = 5.98 \times 10^{24}$ kg, $a = 6.36 \times 10^6$ m (a = equatorial radius), and $C = 0.33\,Ma^2 = 7.98 \times 10^{37}$ kgm^2 one obtains

$$\frac{dE}{dt} = 2.87 \times 10^{12}\,\text{Watt}. \qquad (4.3.3-9)$$

This can now be compared with values obtained in the theory of tidal phase lag (cf. Sect. 4.4.3). One obtains that the latter effect is of the right order of magnitude regarding the energy required to be lost for an explanation of the observed slowing down of the rotation of the Earth. There is a deficiency in the phase lag energy of about 50%, but in view of the many uncertainties in the model and estimates made, this is perhaps not too serious.

4.4 Tidal Effects

4.4.1 Tidal Variations of the Force of Gravity

We have already referred to the existence of tidal effects on several occasions. It is now our aim to analyze these somewhat more accurately.

The tides are due to the varying attraction of the Sun and the Moon on individual points on the Earth. The variations in the vertical gravity acceleration on a perfectly *rigid* Earth due to the varying relative position of the Sun and Moon are of the order of 2×10^{-6} m/s^2; at the same time the variation of the vertical (its tilt) has an amplitude of about $0''.04$. These values can be calculated from astronomical considerations[17].

---

17 Melchior, P.: Earth Tides. Geophys. Surv. 1:275 (1974)

The most simple mathematical expression of the potential of the perturbing body is obtained by using local coordinates of the latter, viz. the zenith distance z and the azimuth. One has for the potential

$$W = -\frac{\varkappa\mu}{2}\frac{a^2}{r^3}(3\cos^2 z - 1) \tag{4.4.1-1}$$

where μ is the ratio of the mass of the perturbing body to that of the Earth, a is the distance to the Earth's center, and r the distance between Earth and perturbing body; $\varkappa$ is the universal gravitational constant.

To make this expression useful for calculations on the Earth, it is convenient to introduce both, the equatorial coordinates (hour-angle H and declination δ) and astronomical coordinates of the observatory point on Earth (latitude φ and longitude λ). Then, already Laplace has shown that one can obtain the following expression for the tidal potential of a perturbing body

$$W = D\left(\frac{c}{r}\right)^3\left[\begin{array}{ll}\cos^2\varphi\cos^2\delta\cos 2H & \text{(A)}\\[2mm] + \sin 2\varphi\sin 2\delta\cos H & \text{(B)}\\[2mm] + 3\left(\sin^2\varphi - \frac{1}{3}\right)\left(\sin^2\delta - \frac{1}{3}\right) & \text{(C)}\end{array}\right] \tag{4.4.1-2}$$

where c is the mean (as compared to r, the instantaneous) Earth-perturbing body distance, furthermore

$H = t - \alpha - \lambda$ (α is a possible phase shift)

t = sidereal time at the observation point

$D = \dfrac{3}{4}\varkappa\mu\dfrac{a^2}{c^3}$ (Doodson's constant)

The three terms (denoted by A, B, C) in the Eq. (4.4.1 – 2) for the perturbation potential represent functions of the sectorial (A), tesseral (B), and zonal (C) type. The three types are illustrated in Fig. 76.

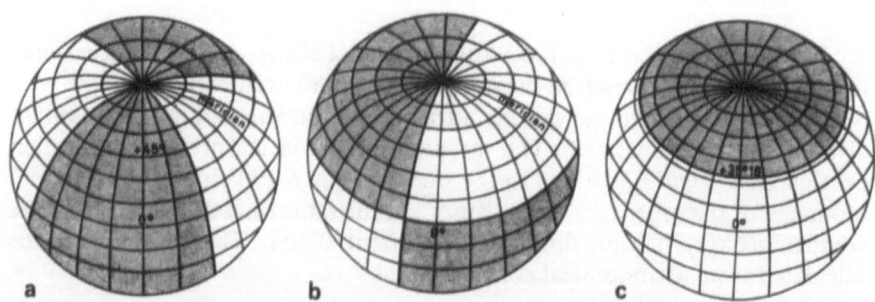

Fig. 76a – c. Geographical distribution of tidal potenial: **a** Sectorial function; **b** tesseral function; **c** zonal function

The perturbing bodies of the gravitational potential on the Earth are the Sun and the Moon. Since it is desirable to obtain a purely harmonic development in terms of the position of these bodies, it is necessary to introduce six independent variables: mean lunar time, mean longitudes of Sun and Moon, longitudes of the lunar perigee, of the ascending node of the Moon, and of the perihelion. These variables are practically linear functions of time, and hence a separation into a great number of periodic functions in time ("waves") is obtained. The most important of these are called M_2 (lunar; period 2 h 25 min 4 s), S_2 (solar; period 12 h) and N_2 (lunar-elliptic, due to the eccentricity of the lunar orbit; period 2 h 39 min 30 s); furthermore K_1 (lunisolar; period 23 h 56 min 4 s = 1 sidereal day), O_1 (lunar, period 25 h 49 min 10 s) and P_1 (solar; period 24 h 4 min). The amplitude of each of these waves can be calculated theoretically for any observation point (primarily a function of latitude). Figure 77 serves as an illustration.

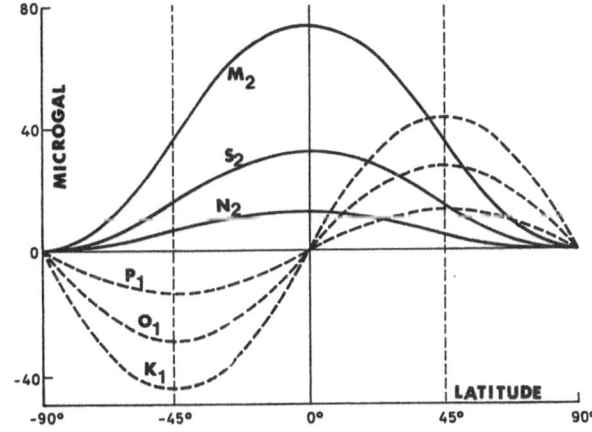

Fig. 77. Amplitude variation of the principal waves as a function of latitude for the vertical component of the tidal force

The theory of the tidal potential can be considerably refined. Thus, e. g., Michelson[18] made corrections for combined rotational and orbital translation.

4.4.2 Earth Tides

The elastic responses of the Earth to the varying gravitational potential due to the varying positions of the Sun and Moon are called "Earth tides"[7].

When the value of gravity diminishes, the Earth's crust rises under the diminishing weight of its mass. Consequently, the gravimeter reading at that point decreases even more, because of the increasing distance to the center of mass. The theoretical tidal effect on a rigid Earth is therefore amplified on an

---

18 Michelson, I.: Can. J. Earth. Sci. 10 (12):1751 (1973)

elastic Earth. Corresponding considerations apply to the deflection of the vertical (here, however, there is a *decrease* in the theoretical effect).

Observations of the Earth's tides are commonly executed by observing the changes in the force of gravity at a series of stations. The common procedure, then, is to analyze the results in the frequency domain, i. e., with regard to spectra. However, Mikumo and Kato[19] have pointed out that a direct analysis in the time domain (i. e., the records considered as time series) may also lead to interesting results. A further different procedure consists in the application of the response method, in which the tide is expressed as a weighted sum of past, present, and future values of a relatively small numer of time-varying input functions[20].

The actual description of Earth tides is best effected by the Love numbers h, k, and l. Of these, h represents the ratio between the height of the Earth tide and the height of the corresponding oceanic tide at the surface; k the ratio between the additional potential generated by this deformation and the distorting potential; l the ratio between the horizontal shift of the crust and the shift of the corresponding oceanic tide. The intensity variations of gravity (vertical component) are amplified in the ratio δ

$$\delta = 1 + h - \frac{3}{2}k \qquad\qquad (4.4.2-1)$$

while oscillations of the vertical are reduced in the ratio γ

$$\gamma = 1 + k - h. \qquad\qquad (4.4.2-2)$$

Observations of gravity along the three components make it therefore possible to calculate the Love numbers h and k; l is derived from direct strain measurements[21].

The fact that the above description is possible shows that the Earth is, in a first approximation, elastic (cf. Sect. 3.4.2). It also stands to reason that there is a regional variation in the Love numbers, and that any found values are, in fact, characteristic for the region in question. In this sense, regional values for the tidal parameters (usually δ and γ rather than the Love numbers) have been determined for various regions[22].

The local variation of the tidal parameters has a bearing upon our knowledge of the internal constitution of the Earth which, accordingly, would also vary regionally[23]. Melchior and Ducarme[24], in fact, have used tidal observations as a method for the prediction of deformations in cavities in the Earth's crust.

---

19 Mikumo, T., Kato, M.: Bull. Disas. Prov. Res. Inst. Kyoto Univ. 26 (2):71 (1976)
20 Lambert, A.: J. Geophys. Res. 79 (32):4952 (1974)
21 Melchior, P.: Geophys. Surv. 1:275 (1974)
22 Melchior, P., Kuo, J. T., Ducarme, B.: Phys. Earth Planet. Inter. 13:184 (1976)
23 Wilhelm, H.: J. Geophys. 44:435 (1978)
24 Melchior, M., Ducarme, B.: Bull. Geodes. 50:137 (1976)

4.4.3 Phase Lag

More important than straightforward tide theory for geodynamic problems are the deviations from Love's elasticity considerations. We have already referred to the fact that there is a phase shift in the response of the Earth to the changes in gravitational potential (Sect. 3.4.3); this problem will be dealt with here in somewhat greater detail.

The phase lag $\delta\omega$ has been experimentally observed. Commonly quoted values run around[25]

$$\delta\omega = 2\varepsilon = 2.16° . \tag{4.4.3-1}$$

There are, in fact, also regional differences (as with the Love numbers) so that one finds values running from a small advance in Europe to a lag of 3.5° in Japan[26].

Any such phase lag $\delta\omega$ automatically implies energy dissipation. First of all, the sectorial tide M_2 produces a torque N acting in a direction opposite to that of the Earth's rotation whose magnitude is

$$N = 5.4 \times 10^{17} \times \sin\delta\omega \text{ Newton meters} \tag{4.4.3-2}$$

so that the energy dissipation rate is

$$\frac{dE}{dt} = N(m-n) = 3.8 \times 10^{13} \sin\delta\omega \text{ Watt} \tag{4.4.3-3}$$

where n is the Moon's and m the Earth's angular velocity. For a phase lag of 2°, this yields

$$\frac{dE}{dt} = 1.35 \times 10^{12} \text{ Watt} . \tag{4.4.3-4}$$

The effect of an energy dissipation of such an order of magnitude on the Earth's rotation has already been discussed in Sect. 4.3.3.

The provenance of the energy dissipation is not entirely clear. Pure oceanic loading[27, 28] or internal friction[29] in the "solid" Earth seem to be insufficient to provide the observed dissipation. Thus, one generally assumes, as suggested by Jeffreys[30], that the energy is primarily dissipated by the friction of tidal currents in the shallow seas. From a consideration of such tidal currents Jeffreys[30] found that an average energy dissipation can be expected which is much of the same order of magnitude as the energy loss suggested by the tidal phase lag.

25 MacDonald, G. J. F.: Rev. Geophys. 2:467 (1964)
26 Melchior, P.: Geophys. Surv. 1:275 (1974)
27 Moens, M.: Phys. Earth Planet. Inter. 13:197 (1976)
28 Groten, E., Brennecke, J.: J. Geophys. Res. 78:8519 (1973)
29 Zschau, J.: In: Tidal Friction and the Earth's Rotation. (eds. P. Brosche, J. Sündermann) Berlin – Heidelberg – New York: Springer, p. 62, 1978
30 Jeffreys, H.: The Earth, 4th edn., Cambridge Univ. Press, p. 230 ff, 1959

As a concluding remark it may be noted that attempts have been made to use tidal energy commercially. Studies of the mathematical problems involved have primarily been made by Duff[31, 32]. The aim is of course, to cause some energy dissipation in the Earth-Moon system for man's own purposes.

4.4.4 Geodynamic Effects

A discussion of tidal forces, thus, is mostly of indirect interest with regard to geodynamic phenomena. However, tidal forces have also been advocated as more directly implicated in geodynamics. Thus, we quote here a calculation of Jeffreys[33] who came up with the conclusion that at the utmost tidal forces might result in a drag at the bottom of the crust of the order of 4 Pa. This force reverses its sign with every tide and can therefore have an orogenetic significance only in such a way that, on the average, the tendency persists to create a bulge at the equator. The long-time effect of the tides is therefore similar to that of the polfluchtkraft (see Sect. 4.6.2), but the magnitude of this tidal force is very small indeed. Similarly, Bostrom[34] and Moore[35] have advocated that tidal drag is the dominant force responsible for lithospheric plate motions on the surface of the Earth. According to these authors, all plates are moving westward. However, an analysis of Jordan[36] showed that tidal torques are far too small to drag the lithosphere at any appreciable velocity.

A somewhat greater effect is caused by the tidal forces in an indirect manner, viz. by their moving the waters in the oceans around. It has been shown by Jobert[37] that the bending of the Earth's crust due to the different amounts of water overlying the crust below the oceans at various times is not negligible. However, the bending of the crust caused in this manner is, of course, again periodic so that no long-term effects (other than possibly due to fatigue fractures) may be expected.

In addition, there is a westward stress due to the secular effect of tidal friction, but its magnitude is only of the order of 10^{-5} Pa. If this stress is assumed to act upon a whole continent, it will cause a compressive stress therein greater than the westward stress roughly in proportion to the ratio of the area of the continent to its cross section[38]. The latter may be approximately equal to $100:1$. This would make the stress equal to about 10^{-3} Pa. This is much less than any stess that could have an orogenetic significance. In fact, it has been

31 Duff, G. F. D.: Mathematical Problems of Tidal Energy. Semin. IRIA, Anal. Controle Syst. 1973:97 – 174 (1973)
32 Duff, G. F. D.: Proc. Int. Congr. Math. Vancouver, p. 87, 1974
33 Jeffreys, H.: The Earth, 4th edn., Cambridge: Univ. Press, 1959
34 Bostrom, R. C.: Nature (London) 234:536 (1971)
35 Moore, G. W.: Geology 1:99 (1973)
36 Jordan, T. H.: J. Geophys. Res. 79 (14):2141 (1974)
37 Jobert, G.: C. R. Acad. Sci. 244:No. 2, 227 (1957)
38 Griggs, D.: Am. J. Sci. 237:611 (1939)

calculated by Jeffreys[33] that, even if there were no Bingham yield stress present in the Earth, to make America move westward to its present distance from the Old World, would take 10^{17} years. To produce the same effect in, say, 3×10^7 years, one would require a tidal friction so high that it would stop the Earth's rotation within a year. In spite of these formidable objections, Nadai[39] has held tidal friction responsible for the drift postulated in the continental drift theory.

A somewhat different mechanism, also connected with tidal drag, was suggested by Knopoff and Leeds[40], which could produce secular motion of the lithospheric plates. It is associated with the loss of kinetic energy resulting from the secular deceleration of the Earth's rotation (cf. Sect. 4.3.3). The torque arising from the Earth's deceleration, when applied over large regions of the tectonosphere, can be shown to be more than sufficient for providing a plate-driving force.

4.5 The Question of Stability of the Earth's Axis of Rotation

4.5.1 The Problem

As outlined in the first two chapters of this book, many geological and geophysical observations point toward the likelihood of a change in the direction of the axis of rotation (polar wandering) during geologic time. The first problem that arises is, in fact, that of defining and verifying directly this indicated polar wandering.

Inasmuch as paleomagnetism indicates that the continents have been drifting, it is necessary to separate the gross polar wandering from the *individual* plate motions. Jurdy and Van der Voo[41] proposed to do this by finding the rigid rotation which best fits statistically in a least squares sense the observed displacements of the crustal plates. If this is done, it is in fact doubtful whether any "gross polar wandering" occurred at all, at least since the early Cretaceous[42,43]. The procedure of Jurdy and Van der Voo was shown by Simpson[44] to be related to equations which Lliboutry[45, 46] and Solomon and Sleep[47] used to calculate absolute plate motions. In these instances, the "absolute" plate motions seem to imply an essentially fixed axis of the Earth's rotation.

39 Nadai, A.: Trans. Am. Geophys. Union 33:247 (1952)
40 Knopoff, L., Leeds, A.: Nature (London) 237:93 (1972)
41 Jurdy, D. M., Van der Voo, R.: J. Geophys. Res. 79:2945 (1974)
42 Jurdy, D. M., Van der Voo, R.: Science 187:1193 (1975)
43 McElhinny, M. W.: Nature (London) 241:523 (1973)
44 Simpson, R. W.: J. Geophys. Res. 80:4823 (1975)
45 Lliboutry, D.: J. Geophys. Res. 79:1230 (1974)
46 Lliboutry, D.: Nature (London) 250:298 (1974)
47 Salomon, S. C., Sleep, N. H.: J. Geophys. Res. 79:2557 (1974)

Inasmuch as there is some doubt regarding the reality of "gross" polar wandering over geologic times, attempts have also been made to determine a possible secular drift of the pole at the present time by direct means. This has been done by a variety of authors since 1922; a summary of the data and results has recently been presented by Poma and Proverbio[48]. The latest of the determinations listed by these authors (for the period 1900 – 1969) yields a displacement velocity of the pole of 0.00307 s/year (or 50 m per million years) along the direction of 69.6°W longitude. This, in geologic times, would certainly make substantial displacements possible. Nevertheless, here, too, the pole shifts are calculated with regard to certain points on Earth assumed as fixed [the positions of the BIH (Bureau International de l'Heure observatories)], so that it is not a priori clear whether the pole shift is, in fact, "gross" or whether the observatories have shifted with regard to the pole.

In spite of some of the uncertainties of the actual existence of polar motion, the occurrence of a change of the position of the pole has to be at least considered as a possibility.

The postulate of such a change poses certain difficulties from a dynamical viewpoint: The Earth is a body rotating around the principal axis of the moment-of-inertia tensor which corresponds to the largest moment of inertia; such an axis is a stable axis of rotation, – at least for a rigid body. In order to justify the heuristic inference of polar wandering dynamically, one must therefore seek to refute the apparent a priori impossibility of its occurrence.

Attempts to do this have been based upon various considerations. We shall deal with these below.

4.5.2 Effects of Circulations on a Rigid Earth

Let us first consider[49] the effects of displacements of matter on the surface of the Earth, in particular of such displacements which have the form of circulations. Such circulations are produced and maintained entirely by forces within the Earth and neither affect its total moment of inertia nor the position of its center of gravity. Circulations can be characterized by their angular momentum vector relative to the rotating Earth. The total angular momentum of the Earth, then, can be separated into the angular momentum without the circulations plus the relative angular momenta of the circulations.

During the motion of the Earth along its orbit, the total angular momentum of the Earth remains a constant vector in space. Thus, during the occurrence of a particular circulation, the instantaneous axis of rotation displaces itself. The point on the Earth's surface where the latter is pierced by the total angular momentum vector is not fixed. This point is not the pole of instantaneous rotation, but would be the pole if the circulation would come to

48 Poma, A., Proverbio, E.: Astron. Astrophys. J. 47:105 (1975)
49 Goguel, J.: Ann. Géophys. 6:139 (1950)

a stop. Goguel calls this the "permanent pole" (designated PP in the accompanying Fig. 78) of the Earth. Since the angular momentum of the circulation is fixed with regard to the Earth (at least as long as the circulation remains stationary), the displacement of the pole from the permanent pole is constant. The angle ε of this displacement (cf. Fig. 78) can be calculated for various cases from the knowledge of the total moment of inertia of the Earth, equal to 6.77×10^{37} m^{2}kg and from the angular momenta of the possible circulations. Goguel considered winds and ocean currents.

With regard to winds, let us represent a cyclonic movement by assuming that, within a radius of 2000 km, a wind of 48 km/h affects the atmosphere to a height of 5000 m. Since the density of air can be, on the average, assumed as approximately equal to 1 kg/m^3, the angular momentum in question turns out to be equal to 12.5×10^{23} m^{2}kg/s. The position of the cyclones and anticyclones is strongly influenced by the distribution of continents and oceans. Goguel assumes that the resulting angular momentum of all the wind movement is, on the average, constant and equal to one-half of the angular momentum of one single cyclone. In this fashion, he obtains an angle ε equal to about 10^{-10} radians.

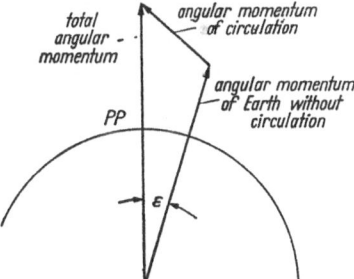

Fig. 78. Goguel's[49] decomposition of the Earth's angular momentum

Now Goguel assumes that an angle ε between the instantaneous and permanent poles would induce a polar wandering of $2\pi R\varepsilon$ per day at right angles to the connecting line between the poles, R being the radius of the Earth. This wandering would be an example of the nutation of a symmetric top as seen in the body-fixed system. It would correspond to about 1.46 m/year, a very large value indeed. However, if the Earth is assumed to be an ellipsoid instead of a sphere, the equatorial bulge brings polar wandering to a halt.

A similar conclusion can be reached if one considers circulations in the sea. Thus, following Goguel, let us represent an ocean current by assuming a circular trajectory of 5000 km diameter, 500 m depth and 100 km width, with a velocity of 3.7 km/hr. Goguel calculates the angular momentum of such a current as equal to 10^{24} m^{2}kg/s, which is of the same order of magnitude as that obtained above for a single cyclone. One can estimate, then, the resulting angular momentum from the known ocean currents; Goguel obtains 3×10^{23}

m^2kg/s, in the direction of the meridian of $7\,°E$. The angle ε of the pole turns out to be about one-quarter of that calculated for the cyclones and the polar wandering induced in this manner in a perfectly spherical Earth should therefore be about one-quarter of that calculated above. The ellipticity of the Earth, however, will again make polar wandering impossible.

The above deductions have been made for *a solid* Earth. If the Earth is able to *yield,* then matters are entirely different. It has been pointed out by Inglis[50] that in this case circulations have a very similar effect as any other asymmetry in the crust of the Earth. The influence of such asymmetries upon the position of the axis of rotation in a *yielding* Earth will be discussed in the Sections just following. However, in spite of this possible effect of circulations, their quantitative influence will in any case be much smaller than that due to geographical asymmetries; – simply because of the orders of magnitude involved.

4.5.3 Polar Wandering in a Yielding Earth

In order to obtain any possibility of polar wandering at all, one has to consider a model of the Earth where it is assumed that the latter is able to *yield*. In such an Earth, "polar wandering" means of course a shift of the Earth with respect to the axis of rotation; the actual motion of the latter in space may thereby remain relatively small.

The first attempt to obtain polar wandering assuming a yielding Earth seems to have been an essay by Darwin[51]. The result was that polar wandering is possible for a "fluid" Earth; – but it has been pointed out by Lambert[52] that there was an algebraic error in Darwin's calculations. Jeffreys[53] rectified the error in Darwin's calculation and obtained as a consequence of Darwin's physical assumptions that polar wandering was impossible.

However, entirely non-analytic reasoning by Gold[54] would make it plausible that significant polar wandering in an Earth which is capable of yielding, should be expected. Munk[55], inspecting the physical basis of Darwin's calculations, noted that one of the assumptions was probably incorrect. This can be demonstrated as follows.

The problem of polar wandering on a plastic Earth involves three types of poles: The pole F of rotation, the pole F' of the moment of inertia tensor (pole of figure) corresponding to the *instantaneous* shape of the Earth at the moment under consideration, and the geoidal pole F'' corresponding to the pole of the ellipsoid approximating the Earth's instantaneous shape.

---

50 Inglis, D. R.: Rev. Mod. Phys. 29:9 (1957)
51 Darwin, G. H.: Philos. Trans. R. Soc. London 167:Pt. 1, 271 (1877)
52 Lambert, W. D.: Bull. U. S. Natl. Res. Counc. No. 78, Chap. 16 (1931)
53 Jeffreys, H.: The Earth, 3rd edn. London: Cambridge Univ. Press, p. 343, 1952
54 Gold, T.: Nature (London) 175:526 (1955)
55 Munk, W. H.: Nature (London) 177:551 (1956)

Darwin now made the following assumptions which are implicit in his equations: (a) the veloctiy of wandering of the rotation pole F is proportional to its separation from the pole of figure F'; (b) the geoidal pole F'' moves at a rate proportional to the separation of the rotation pole F from the pole of figure F'.

According to Munk[55], the assumption (b) does not seem warranted. It seems much more likely that the geoidal pole F'' moves at a rate proportional to its separation from the rotation pole F, for it is the magnitude of this separation which determines the stresses in the Earth and hence the rate of flow altering the shape. It thus appears that assumption (b) should be replaced by the following: (b') the geoidal pole F'' moves at a rate proportional to its separation from the pole of rotation.

It thus turns out, quite intuitively[56], that polar wandering should be quite a rapid process as soon as a small asymmetry has arisen somewhere in the Earth's crust.

The mathematical expression for a special case of the above reasoning has been achieved by Milankovitch[57] long before the intuitive arguments of Gold and Munk became available. Milankovitch considered the special case where the Earth is so quickly adjusting its shape that the rotation pole and the geoidal pole coincide.

Following Milankovitch, we approximate the equilibrium figure of the Earth by an ellipsoid of rotation whose meridian is given by the equation

$$r = a(1 - v \sin^2 \varphi) \qquad (4.5.3-1)$$

where r is the length of the radius vector, a the equatorial radius of the ellipsoid, v the eccentricity and φ the geocentric latitude. The point corresponding to $\varphi = 90°$ is the geoidal pole F''. If we denote the principal moments of inertia of the ellipsoid by A, B, C, then we have because of symmetry properties

$$B = A. \qquad (4.5.3-2)$$

According to general theorems of mechanics, the moment of inertia T with reference to any arbitrary axis ζ through the center of the ellipsoid is given by

$$T = A \cos^2 \alpha + B \cos^2 \beta + C \cos^2 \gamma \qquad (4.5.3-3)$$

where $\cos \alpha$, $\cos \beta$, $\cos \gamma$ are the direction cosines of the axis ζ. In the present case, the last equation reduces to

$$T = A + (C - A) \cos^2 \gamma. \qquad (4.5.3-4)$$

---

56 Gold, T.: Nature (London) 175:526 (1955)

57 Milankovitch, M.: Glas. Acad. R. Serbe 152:39 (1932) – Handbuch der Geophysik, Bd. 1, Abschn. 7, Kap. 25, S. 438. 1933 – Publ. Math. Univ. Belgrade 1:129 (1932) – Glas. Acad. R. Serbe 154:1 (1933) – Milankovitch, M.: Kanon der Erdbestrahlung und seine Anwendung auf das Eiszeitproblem; Ed. Spéc. Acad. R. Serbe Tome 133, Belgrade, 633 pp. 1941

This defines a scalar field $T = T(\alpha, \beta, \gamma)$ describing the dependence of the moment of inertia of the "equilibrium Earth" as a function of the direction of the axis with regard to which it is taken.

However, it must be assumed that the Earth is not in an equilibrium condition. The moment of inertia referring to the axis ζ is therefore not T, but, say, J:

$$J = T + \Omega \tag{4.5.3-5}$$

where now Ω is that part of the moment of inertia which is due to the deviation of the Earth's surface from an equilibrium figure. It must thus be expected that Ω as a function of α, β, γ is not symmetrical with regard to the axis of rotation.

Owing to the asymmetry of Ω, the pole F' of J (i. e., the pole of figure; this corresponds to an extremal value of J) does not coincide with the geoidal pole F'' of the equilibrium figure, but must be somewhere near it. The coordinates ξ, η of F' with respect to the geoidal pole F'' can be found from the equation expressing that there is an extreme value for J for those coordinates:

$$\frac{\partial J}{\partial \xi} = 0; \qquad \frac{\partial J}{\partial \eta} = 0, \tag{4.5.3-6}$$

or

$$\frac{\partial T}{\partial \xi} + \frac{\partial \Omega}{\partial \xi} = 0; \qquad \frac{\partial T}{\partial \eta} + \frac{\partial \Omega}{\partial \eta} = 0. \tag{4.5.3-7}$$

It is convenient to use as coordinates ξ, η orthogonal coordinates in the plane tangent to a unit sphere engendered by the variable axis ζ at the point where it is penetrated by the axis through F''; the origin of these coordinates being at that point of penetration.

Since the pole of figure F' is very near the geoidal pole F'', it is possible to neglect powers higher than the first of ξ and η. We can thus express the last equation as follows:

$$\left.\begin{array}{l} \dfrac{\partial T(0,0)}{\partial \xi} + \xi \dfrac{\partial^2 T(0,0)}{\partial \xi^2} + \dfrac{\partial \Omega(0,0)}{\partial \xi} + \xi \dfrac{\partial^2 \Omega(0,0)}{\partial \xi^2} = 0 \\[2ex] \dfrac{\partial T(0,0)}{\partial \eta} + \eta \dfrac{\partial^2 T(0,0)}{\partial \eta^2} + \dfrac{\partial \Omega(0,0)}{\partial \eta} + \eta \dfrac{\partial^2 \Omega(0,0)}{\partial \eta^2} = 0. \end{array}\right\} \tag{4.5.3-8}$$

Furthermore, we have in virtue of Eq. (4.5.3-4):

$$\frac{\partial T}{\partial \gamma} = -(C - A) \sin 2\gamma, \tag{4.5.3-9a}$$

$$\frac{\partial^2 T}{\partial \gamma^2} = -2(C - A) \cos 2\gamma, \tag{4.5.3-9b}$$

and hence, since $d\xi = d\gamma$, one obtains finally

$$\xi = \frac{1}{2(C-A)} \frac{\partial \Omega(0,0)}{\partial \xi}; \quad \eta = \frac{1}{2(C-A)} \frac{\partial \Omega(0,0)}{\partial \eta}. \quad (4.5.3-10)$$

The vector a of displacement of the pole of figure F' with regard to the geoidal pole F'', in a plane tangent at the penetration point of the axis through F to the sphere engendered by unit vectors along the variable axis ζ, is therefore given by

$$a = \frac{1}{2(C-A)} \operatorname{grad} \Omega. \quad (4.5.3-11)$$

As outlined earlier, we now assume that the Earth can yield so fast that the pole of rotation F and the geoidal pole F'' always coincide. This is a special case of assumption (b') above implying that the constant of proportionality implied by that assumption is very large. Finally, assumption (a) yields for the velocity v by which the pole of rotation F (which is now identical to F'') moves:

$$v = c'a \quad (4.5.3-12)$$

where c' denotes a proportionality constant. It follows that the equation governing polar wandering is

$$v = c \operatorname{grad} \Omega, \quad (4.5.3-13)$$

where c is again a certain constant. The last equation represents what has been known in the German literature for a long time under the name of "Milankovitch's theorem".

The above form of the dynamic condition of polar wandering is obviously well suited to the model of the Earth where one assumes an essentially fluid substratum that can assume an equilibrium position instantly, with all the deviations concentrated in a thin crust. Then, the field Ω is independent of time and the possible polar paths are those corresponding to the last equation. The changing of the elevation of points of the crust owing to the adjustment of shape to the instantaneous equilibrium figure of the Earth is therefore automatically taken into account. The polar paths can be calculated and are as shown in Fig. 79. Here it has been assumed that the axes X, Y, Z correspond to Ω_1, Ω_2, Ω_3, respectively, the latter being the eigenvalues of Ω with $\Omega_1 < \Omega_2 < \Omega_3$.

The next problem is to calculate the moment of inertia for any position of the axis of rotation. Supposing the variation of density with depth to be uniform for all continents [designated by $\varrho'(r)$] and for all oceans [$\varrho(r)$], the excess Q of continental over oceanic inertia equals (ϑ = colatitude, φ = longitude with regard to the axis under consideration)

$$Q = \iiint (\varrho' - \varrho) r^4 (\sin^2 \vartheta \sin \varphi + \cos^2 \vartheta) \sin \vartheta \, dr \, d\vartheta \, d\varphi = Iq \quad (4.5.3-14)$$

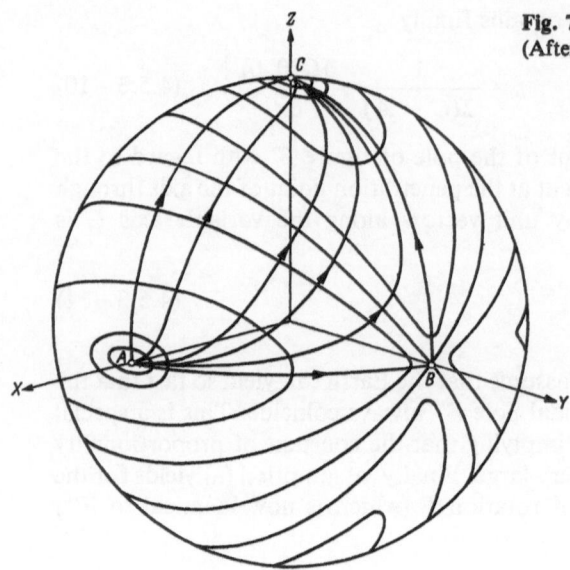

Fig. 79. Possible paths of polar wandering. (After Milankovitch [59])

with

$$I = \int (\varrho' - \varrho) r^4 dr \qquad\qquad (4.5.3-15)$$

and

$$q = \iint (\sin^2 \vartheta \sin \varphi + \cos^2 \vartheta) \sin \vartheta \, d\vartheta \, d\varphi , \qquad\qquad (4.5.3-16)$$

where the integrals over the angles are evaluated over the continental area only. Munk [58] evaluated the values for q for various axes assuming standard crustal sections; the resulting "q-topography" is shown in Fig. 80. The possible polar paths are the orthogonals to the q niveau lines.

Milankovitch did not attempt to give actual values for the integrals, but only took a "different" areal mass density for continents as compared with oceans. He then estimated the path of the pole for the present distribution of continents and obtained the path shown in Fig. 80. Actually, the direction of the motion would be reversed. It had already been pointed out by Gutenberg [60] that, for an isostatically adjusted Earth, the center of gravity of the lighter masses would lie above that of the heavier displaced material, so that the time arrow implied in Fig. 80 should point into the opposite direction. This would put the present pole into a position which is about as far removed from its stable position as it could be. This seems hardly satisfactory. However, it

58 Munk, W. H.: Geophysica 6, No. 3:335 (1959)

59 Milankovitch, M.: Kanon der Erdbestrahlung und seine Anwendung auf das Eiszeitproblem. Ed. Spéc. Acad. R. Serbe Tome 33, Belgrade 1941

60 Gutenberg, B.: In: The Internal Constitution of the Earth, p. 203. New York: Dover 1952

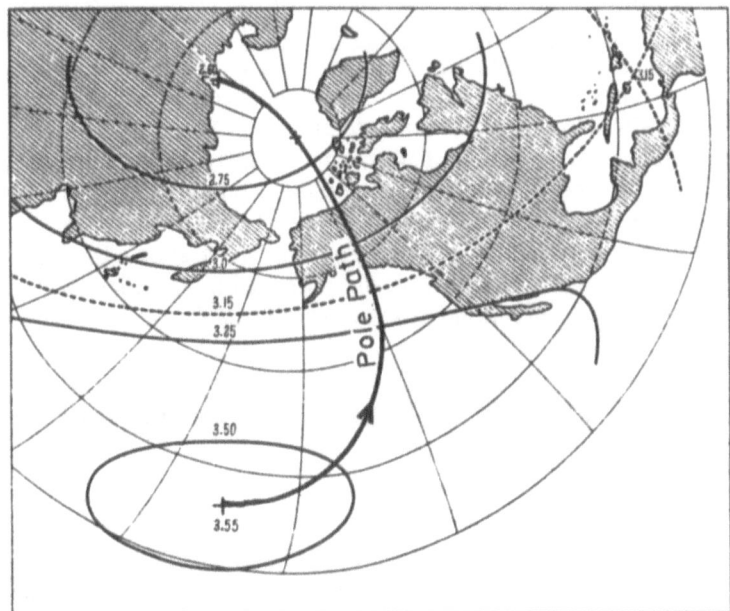

Fig. 80. Munk's[58] q-topography with Milankovich's[59] pole path shown

seems that the assumption of the sign as accepted by Milankovitch would not be as absurd as it appears at first glance. In the isostatic model of the Earth, the inertia of the crust depends on the second order term resulting from the slightly larger radial distance of continents and mountains as compared with oceans from the center of the Earth. If it is supposed that the isostatic balance does not hold precisely, but that there is erosion of continental matter and sedimentation on the ocean floor which is not compensated, then this represents a first order effect which might reverse the sign of the polar wandering so as to be in conformity with Milankovitch's assumption. It has also been pointed out that ridges and trenches may have a greater potential for polar-wandering excitation than continental masses[61].

It may be noted that a displacement of the pole relative to the continents can also be regarded as a shift of the continents in their position, i. e., as "continental drift", rather than as "polar wandering". Assuming that the pole tends to move away from the continental masses is the same as assuming that a "polfluchtkraft" is operative (cf. Sect. 4.6.2). Since the sign of the latter is uncertain, it need not surprise the reader that the same is true for the direction of "polar wandering".

---

61 Jurdy, D.: J. Geophys. Res. 83:4989 (1978)

4.5.4 Convection Currents in the Mantle

Inasmuch as any type of cirulation affects the angular momentum of the Earth, it has also been thought that possible convection currents in the mantle (these will be discussed in more detail later, cf. Sect. 6.2.3) could cause polar wandering. According to Takeuki and Sugi[62], mantle convection may cause secular changes of the Earth's inertia products and lead to secular polar motion of the order of $0.003 \pm 0.009''$/year roughly in the direction of 90°W. Similarly, Goldreich and Toomre[63] have shown that large angular displacements of the Earth's pole are possible on a geologic time scale owing to convection-caused inhomogeneities within the Earth.

4.6 Other Forces Due to the Rotation of the Earth

4.6.1 General Remarks

We have seen that the rotation of the Earth can have geodynamic effects. These are mainly caused in an indirect manner, viz. by the slowing down of the Earth's rotation.

However, direct effects by rotation-caused forces upon the tectonosphere have also been proposed. The first of these is the pole-fleeing force, due to the tendency of any non-rigid rotating body to maximize its moment of inertia. The second is the Coriolis force; this is an inertial force known to act upon any mass that is in motion relative to a rotating frame of reference.

We shall discuss these two forces in turn.

4.6.2 The "Polfluchtkraft"

Inasmuch as there is a general tendency for the moment of inertia of a rotating deformable body to maximize itself, it may be expected that all masses would tend to move as far as possible from the axis of rotation. The force effecting this has been termed "pole-fleeing force", or by its German name, which has also been accepted into the English language, "polfluchtkraft".

The existence of a polfluchtkraft was first postulated by Eötvös[64] who noted that the direction line of the vertical (i. e., the force line of gravity) viewed in a meridional plane is curved in a rotating ellipsoidal Earth, the pole being located on its concave side. Furthermore, the center of gravity of the

62 Takeuki, H., Sugi, N.: Polar wandering and mantle convection. In: Rotation of the Earth (eds. P. Melchior, S. Yumi). Dordrecht: Reidel, p. 212, 1972
63 Goldreich, P., Toomre, A.: J. Geophys. Res. 74:2555 (1969)
64 Eötvös, R. v.: Verh. 17. Allg. Konf. Int. Erdmessung, I. Teil, 1913, p. 111. See also: Eötvös, R. v.: Gesammelte Arbeiten. Budapest: Academy 1953

floating mass in which the weight is acting, must lie higher than the center of gravity of the displaced substratum (metacenter) in which the buoyancy force is acting. The buoyancy force as well as the weight are acting in the direction of the tangent to the corresponding force line of the gravitational field; because of the latter's curvature (mentioned above), Eötvös reasoned that the two forces would not have the same direction and thus could not cancel each other, but would have a small resultant towards the equator. This is the polfluchtkraft. The situation is illustrated in Fig. 81.

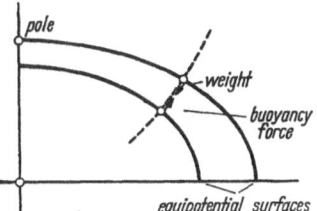

Fig. 81. Origin of the polfluchtkraft

In the wake of Eötvös' qualitative argument, there have been many attempts to calculate the polfluchtkraft analytically. Such attempts have been made notably by Epstein[65], Lambert[66], Ertel[67], and Milankovitch[68].

However, rather severe criticisms against this type of deduction have been voiced. The criticisms are mainly directed against the models which were assumed in the various deductions of the polfluchtkraft and seem to be rather pertinent. The reality of a polfluchtkraft must therefore be severely questioned.

Thus, Prey[69] has voiced a series of pertinent criticisms which can be summarized by stating that one should consider an *extended* floating mass, since a mass with small horizontal dimensions, i. e., a pencil-shaped body as considered in Fig. 81, could never attain equilibrium at all, but would simply tip over. Only a spherical body could attain equilibrium in this fashion. If an extended body is considered, it can attain equilibrium by tilting a little. The geometry assumed is therefore, according to Prey, entirely inadequate.

Prey substantiated his criticisms by explicitly calculating the equilibrium position of an extended "continent" floating upon a denser substratum. Because the geometry of such an arrangement is rather complicated, the calculations are correspondingly lengthy and involved. Starting from a position in which the floating mass is bounded above and below by equipotential surfaces of the gravity field, he showed that equilibrium can be attained by moving it

65 Epstein, P. S.: Naturwissenschaften 9, No. 25:499 (1921)
66 Lambert, W. D.: Am. J. Sci. 2:129 (1921)
67 Ertel, H.: Gerlands Beitr. Geophys. 43:327 (1935)
68 Milankovitch, M.: Kanon der Erdbestrahlung (l. c.)
69 Prey, A.: Gerlands Beitr. Geophys. 48:349 (1936)

by a very small amount *toward* the pole and tilting it. The necessary displacements are extremely small indeed.

Thus, if the continent is rigid, there exists an equilibrium position in which no forces are acting. In this equilibrium position, internal shear stresses will be present because the boundaries of the continent above and below do no longer coincide with the equipotential surfaces of the gravity field. If the continent is not assumed as rigid, there is therefore the possibility that it might yield under those stresses to adapt itself to a condition where its boundaries would again coincide with the equipotential surfaces. Hence it could be driven again into a new equilibrium position by moving slightly further toward the pole and tilting. This process could conceivably repeat itself and result in a net drift of the (deformable) continent *toward* the pole.

The relation of Prey's calculations with respect to those of Eötvös (and Ertel) can therefore be stated as follows. Disconnected floating "pencils" (the model used by Eötvös) are subject to a polfluchtkraft (as long as they stay upright) and *drift toward the equator,* but if a continent is large and strong enough so that tilting can take place, there is a net displacement towards the pole. If, in addition, the continents are able to yield slowly to shear stresses, an actual *drift toward the pole* will be the net result. Thus, if continents are weak enough to yield slowly, but not so weak as to yield instantly, there will be a poleward motion. The whole matter hinges on the question whether one can assume instantaneous isostatic equilibrium everywhere on the Earth (Eötvös) or whether this is not permissible (Prey). It would appear that Prey's model is closer to reality than that of Eötvös and his followers.

4.6.3 The Coriolis Force

The second of the two effects of the rotation of the Earth mentioned in Sect. 4.6.1 is the Coriolis force. The fact that the Earth's axis of rotation is inclined toward the plane of the ecliptic has the effect that the axis precesses. The forces causing this precession have a different action on parts of the Earth of different density, hence a component results that might conceivably make the continents move.

However, Jeffreys[70] analyzed this force, too, and found that it cannot cause stresses greater than 6 Pa. Moreover, it is also mainly alternating in direction and therefore cannot be of any more significance than the tidal forces which have been shown to be negligible as far as geodynamics is concerned.

---

70 Jeffreys, H.: The Earth, 2nd edn., London: Cambridge Univ. Press, p. 304, 1929

5. Planetary Problems

5.1 The Origin of the Earth as a Planet

5.1.1 Origin of the Universe

The Earth is one of many planets and its origin is thus intimately tied up with that of the solar system. The solar system, in turn, is part of the galaxy. Thus, ultimately, the question regarding the origin of the Earth is tied up with that regarding the origin of the Universe, of the galaxy and of the solar system.

The theories of the origin of the Universe can be split into two types: Either there was a "big bang" at the "beginning" in which some "superatom" exploded, or the Universe, in which stars were constantly being formed and destroyed, was at all times in a steady state.

From various sources, the evidence available at present seems to show the picture of a "Universe" which "began" with a "big bang" about 12.5×10^9 years ago[1]. This follows from many results of the study of quasi-stellar objects, galaxies, and clusters of stars, all of which point to an expansion process. The expansion rate is the greater, the farther away the objects in question are from a common center, from which all masses seem to have begun to expand about 12.5×10^9 years ago. In view of all this evidence it is difficult to retain the idea of a stationary state although the latter would perhaps be more satisfactory from a philosophical point of view than the "big bang"-theory.

Of course, it is neither known what the Universe was like before the "big bang" nor what its ultimate fate will be. Indications are that an ever-expanding Universe is unstable[2]. Thus, the present expansion might be part of a pulsation in which phases of expansion alternate with phases of contraction. Thus, some sort of "stationary" state might be present after all in which one "big bang" follows the next at regular intervals.

5.1.2 Origin of the Solar System

With regard to the origin of the solar system, which includes the Earth, one has again essentially two types of theories which can be termed "uniformi-

---

1 Hanes, D.: Mon. Not. R. Astron. Soc. 188:901 (1979). Other authors put the "big bang" further back
2 Barrow, J. D., Tipler, F. J.: Nature (London) 276:453 (1979)

tarian" and "cataclysmic". A useful account of this subject has been given, e. g., by Smart[3] to which the reader is referred for further details.

a) Uniformitarian Theories. The prototype of a uniformitarian theory is the nebular condensation hypothesis of Laplace. In it is assumed that the whole solar system was a gaseous nebula at its beginning. Such a nebula would have a rather dense core with a very thin atmosphere reaching to beyond the present boundaries of the solar system. Owing to gravitational attraction, the nebula would slowly contract which, in turn, would cause any initial rotation to become more rapid. Eventually the rotation would become so fast that at the outer boundary of the nebula a gaseous ring would be thrown off. The latter, in time, would condense to form the first planet. This process would repeat itself until all the planets were formed. The original core of the nebula would form the Sun.

Although the above theory of Laplace has some apparent success, there are in fact many severe difficulties. The most serious one is that the distribution of angular momentum in the solar system is at complete variance with any distribution that would be consistent with the theory: In the solar system, most of the angular momentum is found in the distant planets; Laplace's theory yields the reverse.

Essentially the same ideas as those of Laplace have also been defended by Kant. The latter author, however, did not assume that the nebula was subject to a primeval rotation, but tried to deduce that such a rotation would automatically develop. The difficulties in the theory are thereby not lessened.

A more modern revival of the uniformitarian theory has been proposed by von Weizsäcker[4, 5] who assumed that the Sun was formed from one of the interstellar dust clouds which are fairly common in the Milky Way. The planetary system was thereby formed as part of the process by condensation of the dust particles into the required number of larger masses. The dust cloud is assumed to be in a state of turbulent motion and endowed with a definite angular momentum. At a late stage of the process it would have a disc-like shape. Weizsäcker has shown that certain internal states of motion in the disc are more stable than others; − in fact that a pattern of vortices may develop which would be capable to persist in quasi-stationary motion for a considerable length of time. At the boundaries of the vortices the dust would collect and form the nuclei for the future planets. At this stage of the evolution the quasi-stationary pattern of vortices may disappear as the planets could grow by themselves by further accretion of matter. Weizsäcker estimates that the time required for a planet to grow to its final size was about 10^8 years, whereas the Sun might have been "finished" in 10^7 years. The difference of a factor 10 might just have been sufficient to allow for most of

3 Smart, W. M.: The Origin of the Earth, London: Cambridge Univ. Press 1951
4 Weizsäcker, C. F. v.: Z. Astrophys. 22:319 (1944)
5 Chandrasekhar, S.: Rev. Mod. Phys. 18:94 (1946)

the lighter elements to escape from the planets and thus to account for the different composition of the Sun (consisting mostly of hydrogen) from that of the rest of the solar system.

Finally, Kuiper[6] modified the previous discussions by assuming gravitational instability within a disc-shaped solar nebula as a source of gaseous spheres (protoplanets) which eventually would contract to form planets.

b) *Cataclysmic Theories.* The basis of cataclysmic theories of the origin of the solar system is the assumption of some catastrophe. A good example of such a theory is the hypothesis, due to Hoyle[7], that the Sun was part of a binary system and that the Sun's companion blew up as a supernova. According to Hoyle, a slight eccentricity of the explosion in the Sun's companion would produce the correct distribution of angular momentum in the solar system. Furthermore, the nuclear chemistry of a supernova explosion (being completely unknown) could account for the different composition of Sun and planets.

Other cataclysmic theories include the tidal theory of Jeffreys[8] in which it is envisaged that the Sun was disrupted by a tidal resonance effect with a passing star. The fragments would ultimately form the planets. Difficulties arise in this theory from two sides. First, if it is assumed that the actual fragments of the Sun formed the planets, there is the difference in chemical constitution of the Sun and of the planets to be accounted for. This is only possible by assuming a tremendous thinning out of the fragments just after the catastrophe and a subsequent slow recondensation which would provide time for the light elements to escape. Second, it is again not easy to account for the distribution of angular momentum in the solar system. To avoid this difficulty it has been assumed that the tidal effect occurred with a hypothetical binary companion of the Sun instead of with the Sun itself. Similar problems occur if the catastrophe is assumed to be a head-on collision rather than a tidal resonance effect.

In conclusion, it may be remarked that all theories of the origin of the solar system have two principal difficulties to cope with: First there is the distribution of angular momentum in the solar system which is chiefly (98%) concentrated in the planets and not in the Sun, Jupiter making the biggest single contribution. Second, there is the difference in composition between the Sun and the planets. The former consists chiefly of hydrogen, the latter of heavier elements. The uniformitarian theories differ from the cataclysmic ones mainly by the likelihood of the occurrence of the postulated process. In all uniformitarian theories, the acquiring of a planetary system is part of the normal evolution of any star, whereas in cataclysmic theories this would be an

---

6 Kuiper, G. P.: Chapter 8 in Astrophysics (ed. Hynek). New York: McGraw-Hill Publ. Co. 1951

7 Hoyle, F.: Proc. Cambr. Philos. Soc. 40:265 (1944)

8 Jeffreys, H.: The Earth, 2nd edn. London: Cambridge Univ. Press 1929

extremely rare and unique occurrence. Agreement as to the correct theory has obviously not yet been achieved.

In order to estimate the age of the solar system, one can remark that the Sun, which belongs to the first principal series of "small stars", has to be younger than 7×10^9 years. This fits together well with the value of 12.5×10^9 years given above for the age of the Universe. For the planets, including the Earth, to form, some further billions of years would be required, which fits together well with the age of the crust ("geological zero point") of 4.6×10^9 years (cf. Sect. 2.5.3). At an instant 2×10^9 years ago the tectonic style became prevalent which is effective to the present time.

5.1.3 Birth of the Moon

Another event that has been connected with the Earth's early history is the birth of the Moon. The theories of the origin of the Moon range from those postulating fission from the Earth to those assuming capture by the Earth or simultaneous formation with the Earth.

The theories postulating cataclysmic fission of the Moon from the Earth used to enjoy considerable popularity. The physical possibility of the separation of the Moon from the Earth has been sought in a resonance effect in the oscillations of the Earth with the tidal forces exerted by the Sun[9] or else in an internal explosion[10]. Such a cataclysmic birth of the Moon would have far reaching consequences upon the physiographic appearance of the Earth. These consequences have been discussed for the first time by Fisher[11]; they have been restated more recently, e. g., by Bowie[12] and by Escher[13]. Accordingly, it is assumed that the Earth already had a solid crust when the Moon was torn off. Consequently, a tremendous wound would be left which would be represented now by the Pacific Ocean. The "suction" caused by the wound would help to break up the remaining part of the crust and would cause the pieces to move toward the Pacific.

However, since lunar rock samples have become available through the American Apollo missions, it has become possible to compare the latter with terrestrial crustal and mantle rocks. The weight of the chemical evidence seems to indicate that the terrestrial mantle and the lunar crust differ substantially in composition for many elements[14]; this and the complete absence of water[15] in lunar rocks rules out the possibility that the Moon can ever have been a part of the Earth.

9 Darwin, G. H.: Philos. Trans. R. Soc. London, Part II, p. 532, (1879)
10 Quiring, H. L.: Gerlands Beitr. Geophys. 62:81 (1952). − Z. Dtsch. Geol. Ges. 105:203 (1953). − Neues Jahrb. Geol. Palaenontol. Mineral. 3:140 (1961)
11 Fisher, O.: Nature (London) 25:243 (1882)
12 Bowie, W.: Sci. Mon. 41:444 (1935)
13 Escher, B. G.: Bull. Geol. Soc. Am. 60:352 (1949)
14 Taylor, S. R.: Nature (London) 281:105 (1979)
15 Cochran, W.: Geotimes 24 (9):15 (1979)

Thus, only the theories of capture or of simultaneous formation of the Moon in the neighborhood of the Earth remain as contenders for the origin of the Earth-Moon system. A definite choice between the two possibilities cannot be made as yet.

5.2 The Evolution of the Earth as a Planet

5.2.1 The Problem

We are now about to consider the problem of the evolution of the Earth as a whole. This problem is essentially of a thermodynamic nature, inasmuch as the relation between the fusion curve and the temperature curve determines the evolution of the internal constitution of the Earth.

In conformity with the various theories of the origin of the solar system and therewith of the Earth, one arrives at a series of possibilities for the "initial" course of the temperature in the Earth's interior, i. e., the course at the "geological zero point", 4.6×10^9 years ago. These possibilities are due to various authors and have been compiled by Mayeva[16]. Her compilation is shown in Fig. 82. Based on the possible initial states as shown in Fig. 82, the problem is now that of deducing the changes with time of the pertinent conditions.

5.2.2 Theory of the State of the Earth's Interior

Any discussion of the evolution of the Earth has to start from an analysis of the possible conditions obtaining in the core and mantle during time. Inasmuch as only little is known about these regions even referring to the present time, our knowledge is even less for the geological past. One can, therefore, only deal with general statements about materials under high pressure and make speculations about the evolution of the conditions.

The state of the Earth's interior is determined by the Grüneisen function y defined as follows

$$y = V\alpha K/C_v \tag{5.2.2-1}$$

where α is the volumetric thermal expansion coefficient, K the isothermal bulk modulus, V the molar volume and C_v the molar specific heat at constant volume. The Grüneisen function y is close to proportional to V; since the latter does not vary much, at least in the lower mantle, y itself is almost a constant (Grüneisen's law); it is then called *Grüneisen parameter*[17].

---

16 Mayeva, S. V.: Bull. (Izv.) Acad. Sci. USSR Geophys. Ser. 1971 (1):3 (1971)
17 Anderson, O. L.: J. Geophys. Res. 84 (87):3537 (1979)

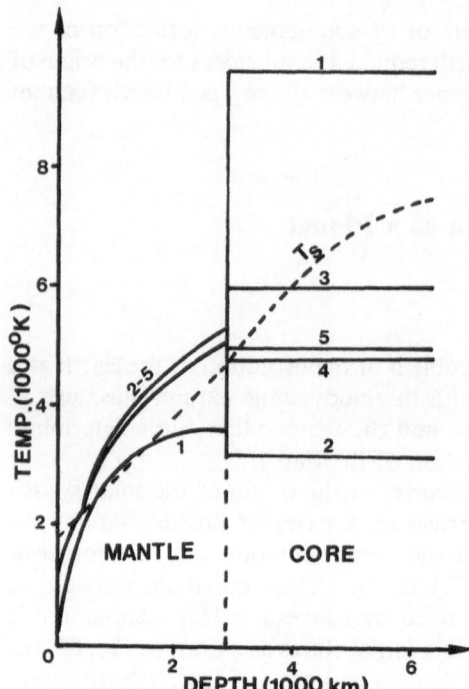

Fig. 82. Possible original temperature distributions (1 – 5) in the interior of the Earth. T_S is the fusion-point curve. (Modified after Mayeva[16])

The Grüneisen parameter has a fundamental significance in the discussion of the state of a material, because it connects the coefficient of thermal expansion of that material to the specific heat. Therefore, many attempts have been made to determine or predict the Grüneisen parameter for the Earth's interior. This bas been done based on experimental (shock wave) and on theoretical considerations. Such studies have been reported for the mantle[18] and for the core[19-22].

It turns out that the results are fairly independent of the materials involved, because general statements can be made about the behavior of materials at high pressure. The details of the investigations, however, are beyond the scope the present study; we shall only refer to the proposed values as the need arises.

Important are also the melting point data: At what temperature and pressure is the material inside the Earth molten. Some time ago, Simon[23] formu-

18 Anderson, O. L.: J. Geophys. Res. 79 (8):1153 (1974); Butler, R., Anderson, D. L.: Phys. Earth Planet. Int. 17(2):147 (1978)
19 Jacobs, J. A.: The Earth's Core. London, New York: Academic Press 1975
20 Boschi, E., Mulargia, F.: Ann. Geofis. (Roma) 30 (1):205 (1977)
21 Mulargia, F., Boschi, E.: Phys. Earth Planet. Int. 18:13 (1979); Geophys. J. R. Astron. Soc. 53:531 (1978)
22 Jamieson, J. C. et al.: J. Geophys. Res. 83 (B12):5929 (1978)
23 Simon, F. E.: Trans. Faraday Soc. 33:65 (1937)

lated an equation relating the melting point temperature T_m to pressure as follows

$$P = A \left[\left(\frac{T_m}{T_{m0}} \right)^C - 1 \right] \qquad (5.2.2-2)$$

where A is a constant and T_{m0} the melting temperature at some standard (atmospheric!) pressure. The exponent C is related to the Grüneisen parameter γ as follows[24]:

$$C = \frac{6\gamma + 1}{6\gamma - 2}. \qquad (5.2.2-3)$$

The Simon equation seems to give, in general, melting temperatures which are too high; thus, modifications have been proposed which are based on the atomic structure of each substance involved. From such relationships, various types of melting point curves have been obtained for the core of the Earth (whose outer part is known to be liquid)[25]. Such curves may cross and recross the adiabatic temperature curve of the Earth, thus giving an explanation for the fact that only part of the core is molten. From such curves, interesting conjectures may also be made regarding the history of melting in the Earth.

5.2.3 The Earth's Thermal History

The various theories of the origin of the Earth discussed in Sect. 5.1.2 of this book are of importance in geodynamics because of their implications regarding the Earth's thermal history. The layered structure of our globe suggests that the Earth might have gone through a molten state at one stage of its life; for in a molten body the differentiation into a dense core and progressively less dense upper strata is most naturally accomplished. Most discussions of the Earth's thermal history, therefore start with a hot Earth which is gradually cooling down.

It is therefore of some interest to investigate whether there are any cosmological indications that the Earth has gone through a molten stage, in the light of the various theories of the origin of the solar system. In the cataclysmic theories, where it is assumed that the planets were formed from the fragments of a star, it would presumably be natural to assume that these fragments were hot. In this instance, it should not be overlooked, however, that the fragments could possibly have been spread out after the explosion into a gas cloud of very low density; the gas cloud, in turn, would condense later to form the planets. Any dispersed gaseous matter in the stellar space would of

---

24 Gilvarry, J. J.: Phys. Rev. 102:308, 317, 325 (1956)
25 For a good summary, see Jacobs, J. A.: The Earth's core. London, New York: Academic Press 1975

necessity be cooled down very rapidly to form "ice crystals" which creates the problem of explaining the subsequent condensation in such a fashion so as to arrive at a hot Earth. In this instance, the problem is the same as that encountered in the uniformitarian theories of creation of the solar system. Here, no cataclysm is assumed in the first place and a way must be found to condense a cold cloud of interstellar matter in such a fashion that it becomes hot; – at least if it is desired to start the development of the Earth from a hot sphere.

There are two ways in which the Earth could have heated up during or immediately after its formation, the latter being assumed to have taken place from a cloud of dispersed material: first by a conversion of mechanical energy into heat during the contraction of the cloud, and second by the effect of radioactivity of the constituent elements in the Earth.

The process of condensation or contraction of a cool gas cloud to a planet has been discussed by Hoyle[26]. In his investigations, Hoyle came to the conclusion that such cold condensation would convert sufficient mechanical energy into heat to melt a planet of the size and composition of the Earth. However, a scrutiny of his argument shows that it depends very sensitively on the speed of condensation, a quantity which is not very well known.

It seems therefore that, except in the case of an extremely rapid condensation, there is not enough heat from mechanical energy available to melt the Earth if it was formed from a cold cloud. We shall therefore investigate the second possibility of heat production, viz. that due to the presence of radioactive matter in rocks. A study of this possibility has been undertaken by Urry[27]. If it is taken into account that radioactive matter decays exponentially, then it is reasonable to expect that the density of heat-generating matter within the Earth must have been much greater during the primeval days than it is at present. Urry's result is that the radioactivity would be sufficient to melt the Earth. The same conclusion was arrived at by Birch[28] using newer values for the radioactive decay constants: an initially cold Earth containing as little as 0.1% of potassium would eventually melt, – at least partially. The time required to reach the molten state might not exceed 10^8 years.

From these investigations it would appear as likely that even had the Earth been formed by cold accretion, it would have melted early in its life. However, it is possible that the above authors overestimated the effect of radioactivity. Jacobs[29] showed that the final solution of the heat conductivity equation is the superposition of the cooling of a non-radioactive Earth from its "initial" temperature (assumed by Jacobs as high) plus the heating-up of an originally cold Earth due to radioactivity. If one inspects Jacob's solution for the latter case, it becomes evident that no very high temperatures are reached (except at the surface; see Fig. 83). For his calculations, Jacobs assumed the present

26 Hoyle, F.: Mon. Not. R. Astron. Soc. 106:406 (1946)
27 Urry, W. D.: Trans Am. Geophys. Union 30:171 (1949)
28 Birch, F.: J. Geophys. Res. 56:107 (1951)
29 Jacobs, J. A.: Publ. Bur. Centr. Séismol. Int. A 19:155 (1956)

layered distribution of elements inside the Earth, but he made allowance for the time variation of radioactivity. The end result of the calculations shows that an initially cold Earth does indeed heat up, but not enough to reach the stage of melting (except at the very surface; cf. Fig. 83). Similar calculations were made by Allan[30] with corresponding results.

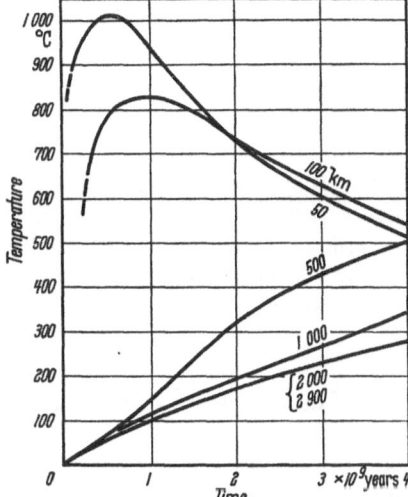

Fig. 83. Temperature distribution at various depths in a radioactive Earth, starting from zero temperature. (After Jacobs[29])

The basic assumption of the heat-conductivity equation as fundamental to the thermal history of the Earth makes no provision for the possible existence of thermal convection currents in the mantle. These, however, would tend to counteract any temperature increase. On the other hand, if chemical differentiation (say, of mantle material into crust and core material) is taken into account, the process could be exothermic leading to liberation of additional heat. However, the proposed mechanisms are highly hypothetical and meet with many difficulties[31]. Therefore, if a molten stage of the Earth is required, it seems that it is unavoidable to start with a "hot" Earth.

Because of the decay of radioactivity, the Precambrian geothermal gradients have generally been assumed to have been steeper than the present-day ones. However, an analysis of rocks in the Canadian Shield does not seem to bear out any evidence for such steep gradients[32]. This fact poses something of a puzzle which one has attempted to solve by postulating that the extra heat was removed from the interior of the Earth through special conditions (extra lengths) of the Precambrian mid-oceanic ridges[32].

30 Allan, D. W.: Endeavour 1954:89 (1954)
31 Vityazev, A. V., Lyustikh, Y. N., Nikolaychik, V. V.: Bull. (Izv.) Acad. Sci. USSR, Earth Phys. 13 (8):3 (1977)
32 Burke, K., Kidd, W. S. F.: Nature (London) 272:240 (1978)

On the other hand, a decrease of the heat flow of continental rocks with the tectonic age of these rocks has commonly been observed[33, 34]. This has usually been ascribed to the effects of erosion[35] which removes the radiogenic contribution from the surface heat flow.

5.3 Continents and Oceans

5.3.1 The Nature of the Problem

The physiographic appearance of continental and oceanic areas is entirely different. It appears that there is also a fundamental chemical difference between the two types of regions of the Earth's crust.

Thus, the question arises how this difference came about in the first place and how it has been maintained since. A discussion of the possible answers to this question is the subject of the present section.

5.3.2 The Formation of Continents by Convection

The concept of Laurasia and Gondwanaland or Pangea as primeval continents outlined in Sect. 1.3.1, immediately poses the question as to how the latter may have originated. One of the possibilities that has been advocated to this end is the hypothesis of the formation of continents by convection[36].

In this hypothesis it is assumed that in the early days of its history, the Earth was well-nigh liquid. The heavier material sank to the center to form the core, and what is now the mantle proceeded to cool (heat being lost into the Universe) by thermal convection. One school of scientists assumes that these convection currents in the mantle of the Earth are still operative to the present day, thereby providing a force for orogenesis (cf. Sect. 6.2.3). An alternative opinion, however, is that such convection currents were possible only in the primeval days of the Earth. Whatever the solution to this question may be, it seems agreed that thermal convection would be a possible means of creating primeval continents.

The creation of a continent by convection can be envisaged to have occurred in one of two ways. First, if one has a rising current, then the material just above it will be brought to a higher elevation than the surrounding material owing to the effects of mechanical dragging. It thus would form a "continent". However, the "continent" would be able to subsist only for so long as the corresponding convection current is operative. In order to adopt

33 Polyak, B. G., Smirnov, Y. B.: Geotektonika (Moscow) 4:205 (1968)
34 Vitorello, I., Pollack, H. N.: J. Geophys. Res. 85:983 (1980)
35 England, P. C., Richardson, S. W.: Geophys. J. R. Astron. Soc. 62:421 (1980)
36 Hills, G. F. S.: The Formation of Continents by Convection. London: E. Arnold & Co. 1947

this view, it must therefore be assumed that rising convection currents are operative everywhere underneath continents up to the present day. A second way by which continents may be formed is by assuming the latter as much lighter than the liquid. The continents would thus correspond to "scum" (in isostatic equilibrium) on a liquid which accumulates over the *descending* branch of a convection current. After the convection stopped, the "scum" (i. e., the continents) would simply remain in its prior position or possibly get dragged around and broken up due to incidental causes (see Fig. 84).

Fig. 84. Formation of a continent (*black*) by "scum" accumulation

The hypothesis of the existence of Laurasia and Gondwanaland as primeval continents implies that the primeval convection currents would have had such a geometrical arrangement that two continents formed at the (then) poles of the Earth. If the scum theory is adopted, this means that the currents must have been descending at the poles, otherwise they would have to be rising at the poles.

However, the primeval existence of Laurasia and Gondwanaland is not at all certain. A different system of primeval continents is arrived at from the observation that four old shields, at present, have a position roughly at the corners of a tetrahedron. If it is not conceded that continents may have moved around much during the history of the Earth, an explanation for the position of these continents may be sought in the assumption of an octahedral system of convection currents. This has been proposed by Vening Meinesz[37]. Figure 85 shows the system of currents. The creation of this system could be made

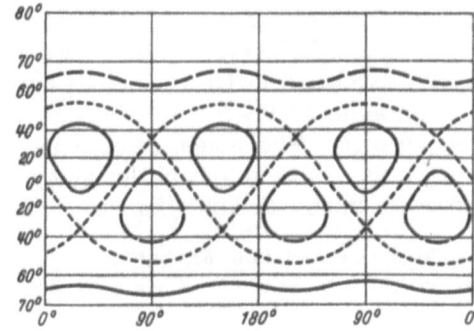

Fig. 85. Octahedral arrangement of convection currents (inward currents *dotted*, outward currents *solid*). (After Vening Meinesz[37])

37 Vening Meinesz, F. A.: Versl. K. Akad. Wet. 53:No. 4 (1944)

plausible by the remark that a regular pattern is most likely to occur. The octahedron is the only regular surface in which an even number of sides touch in one corner, and this is a necessary condition in a convection current system.

Regarding the physical possibility of convection, we recall the fundamental facts of thermohydrodynamics in viscous substances given in Sect. 3.2.6. These show indeed that convection currents are possible in viscous media, provided certain fundamental, rather narrow, requirements are fulfilled. The attempts at elucidating the thermal history of the Earth by diffusive heat transfer alone (cf. Sect. 5.2.3) have, in fact, led to geotherms which may be unstable against convection[38]. The various possible convection patterns can be connected with characteristic parameters of the hydrodynamic regime[39].

A much-quoted attempt at establishing a mechanical scheme of convection currents in a primeval Earth has been made by Pekeris[40]. For his calculation, Pekeris chose a model of the Earth in which there is a temperature variation with depth as well as a zonal temperature variation over the surface of the Earth. The zonal temperature variation is a much more effective cause of convection currents than that presented by a temperature variation with depth. The mean temperature variation (due to exposure to solar radiation) from the equator to the poles is about 60 °C and penetrates to great depth since it is independent of time. Another possible source of zonal temperature variations is that the bottom of the oceans is uniformly at a temperature of about 2 °C and, in addition, that there is only a small crustal layer over the latter.

Pekeris analyzed two particular models. He assumed an axis of rotational symmetry (not necessarily coincident with the axis of rotation of the Earth) and calculated the flow patterns as a function of colatitude Θ (with respect to that axis) and radial distance from the center. The assumed zonal temperature variation in the first model is from 100° to $-100°$ if Θ varies from 0 to 180°, and in the second model from 100° to $-50°$ to 100° if Θ varies from 0° through 90° to 180°. Pekeris found rising convection currents underneath the temperature maxima (i. e., at $\Theta = 0°$ in his first, and at $\Theta = 0°$ and 180° in his second model) and descending currents underneath the temperature minima (i. e., $\Theta = 180°$ in the first and $\Theta = 90°$ in the second model). The result for the second model is shown in Fig. 86.

Similar calculations as those just mentioned have also been made by Chandrasekhar[41], by Urey[42], and by Latynina[43] with corresponding results.

According to Pekeris, a rising current creates a continent, a descending one an ocean. The system of currents, thus, would be self-perpetuating if continents would automatically always stay hotter than oceans. This is, however, an unsettled question.

38 Tozer, D. C.: Sci. Prog. 64:1 (1977)
39 Sharpe, H. N., Peltier, W. R.: Geophys. Res. Lett. 5:737 (1978)
40 Pekeris, C. L.: Mon. Not. R. Astron. Soc. Geophys. Suppl. 3:343 (1935)
41 Chandrasekhar, S.: Philos. Mag. 43:1317 (1952)
42 Urey, H. C.: Philos. Mag. 44:227 (1953)
43 Latynina, L. A.: Bull. (Izv.) Akad. Nauk SSSR. Ser. Geofiz. 1958:1085 (1958)

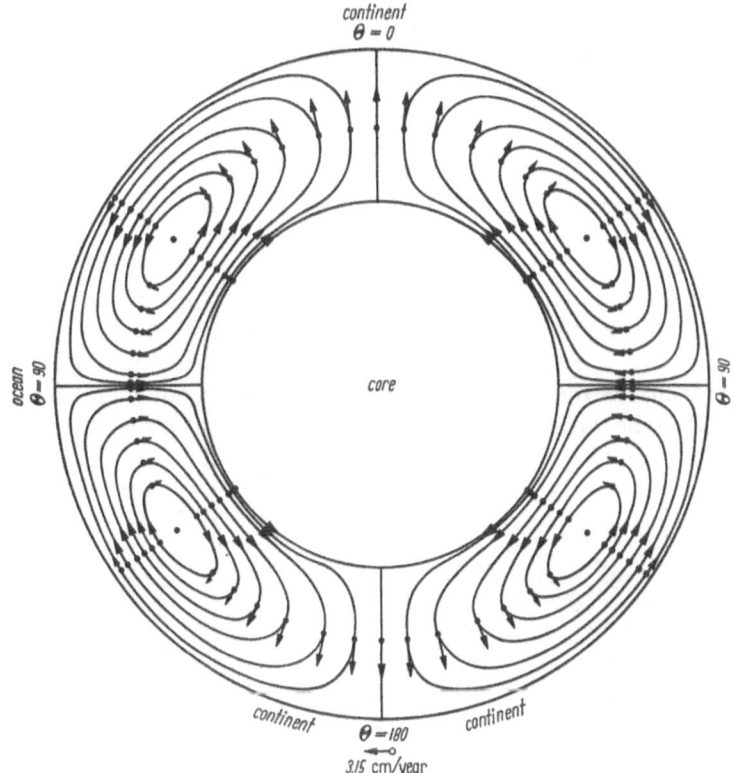

Fig. 86. Convection streamlines and flow velocities in one of Pekeris'[40] models

5.3.3 Possibility of Present-Day Convection

The average heat flow values observed today can be reconciled with the existence of present-day convection currents. The actual thermodynamics of such currents, if they exist at the present time, is, however, not yet quite clear. In this instance, all that was said in Sect. 5.3.2 for primeval currents, also applies to modern currents, except that one has less latitude for adjusting the required numerical parameters (see also Sect. 3.2.6).

Nevertheless, because of the possibility of the existence of such currents, one might want to speculate regarding their present-day effects. Most of these effects will bear upon the theories of orogenesis, and a detailed discussion will therefore be relegated to Chap. 6. However, there might be some continental effects that should be properly dealt with here.

Thus, an important investigation by Rikitake and Horai[44] deals with a possible explanation of the heat flow anomalies observed over mid-ocean

44 Rikitake, T., Horai, K.: Bull. Earthquake Res. Inst. 38:403 (1960)

ridges (see Sect. 2.6.2) in terms of present-day continental convection currents. The result of this investigation is that the larger the diameter of a current, the greater should be the heat flow through the crust above the ascending branch. Thus, the Pacific being a large feature presumably associated with a large convection current, its heat flow anomalies should be high.

Another possible effect of present day convection currents, if they exist, has been pointed out by Licht[45]. We have noted in Sect. 4.2.1 that there are indications that the Earth might be slightly pear-shaped. Such a pear shape could easily be associated with a global system of convection currents as shown in Fig. 87. Licht calculated the perturbation of the gravitational field due to such a system (note that the density of the material in the ascending branches of the currents is less than in the descending branches) and came up with some values that are not inconsistent with those obtained from satellite observations.

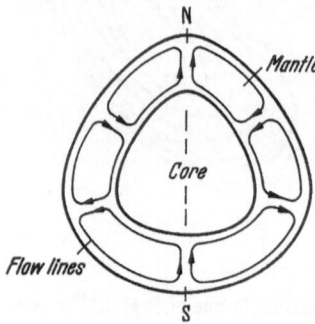

Fig. 87. A global system of convection currents (the *NS axis* is an axis of symmetry) that could cause the Earth to be pear-shaped. (After Licht[45])

However, abundances of lithophile elements in accessible mantle (mid-oceanic ridges) and cosmic materials (presumably indicative of the lower mantle) seem to preclude a convection current system operating through the *whole* mantle: rather, there must be a decoupling at about 500 km depth. [46]

5.4 Historical Remarks

5.4.1 Introduction

The ideas regarding the problem of evolution of continents and oceans given above are those that are current today: They are mostly based on some form of convection in the mantle.

45 Licht, A. L.: J. Geophys. Res. 65:349 (1960)
46 O'Nions, R. K., Hamilton, P. J., Evison, N. M.: Sci. Am. 242 (5):120 (1980)

However, it is only relatively recently that the evidence has become over-whelmingly in favor of some sort of primeval formation and subsequent drift of the continents. Earlier ideas were based on some "fixistic" concept regarding the evolution of the Earth. For historical interest, we shall mention some of these ideas here. It is also possible that some of their premises may in fact be applicable in a minor way. The expansion theory still has quite a number of adherents and may prove some day to hold some truths after all.

5.4.2 Tetrahedral Shrinkage

A theory to account for the morphological facts about continents and oceans which, at the same time, aimed at an explanation of the tetrahedral arrangement of the former, was a theory assuming a particular type of shrinkage. It was based upon the assumption that a tendency exists for a contracting sphere to shrink tetrahedrally, simply because the tetrahedron has minimum volume for a given surface of all *regular* bodies. [47]

Thus, if it be assumed that the Earth had at one time cooled enough so that the outermost layer had become a solid skin incapable of changing its area, then a tetrahedral shape might be considered as the logical outcome of such a process. The corners of the tetrahedron would correspond to the continents, the faces to the ocean basins. A proper arrangement concerning the size of the tetrahedron would also explain the ratio 1 : 2 occupied by continents and by oceans.

An idea very similar to that presented above had been suggested long ago by Davison [48]. Accordingly, the Earth is contracting in its upper layers only (due to cooling) which are therefore in a state of internal tension. Owing to the pressure of the continents, Davison assumed that the amount of stretching under them must have been very much less than under the great oceanic areas. This would tend to make the ocean basins subside even further and present a physical cause for their permanence. Any orogenetic effects would be most pronounced at the junction of the oceans with the continents, thereby leading to the idea of continental growth.

The chief criticism of the tetrahedral shrinkage theory is [49] that the topmost "skin" of the Earth simply does not have such properties as would prevent it from changing its area under the action of tangential forces. It is thus quite inconceivable that it would retain its area upon a shrinking interior; at the very least it would either thicken in spots or else become folded over in the manner of nappes. The evidence of folding seems to show that adjustment of an outer shell to a collapsing interior would take place continually or in a

47 Woolnough, W. G.: Bull. Am. Assoc. Petrol. Geol. 30:1981 (1946)
48 Davison, C.: Philos. Trans. R. Soc. London Ser. A 178:240 (1888)
49 Scheidegger, A. E.: Handbuch der Physik, vol. 47, p. 283. Berlin-Göttingen-Heidelberg: Springer 1956

rapid sequence of diastrophisms rather than in a slow settling to the from of a tetrahedron. Furthermore, the theories of deformation of such an outer shell seem to indicate that buckling would be the mechanism determining the adjustment of a *rigid* shell to a collapsing interior. It has been shown[50, 51] that the deformation of a buckling sphere is symmetrical about a diameter and that the deviations of the shape are given by a series of spherical harmonics along parallels of latitude associated with the diameter of symmetry. This obviates the postulate of tetrahedral shrinkage.

5.4.3 Formation of Continents by Expansion

Several theories have been proposed in which it has been assumed that the Earth is subject to *expansion*. The idea that there is some expansion of the Earth is not new[52]; it has currently been revived because of the recognition that many tensional features, such as the mid-oceanic rifts, are present on the Earth's surface[53-58]. As far as these latter features are concerned, the expansion hypothesis is a hypothesis of orogenesis and, in this context, will be treated in the next chapter.

Expansion, however, has also been advocated as a cause of the origin of continents[59, 60]. Thus, it has been assumed that the Earth was much smaller at the beginning than it is now, having a diameter of about one-half of its present one. Somehow, a "crust" was formed on it, which was everywhere some 30 km thick. Then, as the diameter grew, the original crust broke up and its remnants are the present continents (see Fig. 88). Expansion was supposed to have started by ocean "cracks" like the Mid-Atlantic Rift. An increase by a factor 2 in diameter represents a surface increase by a factor 4, which produces about the right order of relative area occupied by present-day continents. However, such a radius increase also produces a volume increase, and a corresponding density decrease, by a factor 8. Since the present-day average density of the Earth is about 5,500 kg/m³, the average density before the start of the expansion must have been about 44,000 kg/m³.

50 Zoelly, R.: Über ein Knickungsproblem an der Kugelschale. Diss. Eidg. Tech. Hochsch. Zürich 1915
51 Leutert, W.: Die erste und zweite Randwertaufgabe der linearen Elastizitätstheorie für die Kugelschale. Diss. E. T. H. Zürich, 1948
52 Halm, J. K. E.: J. Astron. Soc. S. Afr. 4:1 (1935). − Hilgenberg, O. C.: Vom wachsenden Erdball. Berlin, 1933
53 Heezen, B. C.: Preprints Int. Oceanogr. Crongr. 26 (1959)
54 Egyed, L.: Geofis. Pura Appl. 33:42 (1956)
55 Carey, S. W.: J. Alberta Soc. Pet. Geol. 10:95 (1962)
56 Carey, S. W., O'Keefe, J. A.: Science 130:978 (1959)
57 Groeber, P.: Bol. Inf. Petrolif. No. 311:101; No. 312:181 (1959)
58 Wilson, J. T.: Nature (London) 185:880 (1960)
59 Hilgenberg, O. C.: Vom wachsenden Erdball. Berlin 1933
60 Egyed, L.: Geofis. Pura Appl 45:115 (1960)

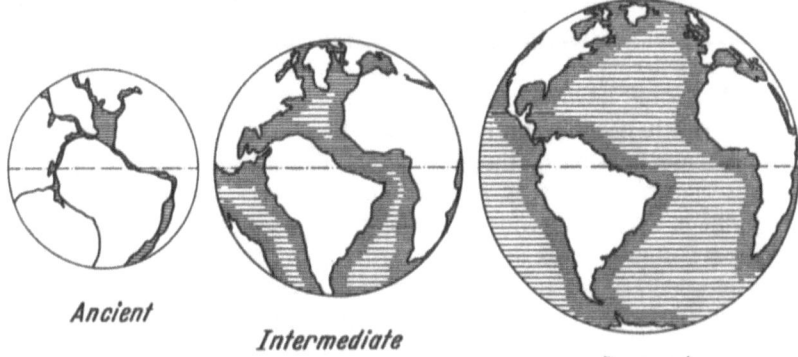

Ancient

Intermediate

Present

Fig. 88. How the Earth's oceans could have developed by global expansion. (After Hilgenberg[59])

Such a great change in density would, at least if the present-day physical laws are assumed to have been valid throughout the Earth's history, cause unsurmountable difficulties. Thus, the gravitational acceleration at the surface of the Earth would have to have been four times bigger at the beginning of the expansion (Paleozoic?) than at the present time and the moment of inertia four times smaller. There is no evidence from fossils for a such a large gravity acceleration in the Paleozoic (plants and animals had about the same stature then as at present), nor for a smaller moment of inertia which would have caused a much more rapid rotation of the Earth than at the present[61] (as shown in Sect. 4.3.3, the length of the day was only some 10% less than at present).

Furthermore, it is difficult to envisage mechanisms that could have caused the expansion.

The most obvious effect that could have caused expansion is a thermal one. One could imagine that, if the Earth started out as a hot liquid sphere, it might form a solid layer on the top like ice on water. This layer might in a way serve as an insulator so that immediately after its formation the interior would heat up again since the heat created by radioactivity could now no longer escape. Although the above idea seems appealing at first glance, it is in fact quite untenable because (unlike ice and water) most rocks are more dense in the solid state than in the molten state. If it is to stay on top, the solid crust cannot therefore simply be the solid phase of the substratum but must be composed of a different substance. If this be assumed, however, then the thermal history for various Earth models can be calculated, as has been reported in Sect. 5.2.3. Accordingly, it is possible that the whole Earth was remelted early in its history but it does not seem reasonable that it would have melted later on. Looking at the values given earlier in Fig. 83, one must admit, however, that the possibility exists that there is a net gain in heat and hence a

---

61 Runcorn, S. K.: Nature (London) 204:823 (1964)

slight expansion of the Earth as a whole during its history. It does not seem possible, though, that this could effect more than the formation of e. g., deep ocean trenches, − although admittedly no exact calculations exist.

Calculations have also been made to investigate whether a chemical change or a phase change in the interior of the Earth could be held responsible for a large-scale expansion of the globe. Such calculations are simply based on a comparison of the energy required to cause the expansion with the energy available in chemical bonds. The energy required for expansion is simply the difference in gravitational potential energy in the small and in the big Earth.

The gravitational energy difference between a small Earth and a big one depends somewhat on the density distribution in the interior. Beck[62] found that for any reasonable density distribution an expansion of the radius of about 100 km is possible, but for expansions of 1,000 km or more this is not so. Similarly, Cook and Eardley[63] estimated that a uniform expansion of the radius of the Earth by 20% would require an amount of energy equal to that required to dissociate almost all the chemical bonds of the molecules constituting the Earth.

It appears therefore that no chemical or similar sources could supply enough energy to cause an expansion of the Earth that would be required to create continents by this mechanism.

Expansion of the Earth has been attributed not only to thermal causes, but also to a slow change of the value of the gravitational "constant" postulated in some cosmological speculations. Jordan[64], in a discussion of projective relativity theory, came up with the conclusion that the quantity x in Newton's law of gravitational attraction

$$F = x\frac{m_1 m_2}{r^2} \qquad\qquad (5.4.3-1)$$

(where F is the force; m_1 and m_2 are the two masses involved; r is the distance between them) should not be a fundamental constant as commonly assumed, but in fact should be a variable which has been slowly decreasing since the beginning of the universe. Similar postulates have also been made by Dicke[65,66].

Beck[67] has estimated the energy available from such a decrease of the gravitational constant and again came to the conclusion that not more than an increase of 100 km in the Earth's radius can be accounted for in this fashion. Thus, if expansion on the postulated scale occurred at all, a completely unknown energy source must be found.

62 Beck, A. E.: J. Geophys. Res. 66:1485 (1961)
63 Cook, M. A., Eardley, A. J.: J. Geophys. Res. 66:3907 (1961)
64 Jordan, P.: Schwerkraft und Weltall. Braunschweig 1952. − Naturwissenschaften 48:417 (1961)
65 Dicke, R. H.: Rev. Mod. Phys. 29:355 (1957)
66 Dicke, R. H.: J. Wash. Acad. Sci 48:213 (1958)
67 Beck, A. E.: J. Geophys. Res. 66:1485 (1961)

If there has been a universal decrease of the gravitational constant, other celestial bodies must have also been affected by it. Thus, the available evidence regarding Mercury has been analyzed by Crossley and Stevens[68] who showed that the required expansion is in conflict with observations.

A yet further modification of the physical laws has been the supposition that *all* mass (but not necessarily the density) increases with time according to the law[69]

$$M = m_0 e^{\alpha t} . \qquad (5.4.3-2)$$

If the mass of the Earth has increased eightfold in 300 million years, one obtains

$$\alpha = 0.00994 , \qquad (5.4.3-3)$$

viz. the condition that *all* mass increases by 1% in 1 million years.

If such speculations are allowed, it is, of course, very difficult, to remain on a known basis. If only the mass increases, but all other physical laws are retained, one again ends up with unsurmountable difficulties (the orbital velocities would change in an unacceptable manner, etc.). However, if all physical laws are questioned and changed, it is, of course, possible to obtain any desired result.

68 Crossley, D. J., Stevens, R. K.: Can. J. Earth Sci. 13(12):1723 (1976)
69 Carey, S. W.: Papers and Proc. R. Soc. Tasmania 112:5 (1978)

6. Orogenesis

6.1 Fundamental Global Tectonic Relationships

6.1.1 Introduction

The present chapter concerns itself with the mechanical explanation of the evolution of the large-scale geological features upon the Earth, i. e., of the mountain ranges, ocean trenches, etc. The genesis of such features is commonly referred to as "orogenesis". The smaller features, such as faults, folds, etc. will be relegated to the chapter on "tectonophysics".

Before starting our exposition it is, however, necessary to recapitulate some fundamental global relationships which must not be violated in the mechanical models proposed. First of all, there are the facts of "plate tectonics". Next, there are some fundamental mass balance relationships concerning geosynclines and orogens. Furthermore, global mass balance relationships also apply to each plate-tectonic cycle involving magma production and subduction. In addition to a mass balance, there must also be a corresponding energy balance. Finally, last but not least, we shall consider the role of rare events in geology.

6.1.2 Plate Tectonics

6.1.2.1 Principles

The physiographic evidence adduced in Chapters 1 and 2 has led of late to the postulation of a global scheme which has been termed "plate tectonics". The literature on the subject is vast and the applications of the model to individual areas numerous.

In a treatise on geodynamics, it cannot be the task to present the various models of plate tectonics as applied to individual geological features, particularly since there is by no means a consensus regarding the conditions obtaining in each individual area. For some general reviews of the geological aspects of the subject, the reader is referred, e. g., to the works of Glen[1] or Smith[2].

---

1 Glen, W.: Continental Drift and Plate Tectonics. Columbus: Merrill 1974
2 Smith, A. G.: Tectonophysics 33:215 (1978)

In connection with geodynamics, it is the general global features of "plate tectonics" which are of importance. These can be briefly stated as follows

1. The crust and upper mantle of the Earth form a relatively brittle "tectono-sphere" (or "lithosphere") below which there is a much weaker "astheno-sphere".
2. The tectonosphere is divided into a relatively small number of plates (hence the name plate tectonics) at whose boundaries almost all tectonic, seismic, and volcanic activity occurs. The plates undergo displacements with regard to each other; these displacements cause tension zones (accreting plate margins), over- and underthrusts (consuming plate margins) and shear faults (conservative plate margins).
3. Hot material rises from the asthenosphere in the tension zones existing in mid-oceanic ridges and "hot spots" and then spreads toward the adjacent continental margins.
4. The spread of the ocean bottom from the mid-oceanic ridges is com-pensated by a subduction in the deep-ocean trenches at the continental margins whereby the material of the ocean bottom is resorbed into the mantle material.

The four propositions delineated above can be restated in the form of the postulation of a three-stage "plate-tectonic cycle", consisting in the processes of (a) rifting, (b) drifting, and (c) collision-subduction. As with all cycles, these processes may shift their positions during geologic time so that remnants of former cycles may be sought upon Earth. Inasmuch as during subduction, parts of the oceanic crust and mantle may be left near the surface, the presence of the corresponding rock types, termed "ophiolites" (petrologically very basic igneous rocks such as basalt, gabbro, and olivine-rich materials in typical sequences), has been taken as characteristic of old subduction zones. Much geological effort has been expended in locating such old subduction zones, i.e., ophiolite belts.

The above picture has generally been agreed upon by the scientific com-munity today, although Wesson[3] has pointed out a number of inconsistencies therein. Less agreement, however, exists regarding the exact delineation of the crustal plates. A broad generalization, given by Le Pichon[4] is commonly used as a base by individual researchers, but the details vary considerably. In addition to the six large plates of Le Pichon, many more smaller ones have also been postulated. Thus, the picture shown in Fig. 6 may be taken as representative today.

For geodynamics, the mechanical conditions obtaining at each stage of the plate-tectonic cycle are of importance. These will now be described in some detail.

---

3 Wesson, P. E.: J. Geol. 80:185 (1972)
4 Le Pichon, X.: J. Geophys. Res. 73:3675 (1968)

6.1.2.2 Rifting

This is, so to speak, the "beginning" of the cycle. Material wells up from the mantle, either in single "mantle plumes" giving rise to individual "hot spots"[5] which "burn holes" into the crust[6], or in linear features, giving rise to mid-ocean ridges[7] and rift valleys[8]. In connection with mid-oceanic ridges, the phenomenon of transform faulting (cf. Sect. 1.6.2) poses some problems. In particular, the persistent near-orthogonal pattern formed by the ridges and the faults defies explanation in terms of rigid plates. Lachenbruch and Thompson[9] proposed a model which is based on energy considerations: The work required to spread a given distance of ridge is much greater than that required to slip that distance of transform fault; the work becomes a minimum if the fault is normal to the ridge. Thus, the observed pattern is explained at once.

6.1.2.3 Drifting

The drifting of a plate can be described instantaneously as a rotation, in conformity with the kinematics of rigid bodies. Thus, for each plate, its (present-day) instantaneous "rotation pole" can be sought. Inasmuch as the surface of the Earth is not a perfect sphere, internal strains and stresses might be expected to arise for geometrical reasons alone. However, the stresses originating from geometry appear to be small in comparison with those caused by dynamic factors. Thus, the origin of the neotectonic stress field determined from various sources, which shows a uniformity over large areas (cf. Sect. 2.4.8), must be ascribed to prevailing dynamic conditions[10]. As these, thermal stresses due to the cooling of the lithosphere or heating in "hot spots"[11], as well as differential effects in the forces which cause the motion of the plates as a whole[12], come to one's mind.

6.1.2.4 Collision-subduction

The final stage of the plate-tectonic cycle requires the material that has welled up at the rifts to be subducted again into the mantle. Subduction takes place at the plate boundaries where a collision occurs: One of the plates descends under the other. The first question that arises here is why in some cases a big mountain range is the result of this process, in others merely a deep trench,

5 Wright, J. B. Nature (London) 244:565 (1973)
6 Delrick, R. S., Crough, S. T.: J. Geophys. Res. 83:1236 (1978)
7 Moore, J. G., Fleming, H. S., Phillips, J. D.: Geology 1974(9):437 (1974)
8 Girdler, R. W., Styles, P.: Earth Planet. Sci. Lett. 33:169 (1976)
9 Lachenbruch, A. H., Thompson, G. A.: Earth Planet. Sci. Lett. 15:116 (1976)
10 Turcotte, D. L., Oxburgh, E. R.: Nature (London) 244:337 (1973)
11 Green, H. G., Dalrymple, G. B., Clague, D. A.: Geology 6:70 (1978)
12 Illies, J. H.: Tectonophysics 29:251 (1975)

and in some, both. Le Pichon[13] ascribed the deciding factor to the *rate* of plate motion (relative velocity greater than 5 cm/year causes a trench, a smaller velocity no trench), Glen[14] to the *density* of the material involved (dense oceanic crust descends to form a trench, light continental crust is piled up into a mountain range). The problem has evidently not yet been settled.

The collision in purely continental areas evidently always gives rise to high mountain ranges, such as the Alps and the Himalaya; the observed crustal shortening values can easily be explained by such a mechanism. In the adjacent plates, the marginal stress field must have an analytic continuation which can be seen in slip lines[15] and joint orientations[16], which might explain the large-scale features of the intraplate tectonic stress field (see above).

More difficult to understand are collisions (and subduction) between an oceanic and continental plate, such as the coast of Peru[17,18]. Even the geometry (dip angle of the subducting plate 10° or 30°) is not clear here. In a descending plate, the strain must change in a certain way. The geometry on a sphere requires a compression which can be calculated[19].

Finally, a word may be added regarding the problem of displacement *rates*. Of greatest interest are the "absolute" plate motion rates (where there is some problem of even defining a fixed reference system, cf. Sect. 2.7.2) as well as the relative motion rates of the plates with regard to each other at their boundaries. Data regarding displacement rates come basically from three sources: (1) geodetic measurements, (2) magnetic anomalies, and (3) seismic strain release.

The geodetic measurements have already been discussed in Sect. 1.7.6. Accordingly, slip rates along faults are commonly of the order of from 10 mm/year to a maximum of 70 mm/year (see Sect. 1.7.6). The data on magnetic anomalies confirm the above values. Thus, a comparison of polarity epochs with magnetic bands parallel to the mid-Atlantic ridge yield spreading rates of 10 – 50 mm/year (cf. Sect. 2.7.2).

Data from earthquake strain release rates again confirm the above values[20]. Here, slip values of 8 – 20 mm/year are common; an exception is the Aleutians-Alaska region with 643 mm/year.

Finally, attempts have been made to define "absolute" values for the plate motion rates. As noted, this poses the problem of defining an appropriate reference system whose solution can be attempted by performing a boundary-velocity minimization or by considering hot-spot traces (the latter being assumed as fixed). The rms absolute translational velocities then turn out to

13 Le Pichon, X.: J. Geophys. Res. 73:3661 (1968)
14 Glen, W.: Continental Drift and Plate Tectonics. Columbus: Merrill 1974
15 Molnar, P. Tapponnier, P.: Geology 5:212 (1977)
16 Scheidegger, A. E.: Geophys. Surveys 4:233 (1981)
17 James, D. E.: Geology 6:174 (1978)
18 Berg, E., Sutton, G. H.: Phys. Earth Planet. Int. 9:175 (1974)
19 Laravie, J. A.: Geology 9:484 (1975)
20 Berg, E., Sutton, G. H., Walker, D. A.: Tectonophysics 39:559 (1977)

be about 20 mm/year for ocean – ocean plate boundaries and 15 mm/year for ocean – continent plate boundaries[21]. Other studies, based on earthquake slip vectors, lead to somewhat smaller values, i.e., 10 mm/year[22].

6.1.3 Geosynclines

As noted in the chapter on physiography, orogenetic processes have often been associated with the notion of a *geosyncline*. The term seems to have been introduced by Dana[23] when he was investigating the Apallachians, and indicates a great thickening of the sedimentary layers in a trough in the Earth's crust which is destined to become a mountain chain. Many types of geosynclines have been discerned, particularly by Stille[24], since the creation of the term. However, it should be noted that the one-time existence of a trough in the Earth's crust in places where there are now mountains is quite hypothetical although it is beyond question that, in orogenetic belts, tremendous thicknesses of sediments are present. As possible present-day geosynclines, the Adriatic Sea[25], the Timor Trough[26], and the Gulf Coast Region[27] have been quoted. Much of the *pro* and *con* for the very concept of "geosyncline" has been reviewed by Knopf[28].

Nevertheless, if it be assumed that geosynclines have existed, one immediately must ask himself what their significance is with regard to the physics of orogenesis.

One hypothesis advanced in the literature assumes that the geosynclines are *caused* by the weight of the sediments that are deposited. However, it can easily be shown that such a hypothesis is in contradiction with the principle of isostasy.

This point has been particularly emphasized by Holmes[29]. Thus, let us assume that the deposition of sediments (density $\varrho_s = 2.4$) took place in a water (density $\varrho_w = 1$) depth of $h_w = 30$ meters, and proceeded until the water depth was completely filled in. If the maximum thickness of sediments deposited in this manner be h meters, then the ultimate amount of the depression of the crust into the mantle ($\varrho_m = 3.4$) is ($h - h_w$). Isostasy then requires

$$h\varrho_s = h_w\varrho_w + \varrho_m(h - h_w) \tag{6.1.3-1}$$

21 Kaula, W. M.: J. Geophys. Res. 80:244 (1975)
22 Minister, J. B., Jordan, T. H.: J. Geophys. Res. 83:5331 (1978)
23 Dana, J. D.: Am. J. Sci. (3) 5:423 (1876)
24 Stille, H.: Einführung in den Bau Amerikas. Berlin: Bornträger 1940
25 Kossmat, E.: Paläogeographie und Tektonik. Berlin: Bornträger 1936
26 Kuenen, P. H.: Sci. Res. Snellius Exp. 5:54 (1935)
27 Bucher, W. H.: Trans. Am. Geophys. Union 32:514 (1951)
28 Knopf, A.: Am. J. Sci. 258 A:126 (1960)
29 Holmes, A.: Principles of Physical Geology. London: Nelson 1944, see p. 380 therein

or

$$h = h_w \frac{\varrho_m - \varrho_w}{\varrho_m - \varrho_s} \tag{6.1.3 – 2}$$

which yields with the above values for the constants

$$h = 30 \times 2.4\,\mathrm{m} = 72\,\mathrm{m}. \tag{6.1.3 – 3}$$

It is seen that the ultimate thickness of sediments that can be depressed in this fashion is but a tiny fraction of the sediment thicknesses surmised for the hypothetical geosynclines. Thus, the weight of the sediments cannot possibly be the *cause* of the formation of a trough.

The relation between geosynclines and isostasy has more recently been investigated by Hsu[30]. Holmes had assumed a constant thickness of the crust under a geosyncline, but it may be assumed that the thickness of the crust h_c (density ϱ_c) may vary. The isostatic relationship then requires, assuming that the deposition takes place in very shallow water ($h_w \sim 0$):

$$h_c^{\mathrm{normal}} \varrho_c = h_s \varrho_s + h_c^{\mathrm{geosyncline}} \varrho_c + h_m \varrho_m \tag{6.1.3 – 4}$$

where h_m denotes the deviation of the mantle – crust interface from the "normal" position; h_m is positive if the deflection is up, negative if it is down (for a mass deficiency above, h_m is positive). Assuming the "normal" crust to be 33 km thick, Hsu obtained values for the sediment thicknesses in relation to crustal thickness, some of which are shown in Table 10.

Table 10. Hsu's[30] calculation of isostasy in geosynclines

$h_c^{\mathrm{geosyncl.}}$ (km)	h_s (km) (for $\varrho_s = 2.4$)
5	13.8
10	11.4
15	8.9
20	6.4
25	3.95
30	1.48
33 $= h_c^{\mathrm{normal}}$	0

The above argument was further followed up by Watts and Ryan[31] who confirmed that the flexure of the lithosphere in response to a sediment load cannot possibly account for the observed sediment thicknesses, even if other than isostatic models are taken as the base for the analysis. One can say,

30 Hsu, K. J.: Am. J. Sci. 256:305 (1958)
31 Watts, A. B., Ryan, W. B. F.: Tectonophysics 36:25 (1976)

therefore, that a process other than loading must be involved in the formation of geosynclines. A review of the possibilities has been given by Bott[32].

Accordingly, some hypotheses attribute subsidence to initial uplift of the lithosphere by thermal expansion, followed by erosion and subsequent cooling causing subsidence[33]. Another suggestion is that subsidence may occur in response to increased density of basic or ultrabasic intrusives[34,35]. In a similar vein, it has been suggested that metamorphism can produce a lower density of the crustal material and thereby subsidence[36].

Finally, hypotheses of trench formation are based on the action of the tectonic stress field. There is the old idea of a tensional state in some parts of the world, which might be in conformity with the mechanics of the formation of the mid-Atlantic and East African rifts[37]. However, on continental margins a mechanism which envisages a down-bending oceanic plate dragging crustal material along to subduction would appear more likely. Thus, Helwig and Hall[38] proposed a mechanism in which a trench results from a stationary state in the subduction process; Dennis and Jacoby[39] assumed a model in which rupture occurs in a subducting plate.

6.1.4 Volume Relationships in Orogenesis

6.1.4.1 General Remarks

The plate-tectonic cycle involves mass "production" at the mid-ocean ridges, subduction at the collision regions. The first process is characteristic of "oceanic" orogenesis, the second of "continental" orogenesis.

6.1.4.2 Continental Orogenesis

The *continental* orogenetic activity is at any one time concentrated in narrow belts that form a worldwide pattern which nearly follow two circles (cf. Sect. 1.4.2). Thus, let us assume that in a single orogenetic cycle two-thirds of two great circles about the Earth are folded into mountains 2 km high and 300 km wide. The length L of a complete orogenetic system is thus

$$L = 5.3 \times 10^4 \, \text{km}. \tag{6.1.4-1}$$

The volume V of an orogenetic system is thus

$$V = 32 \times 10^6 \, \text{km}^3. \tag{6.1.4-2}$$

32 Bott, M. H. P.: Tectonophysics 36:1 (1976)
33 Sleep, N. H.: Geophys. J. R. Astron. Soc. 24:325 (1971)
34 Belousov, V. V.: J. Geophys. Res. 65:4127 (1960)
35 Sheridan, R. E.: Tectonophysics 7:219 (1969)
36 Falvey, D. A.: J. Aust. Pet. Explor. Assoc. 14:95 (1974)
37 Vening Meinesz, F. A.: Bull. Inst. R. Colonial Belge 21:539 (1950)
38 Helwig, J., Hall, G. A.: Geology 2:309 (1974)
39 Dennis, J. G., Jacoby, W. R.: Tectonophysics 63:261 (1980)

If the great mountain ranges of the Earth are assumed to be recreated every 2 million years (this corresponds roughly to the observed erosion and uplift rates), then one would have a rate of mountain building of 16 km^3/ year. One of the assumptions which is often at the basis of a theory of oro- genesis, is that mountain building is due to crustal shortening. There is no doubt that at least an apparent crustal shortening occurred in *some* places. We have stated (Sect. 1.4.2) that for most continental mountain ranges, geological estimates of shortening are of the order of

$$s_A = 50 \text{ km} . \qquad (6.1.4-3)$$

We denote this value of shortening by the subscript "A" to indicate that this is the geologically "apparent" shortening. The Alps are an exception; the observed values of crustal shortening are up to 320 km.

The crustal shortening cannot be entirely independent of the volumes in orogenesis. In fact, there must be a connection with the apparent shortening s_A across the orogenetic system. This connection is a most basic relationship in geodynamics.

By the term "apparent" it is already implied that there also should be a "true" shortening s_T. The apparent shortening is obtained by assuming that in a normal cross-section of a mountain range the length of a stratum (which is a curved line) is equal to the length of that section before it was folded, i.e., when it was flat on the ground, and comparing it with the width of the moun- tain range. The difference is the "apparent shortening" s_A. It is, however, not a foregone conclusion that the strata did not undergo an extension of their length during folding. The "true" shortening may therefore have been less than the "apparent" shortening. Let us assume that the extension of length was by the "extension" factor γ, then we have

$$s_A = \gamma s_T . \qquad (6.1.4-4)$$

Furthermore, during an orogenetic diastrophism, only the surface of the Earth is affected. Let us denote the (hypothetical) depth to which the short- ening is felt by h. Then, if the total length of the orogenetic system is again denoted by L, the volume that appears as mountains is given by

$$s_T L h = V . \qquad (6.1.4-5)$$

Replacing the hypothetical true shortening by the measurable apparent short- ening, and putting all the hypothetical quantities on one side of the equation, we obtain:

$$\frac{h}{\gamma} = \frac{V}{L s_A} . \qquad (6.1.4-6)$$

This is a basic relationship which every theory of orogenesis must fulfill. Such theories yield values for the hypothetical constants; the fact that these are not independent has usually been overlooked.

An interesting outcome is observed if the numerical values obtained earlier are inserted into the basic relationship [Eq. (6.1.4 – 6)]. One then obtains:

$$h/\gamma \cong 12 \text{ km} \tag{6.1.4 – 7}$$

which is of the order of the thickness of the crust (as defined by the Mohorovičić discontinuity); in fact, it is only a little less than the weighted mean thickness of an oceanic (5 km) and continental (35 km) (of frequency 2 : 1) crust (which would yield about 15 km). Thus, if it is assumed that γ is of the order of 1 (no significant extension of the strata), one can explain the geologically observed shortening and the volume of mountains by postulating that the apparent shortening approximately equals the true shortening and that the depth to which orogenesis is felt is determined by the Mohorovičić discontinuity. This leaves one with the difficulty of finding forces that can produce the required large shifts.

On the other hand, if γ is assumed to be significantly larger than 1 (of the order up to 10), then it is easy to find possible forces to produce the required small shortening, but the depth to which orogenesis is felt becomes much larger and the explanation of large extension factors γ itself becomes problematic.

The above argument assumes that there is no density reduction in the material affected by orogenesis. If there is such a density reduction, possibly due to rock metamorphism, say by the *metamorphosis factor* ζ, the basic equation reads [40]

$$\frac{h\zeta}{\gamma} = \frac{V}{Ls_A}. \tag{6.1.4 – 8}$$

A further interesting remark can be made with regard to the *maximum speed* with which crustal shortening can take place. If crustal shortening is assumed to be due to the sliding of the crustal parts in question over the substratum, the work necessary to produce the motion is expended against the frictional resistance occurring at the sliding surface. The resistance W to the edgewise motion (with velocity v) of a circular disc of radius c in a viscous liquid (of viscosity η) has been calculated by Lamb [41]; it is given by the following expression

$$W = 6\pi\eta R v \tag{6.1.4 – 9}$$

with

$$R = \frac{16c}{9\pi} = 0.566 c. \tag{6.1.4 – 10}$$

---

40 It may be noted, however, that rock metamorphism is generally connected with an *increase* in density; hence we have in general, $\zeta < 1$. This is generally ignored in theories of orogensis where, if metamorphism is considered at all, it is always assumed that $\zeta > 1$

41 Lamb, H.: Hydrodynamics, p. 605. New York: Dover Publ. Co. 1945

A *floating* disc experiences only half of this resistance, hence

$$W = 3\pi\eta R v = \frac{16}{3} c\eta v .$$ (6.1.4–11)

If the crustal parts are *sliding* over the substratum, a force as given by the last equation must act on these parts. This introduces stresses τ in the latter whose order of magnitude is

$$\tau = W/(2cH)$$ (6.1.4–12)

where H is the thickness of the crustal part in question. The stresses τ obviously cannot exceed the yield stress ϑ of the surface material:

$$\tau \leq \vartheta$$ (6.1.4–13)

which, in turn, imposes a limit on the speed v with which the crustal short-ening can proceed. Using the appropriate values for η, ϑ etc., one obtains (with $H = 40$ km corresponding to the depth of the Mohorovičić discontinuity in mountainous areas) for the maximum speed at which crustal parts can slide over the substratum

$$v = \frac{3W}{16c\eta} = \frac{3\vartheta 2cH}{16c\eta} = 6 \times 10^{-9} \text{m/s} = 0.18 \text{ m/year} .$$ (6.1.4–14)

Thus, in order to create crustal shortening of the order of 40 km (Rocky Mountains, cf. Sect. 1.4.2), at least about 200,000 years would be required; in order to produce the shortening of 320 km quoted for the Alps, at least about 1.8 million years are necessary. These values constitute the absolute minima of the time necessary to produce the mountain ranges in question. These values are reasonable; they rule out any speculations that mountain building might have occurred by instantaneous catastrophes. It should be noted, however, that the above argument does *not* hold if it is assumed that the substratum is moving in unison with the crust. In that case, speeds faster than those calculated above might be possible.

6.1.4.3 Oceanic Orogenesis

Second, we turn to a discussion of the possible origin of the various structural elements of the *ocean bottom*. The most prominent features, as we have mentioned earlier (Sect. 1.5.3), are the mid-ocean ridges. The volume V of the presently known system of ridges can be estimated as follows.

We take the height H of the ridges as being 3 km (above the abyssal plains):

$$H = 3 \text{ km} .$$ (6.1.4–15)

The width, d, has been measured as being on the average some 1,600 km. If the cross-section A be assumed to be triangular, one obtains:

$$A = \tfrac{1}{2} \times 3 \times 1,600 \, \text{km}^2 = 2,400 \, \text{km}^2 . \tag{6.1.4 – 16}$$

The length L of the system is approximately equal to the circumference of the Earth; hence

$$L = 40,000 \, \text{km} . \tag{6.1.4 – 17}$$

This yields for the volume V

$$V = 96 \times 10^6 \, \text{km}^3 . \tag{6.1.4 – 18}$$

If this be compared with the standard value of V for a continental orogenetic system [cf. Eq. (6.1.4 – 2)], one observes that the oceanic ridge system is roughly three times larger than the former.

As noted, "oceanic orogenesis" is connected with magma production of the mid-oceanic rifts. Present-day estimates are that basalt production of the ocean ridge systems is about 10 km³/year. Of this, 8 km³/year is intrusive. This material must again be resorbed at the subduction side of the tectonic plates.

One can make some estimates regarding the chemistry changes involved with the magma production-subduction process[42]. The heat of fusion of basalt is $4.2 \times 10^5 \, \text{J/kg}^{-1}$. The material must yield its heat content to the ocean water; a simple estimate shows that, at the given magma production rates, enough heat is liberated to boil the whole ocean every 10 million years. Since this does not occur, there must evidently be some chemical changes taking place. One generally assumes that this is a hydration; the seawater-basalt product is *spilite*. Thus, basalt is produced at the ridges, spilite resorbed at the subduction zones.

The subducted material carries with it an amount of seawater. Spilite production is 30×10^{15} g/year containing 5% of water. Hence, the water subduction is 150×10^{13} g/year. The ocean has 1.4×10^{24} g of water; – so the ocean should disappear in 10^9 years.

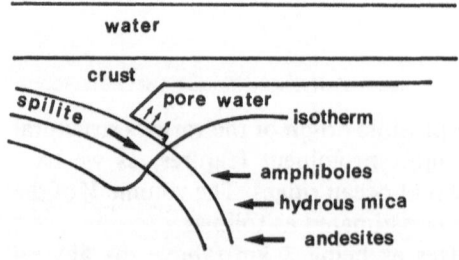

Fig. 89. Schematic view of subduction chemistry

42 Fyfe, W. S.: Magma Production, Influence of the Hydrosphere. R. Soc. Geodyn. Today 1975:25 (1975)

Since this does not happen, water must be regenerated. Figure 89 shows schematically the conditions in a subduction zone. It this zone, the water must be regenerated by dehydration. It reappears again in the andesite-type volcanism. One thus has a type of "hydrological cycle" associated with the plate-tectonic cycle: Water is absorbed and desorbed in the mantle[43].

6.1.5 Global Energetics

We have seen in Sect. 2.6.2 that the heat-flow values on the surface of the Earth are fairly constant. If a mean value of 59 mW/m² is taken to be valid for the whole surface of the Earth, one obtains an energy production of the Earth of 30.09 × 10¹² W. This energy is lost from the Earth into the Universe.

In effect, a more accurate estimate can be made if continental and oceanic areas are considered separately[44]. Assuming the average heat flux on continents of an area of 1.48 × 10¹⁴ m² to be 61.09 mW/m², one obtains an energy loss of 9.04 × 10¹² W. On the oceans of 3.62 × 10¹⁴ m² area one assumes a heat flux of 50.63 mW/m² and obtains, thus, an energy loss of 17.00 × 10¹² W. The total, then, would yield 26.04 × 10¹² W for the whole Earth.

The heat loss from volcanic effects, although spectacular on a local scale, is estimated[45] as only 20.9 × 10¹⁰ W, which is insignificant (1%) in comparison with the normal heat loss of the Earth.

Furthermore, the total tectonic energy of the whole Earth which causes measurable geodynamic displacements and which includes all of the seismic energy, is only about 0.1% of the thermal energy liberated by the Earth in a time unit. Therefore, there is no difficulty in principle in explaining the tectonic motions occurring upon the Earth as effects of the Earth acting as a heat engine[46]; the "efficiency" of that "engine" need only be 0.1%.

The above procedure only takes thermal and mechanical energy into consideration. Evidently, chemical effects can have a great influence on balance calculations. Much of the matter regarding geochemical energy relations has been reviewed in a book by Saxena and Bhattacharji[47]. However, these relations are rarely of a global nature; rather they concern individual geological phenomena, mostly those involving metamorphism. One exception to this is contained in a paper by Shimazu[48] who made an attempt to estimate the

43 For a detailed discussion of the problems involved see also: Fyfe, W. S., Price, N. J., Thompson, A. B.: Fluids in the Earth's Crust. (Developments in Geochemistry, Vol. 1). Amsterdam: Elsevier 1978
44 Williams, D. L., von Herzen, R. P.: Geology 2:327 (1974)
45 Elder, J. W.: Am. Geophys. Monogr. 8:211 (1965)
46 Goguel, J.: Philos. Trans. R. Soc. London 258:194 (1965)
47 Saxena, S., Bhattacharji, S. (eds.): Energetics of geological processes. Berlin-Heidelberg-New York: Springer 1977
48 Shimazu, Y.: J. Earth. Sci. Nagoya Univ. 12(1):85 (1977)

global energy balance including mechanical, metamorphic (chemical), gravitational (uplifting!) and thermal energies. The main conclusion of Shimazu is that, if the problem of orogenesis is regarded as that of a heat engine, the efficiency becomes several per cent and the rate of energy flux from the interior must be several times that of normal heat flow. In this, it must be noted that the chemical changes in metamorphism and volcanism account for a substantial amount of the energy requirements.

6.1.6 Rare Events

In the evolution of the Earth, some evidently abrupt events have taken place. The most impressive of these are faunal extinctions[49]. These are not necessarily of geodynamic significance, but rather (perhaps) the consequence of geodynamic rare events. Thus, the question as to be possible significance of rare events in geodynamics may be posed.

It should be noted that the time spans involved in geodynamics, compared with the time span of the human experience, are extremely large. Thus, the occurrence of an event which is improbable on a human scale, can become a near certainty in geologic time[50]. This may not only apply to faunal extinctions, but also to meteor impacts, land slides, floods, etc.

The effect of the above phenomena in geodynamics has generally been discounted, inasmuch as an "uniformitarian" view is commonly taken of the evolution of the Earth. However, with the appearance of a book by Thom[51] on the mathematical aspects of catastrophe theory, the whole problem has been taken more seriously.

Catastrophe theory proceeds from the assumption that the forces (the independent variables) governing a natural process (the dependent variable) are normally in a stationary dynamic equilibrium which can be described by a smooth surface in function space. However, an abrupt change can occur if the function describing the dependent variable becomes multivalued: Thom has shown that for processes controlled by four independent variables a maximum of seven possible surfaces of equilibrium exist; for two independent variables, only one two surfaces (a fold or a cusp) can exist. The sudden transition form one level to the other represents a "catastrophe".

Zeeman[52] has applied the ideas of catastrophe theory to a variety of processes in the physical and social sciences. Thom[51] himself suggested that the theory might also be applicable to the generation of a fault, the formation of a salt dome and the breaking of very tightly folded strata. A theory of terrestrial catastrophism as governed by underlying galactic phenomena has been proposed by Napier and Clube[53].

49 Simpson, G. G.: Bull. Am. Mus. Nat. Hist. 99:163 (1952)
50 Gretener, P. E.: Bull. Am. Assoc. Pet. Geol. 51(11):2197 (1967)
51 Thom, R.: Stabilité structurelle et morphogènese. Reading (Mass.): Benjamin 1972
52 Zeeman, E. C.: Sci. Am. 234:65 (1976)
53 Napier, W. M., Clube, S. V. M.: Nature (London) 282:453 (1979)

6.2 Theory of Plate Motions

6.2.1 General Remarks

There is little doubt that parts of the Earth's lithosphere ("plates") have undergone large displacements. Before investigating the question regarding the mechanical process which would provide the forces necessary to produce plate motions, we may note that certain general observations can be made regarding this process.

The basic data that are to be explained are as follows:

1. Drift of India since the Eocene (60 million years ago) has been 6,000 km.
2. Drift of India since the Permian (265 million years ago) has been 14,000 km.

It is obvious that the above data cannot be explained by the assumption of uniform motion, since the "short-term velocity", v:

$$v = \frac{6,000 \text{ km}}{60 \times 10^6 \text{ yrs.}} = 100 \text{ mm/year}$$

is much greater than the "long-term velocity", V:

$$V = \frac{14,000 \text{ km}}{265 \times 10^6 \text{ yrs.}} = 50 \text{ mm/year .}$$

It turns out that the actual values of the continental drifts as suggested by the factual (although still somewhat doubtful) evidence can be explained by the assumption of a *random* force[54]. This implies that the continents are subject to random drifting.

Since geological investigations lead to a *continental path*, the logical type of analysis of continental drift is the Lagrangian analysis. We assume that the mean velocity $\bar{v}$ of all the continents is zero; for the analysis, the Earth can of course be considered as flat (coordinates of a continent are then Cartesian; they may be denoted by x and y). Thus we have

$$\bar{v}_x = \bar{v}_y = 0 , \tag{6.2.1 – 1}$$

$$\overline{v_x^2} = \overline{v_y^2} = \tfrac{1}{2}\overline{v^2} = \text{const.} \tag{6.2.1 – 2}$$

A value for the velocity-square $\overline{v^2}$ can be obtained from the short-term velocity v calculated above; one obtains

$$\overline{v^2} = 10^4 \text{ mm}^2/\text{year}^2 . \tag{6.2.1 – 3}$$

Then the total displacement of a continent evolving in time is

$$x = \int_0^t v_x(\tau)\,d\tau , \tag{6.2.1 – 4}$$

---

54 Scheidegger, A. E.: Can. J. Phys. 35:1380 (1957). J. Alberta Soc. Pet. Geol. 6:170 (1958)

$$y = \int_0^t v_y(\tau) d\tau. \tag{6.2.1-5}$$

This leads to[55]

$$\overline{x^2} = \overline{y^2} = \Phi(t) \tag{6.2.1-6}$$

where $\Phi(t)$ can be calculated as follows:

$$\overline{x^2} = \overline{\left[\int_0^t v_x(\tau) d\tau\right]^2} = \overline{\int_0^t \int_0^t v_x(\tau_1) v_x(\tau_2) d\tau_1 d\tau_2}. \tag{6.2.1-7}$$

If we introduce the *Lagrangian correlation coëfficient:*

$$R_x(\tau) = \overline{v_x(t) v_x(t + \tau)}/\overline{v_x^2}, \tag{6.2.1-8}$$

relation $(6.2.1-7)$ can be written as follows:

$$\overline{x^2} = \overline{v_x^2} \int_0^t \int_0^t R(\tau_1 - \tau_2) d\tau_1 d\tau_2 \tag{6.2.1-9}$$

and, upon making a minor transformation[55]

$$\overline{x^2} = 2\overline{v_x^2} \int_0^t (t - \tau) R_x(\tau) d\tau. \tag{6.2.1-10}$$

The last relation is useful to investigate limit cases. Thus, introducing the Lagrangian autocorrelation time L_t

$$L_t = \int_0^\infty R(\tau) d\tau \tag{6.2.1-11}$$

one can for instance investigate the case

$$t \gg L_t. \tag{6.2.1-12}$$

One then obtains from Eq. $(6.2.1-10)$

$$\overline{x^2} \simeq 2\overline{v_x^2} L_t t. \tag{6.2.1-13}$$

With

$$\overline{r^2} = \overline{x^2} + \overline{y^2}, \tag{6.2.1-14}$$

this yields

$$\overline{r^2} = 2\overline{v^2} L_t t. \tag{6.2.1-15}$$

This formula for the average square of the displacement is valid for time intervals which are long compared with L_t. If we insert the estimates for the displacements and the velocity, we obtain a value for the autocorrelation time

---

55 Kampé de Fériet, J.: Ann. Soc. Sci. Bruxelles. Ser. I 59:145 (1939). See also Pai, S.: Viscous Flow Theory, Vol. II, p. 174. New York: D. van Nostrand Inc. 1957

L_t. Recalling that the displacements of the pieces of "Gondwanaland" since the Permian epoch ($t = 2.65 \times 10^8$ years ago) are on the average 14,000 km, and using the earlier value of 10^4 (mm/year)^2 for v^2, we obtain

$$L_t = \frac{\overline{r^2}}{2\overline{v^2}t} = \frac{t}{2}\frac{\overline{V^2}}{\overline{v^2}} \simeq 30 \times 10^6 \text{ years} . \tag{6.2.1-16}$$

It may be noted that this is indeed small compared with the time since the Permian (as was supposed when making the calculation). On the other hand, for very short time intervals, one obtains

$$\overline{r^2} = \overline{v^2}t^2 . \tag{6.2.1-17}$$

Therefore, if the time since the Eocene (60 million years) can be regarded as "short", the time since the Permian (265 million years) as "long", one can explain the order of magnitude of the drift of India and of the drift of the pieces of Gondwanaland simply by assuming a random drift with a velocity square of, on the average, 10^4 mm^2/year^2 and an autocorrelation time of 30 million years. It is worthy of note that the autocorrelation time is precisely of the minimum order of magnitude to divide "short" from "long" time intervals to make the theory consistent.

The above kinematical discussion of plate drift does not elucidate the nature of the random forces which are supposed to cause the drifting. For explaining the existence of such forces, several proposals have been made. A group of these are based upon the occurrence of thermal instability in some form in the mantle. There is, of course, no agreement regarding whether such a thermal instability is possible at all; − thus the problem of thermodynamical stability conditions in the Earth's mantle is the first to be discussed.

If it assumed that instability is possible, two types of mechanisms have been proposed for shifting the plates around. The first of these is based on the assumption of regular patterns of toroidal convection currents, existing either on a global or else on a more local scale. The second mechanism is based on the hypothetical existence of more or less randomly arising "plumes" in the mantle of ascending hot materials. In all these "thermal instability" theories, the lithospheric plates are envisaged as being dragged around passively by the moving material below.

In contrast to the theories based on the hypothetical existence of a thermal instability, it has been thought that the plate-tectonic cycle might be self-exciting, inasmuch as a descending slab in the subduction zones might be denser (while cooler) than the surrounding material and would thus actively descend by gravity, and pull the plate along behind it. This possibility will be discussed in detail.

Finally, it is clear that the lithosphere is affected passively by the moving plates. The "orogenesis" in the continental collision zones has been claimed to be the result of active thrusting, as well as the result of secondary gravity sliding after a pile-up of material. The last paragraphs of this section will be devoted to this problem.

6.2.2 The Problem of Thermal Stability

The first problem to be considered is that of whether a thermal instability (causing convective motion) is possible today in the mantle at all.

In this instance, we recall the basic condition mentioned in Sect. 3.2.6 which states that, in a viscous liquid heated from below, convection cells arise if the product λ of Grashoff and Prandtl numbers is at least 1709. If this number increases, the rate of fluid circulation increases until finally a state of turbulence ensues which is governed by the Reynolds number reaching 2,000. A simple calculation shows that thermal instability is indeed possible in the Earth: for $D = 500$ km, $\varrho = 3,000$ kg/m^3, $\beta = 5 \times 10^{-6}$ deg^{-1}, $\Delta T = 1,000$ deg, $c = 1.05 \times 10^3$ J/kg deg, $\eta = 10^{22}$ Pa $\cdot$ s and $k = 2.09$ W/mdeg one obtains $\lambda = 2,759$. The range of admissible values, however, is very small and the parameters used above are very uncertain.

However, it has been noted that the above argument relates only to a viscous flow law. If a creep law (e. g., of modified Lomnitz type) is used, Jeffreys[56] affirms that by and large a thermal instability cannot arise in the mantle. However, Birger[57] has shown that Jeffreys has studied only mono-tonic instability and that an oscillating thermal instability is indeed possible in a Lomnitz-type mantle. His most important result is that a perturbation in the form of a thermoconvective wave propagating with a speed of 0.1 m/year is possible; this wave corresponds to the threshold of oscillating thermal instability. These thermoconvective waves could, in effect, be the reason for the magnetic anomaly bands paralleling the mid-oceanic ridges which would, thus, *not* be indicative of the motion of the *plates*, but rather of the motion of the heat sources below.

The problem becomes even more complicated if a multicomponent fluid is considered. In this vein, Bulashevich and Khachay[58] studied the convective stability of a plane layer of a two-component viscous fluid one of whose com-ponents is a radioactive heat source. They showed that the stability of the system increases if the heat sources are moved toward the upper boundary. The system is destabilized when heavy radioactive components move toward the lower (hot) boundary. Calculations of the stability of multilayered fluids have also been made by Birger and Shlesberg[59] who showed that the critical characteristic for the onset is the Rayleigh number, and, for typical values of the parameters of the Earth's mantle, found that the dimensions of the con-vection cells must be maximally of the order of 330 km. This is less that the characteristic dimensions of the lithospheric plates (5,000 – 6,000 km). A comprehensive study of the problem of convection through a phase boundary was made by Richter[60] who found it to be possible, but noted that vertical

56 Jeffreys, H.: The Earth. 6th edn. Cambridge: Univ. Press 1975
57 Birger, B. I.: Bull. (Izv.) Acad. USSR Earth Phys. 12(3):15 (1976)
58 Bulashevich, Yu. P., Khachay, Yu. V.: Bull. (Izv.) Acad. Sci. USSR 11(12):13 (1975)
59 Birger, B. I., Shlesberg, S. G.: Bull. (Izv.) Acad. Sci. USSR 13(6):3 (1977)
60 Richter, F. M.: Rev. Geophys. Space Phys. 11:223 (1973)

temperature gradients in the Earth are incapable of causing convection in the mantle, so that horizontal ones have to be advocated as well.

The above calculations all refer to some type of cellular model of insta-bility. There is also the possibility of local "runaway" conditions which, in some way, could be linked with hypothetical mantle plumes or periodic oro-genetic "pulsations". This problem has been studied (among others) by Rice and Fairbridge[61]. By including viscous dissipation in the flow equations one obtains a cycle mechanism that could create relaxation oscillations. In this way, single recurring runaway pulses can be obtained.

6.2.3 Convection Currents

The possibility of thermoconvective instability within the Earth's mantle has given rise to the idea that regular convection currents might exist to the present day which would drag the continents around.

Convection on a global scale in a primeval Earth has already been dis-cussed in Sect. 5.3.2. With regard to plate tectonics, the idea is that such con-vection currents also exist to the present day. Gough[62] believes to see evidence for present-day single-cell mantle convection in the antisymmetric departure of the geoid from the spheroid of best fit. He argues that an up-current underlies the low-geoid and vice versa. He argues that the proposed whole mantle convective system is consistent with the rapid northwestward motion of the Pacific plate, with fast spreading of the East Pacific rise and with slow spreading of the North Atlantic Ridge. Similar claims have been made by McKenzie et al.[63].

A somewhat different view of global convection has been taken by Run-corn[64]. Following upon the ideas presented in Sect. 5.3.2, Runcorn has postu-lated a continual existence of global convection current patterns from prime-val times to the present day. Herein, the patterns are supposed to change in an intermittent way in response to a slow decrease of the mantle region. The idea is that the mantle slowly differentiates into heavy and lighter materials, the heavy substances being incorporated into the core which grows thereby. As the core grows, the mantle layer becomes thinner and the global convection currents become arranged into a greater and greater number of cells. Each time a higher mode of arrangement involving more cells is attained, the litho-spheric plates would be induced to seek new equilibrium positions so as to cor-respond to the new cellular pattern below. A new orogenetic cycle would be the consequence of this process.

---

61 Rice, A., Fairbridge, R. W.: Tectonophysics 29:59 (1975)
62 Gough, D. I.: Earth Planet. Sci. Lett. 34(3):360 (1977)
63 McKenzie, D. et al.: Nature (London) 288:442 (1980)
64 Runcorn, S. K.: Tectonophysics 63:297 (1980)

The above theories assume a uniform convection pattern on a global scale. However, the scale of the convection[65] is by no means agreed upon; the convection may or may not extend into the lower mantle. In this vein, individual convection currents (without composing them into a global pattern) have been proposed[66]. De Bremaecker[67] made some numerical calculations of pseudo steady-state models of convection in the mantle, but found that, thus far, no model satisfies the heat-flow data or yields a constant surface velocity. A more analytical approach to the problem was made by Walzer[68] who was able to show that convection "rolls" can develop below and above a heat-generating plane. Numerical calculations were also made by others[69-72].

The above models were modified by Richter and Parsons[73] who assumed two systems of convection currents of different scales superposed upon one another. Jacoby and Ranalli[74] assumed a non-linear rheology.

Finally, one may mention that laboratory models of convection currents and their effects have been made for a long time. Thus, Griggs[75] has already performed model experiments in which the currents were represented by rotating drums in a viscous liquid. A typical result, meant to represent the development of geosynclines and an accumulation of crustal material is shown in Fig. 90.

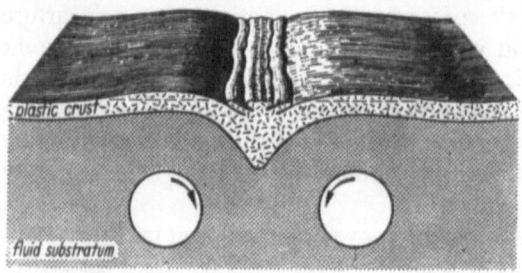

Fig. 90. Stereogram of Griggs'[75] model of convection currents showing the development of an orogenetic system

Although the outcome of the experiments has a striking resemblance with certain orogenetic phenomena, it should be noted that the thermal similarity

65 O'Connell, R. J.: Tectonophysics 38:119 (1977)
66 Torrance, K. E., Turcotte, D. L.: J. Geophys. Res. 76:1154 (1971)
67 De Bremaecker, J. C.: Tectonophysics 41:195 (1977)
68 Walzer, U.: Pure Appl. Geophys. 112:96 (1974)
69 Gebrande, H.: Ein Beitrag zur Theorie thermischer Konvektion im Erdmantel mit besonderer Berücksichtigung der Möglichkeit eines Nachweises mit Methoden der Seismologie. 159 pp. Ph. D-Diss. Univ. Munich 1975
70 Liu, H. S.: Tectonophysics 65:225 (1980)
71 Hsui, A. T.: Tectonophysics 50:147 (1978)
72 Kopitzke, U.: J. Geophys. 46:97 (1979)
73 Richter, F. M., Parsons, B.: J. Geophys. Res. 80:2529 (1975)
74 Jacoby, W. R., Ranalli, G.: J. Geophys. 45:299 (1979)
75 Griggs, D.: Am. J. Sci. 237:611 (1939)

conditions had been completely neglected. More recently, laboratory models have been made by Whitehead[76], by Davis and Walin[77] and by Jacoby[78]. However, all these, partly very suggestive, experiments suffer from the difficulty as to how far the laboratory results can be compared with nature (cf. the discussion on mechanical scaling in Sect. 7.2.3).

Against all the above theories of convection stands the very basic objection of Birger and Shlesberg[79] already mentioned in Sect. 6.2.2, according to which convection cells cannot be larger than about 330 km in diameter. For an explanation of plate tectonics, one would expect the cells to be of a size corresponding to the size of the plates; the latter, however, is one order of magnitude larger.

6.2.4 Mantle Plumes and Hot Spots

The possible thermal instability in the Earth has not only been thought to have given rise to extended convection cells, but also to single "plumes" of hot materials rising in the mantle. Such plumes would be felt as "hot spots" on the surface of the Earth. Plumes have been advocated either as the cause of linear traces of volcanism[80-82] on the Earth's surface, or as actually providing a plate-driving mechanism. Thus, Morgan[83] has suggested that a plume flow ascending at the rate of 2 m/year and then radiating out from under the hot spot could provide sufficient viscous drag (shear stresses of some ten MPa) to drive the plates.

In an analysis of the above ideas, the first problem is that of the possibility of the proposed plumes. These plumes should originate deep in the mantle inasmuch as they are thought to be quasi-stationary with regard to the lower mantle, thus leaving a "hot-spot trace" on the moving plates.

Of the theoretical attempts at elucidating (or at least modelling) the mechanism of plumes, the study of Parmentier et al.[84] is worthy of note. These authors made numerical (computer) experiments which yielded plume-like structures in cylindrical (axi-symmetric), base-heated, variable viscosity Newtonian flows. The driving mechanism is provided by density differences. The problem was formulated in terms of a pair of transport equations for vorticity and thermal energy; the Boussinesq approximation was adopted. Steady-state solutions of the transport equations were obtained by using

76 Whitehead, J. A.: Tectonophysics 35:215 (1976)
77 Davis, P. A., Walin, G.: Tellus 29:161 (1977)
78 Jacoby, W.: Tectonophysics 35:103 (1976)
79 Birger, B. I., Shlesberg, S. G.: Bull. (Izv.) Acad. Sci. USSR 13(6):3 (1977)
80 Wilson, J. T.: Philos. Trans. R. Soc. London Ser. A258:145 (1965)
81 Wilson, J. T.: Tectonophysics 19(2):149 (1973)
82 McDougall, I., Duncan, R. A.: Tectonophysics 63:275 (1980)
83 Morgan, W. J.: Bull. Am. Assoc. Pet. Geol. 56:203 (1972)
84 Parmentier, E. M., Turcotte, D. L., Torrance, K. E.: J. Geophys. Res. 80(32):4417 (1975)

finite-difference approximations. Of course, the results depend in some measure on the parameters (like viscosity and viscosity dependence on temperature) chosen. A typical result of the calculations is shown in Fig. 91; here the stream lines are given as fractions of ψ_{max} which represents the volume flow per unit volume of the convection cell. This flow pattern was then used to determine the average shear stress on the base of the lithosphere due to plume flow. The average stress (the magnitude of the radial shear stress averaged over the area) for reasonable parameters turned out to be 9×10^5 Pa and the maximum radial stress 12.8×10^5 Pa, which is by an order of magnitude smaller than the stresses required to drive the plates. Similar results were obtained in an analytical boundary-layer approach to the plume problem by Yuen and co-workers[85,86], who made a study allowing for both Newtonian and non-Newtonian temperature-dependent rheologies. Again, stresses in the upwelling flows of $10^5 - 10^6$ Pa were found.

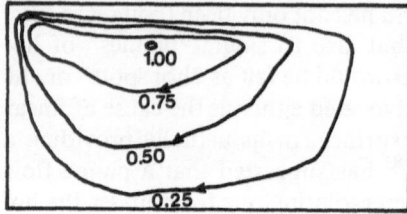

Fig. 91. Structure of base-heated thermal convection with variable viscosity. Streamlines given as fractions of ψ_{max} which represents the volume flow per unit volume of the convection cell. (After Parmentier et al.[84])

The problem of the existence of plumes has also been studied in a much more general fashion by Runcorn[87] who started from the simple assumption that, in a steady-state situation, an upward moving cylindrical plume would be isolated from the rest of the mantle by a film of negligible viscosity. If this film is removed, then the movement of the plume will be communicated to the rest of the mantle by the diffusivity equation and the characteristic diffusion time τ is

$$\tau = \varrho L^2/\eta \tag{6.2.4-1}$$

where ρ is density, η the viscosity and L the diameter of the plume. Taking $\eta \sim 10^{20}$ Pa $\cdot$ s and $\varrho = 3,000 - 4,000$ kg/m^3 shows that the plume will double its diameter in less than a millisecond. A plume *not* isolated from the rest of the mantle, thus, could not possibly exist. The analogy between plumes in the Earth's mantle and in the atmosphere is wrong[88]. In the atmosphere the velocity terms, in the mantle the viscosity terms are the significant ones in the Navier-Stokes equation. It is clear, therefore, that plumes cannot be advocated as plate-driving mechanisms.

85 Yuen, D. A., Schubert, G.: J. Geophys. Res. 81(14):2499 (1976)
86 Yuen, D. A., Peltier, W. R.: Geophys. Res. Lett. 7(9):625 (1980)
87 Runcorn, S. K.: Tectonophysics 21:197 (1974)
88 Runcorn, S. K.: Tectonophysics 63:297 (1980)

A different matter, of course, is the hypothesis of plumes (which are "slower" than those advocated as plate motors) causing passive effects on the Earth's crust. Thus, hot spots are held responsible for volcanism, uplift and plate breakings. For this purpose, studies have been made[89] of the temperature effect of a stationary plume on a moving plate above. The study of Parmentier et al. mentioned earlier was extended by these authors to an investigation whether the conditions around the islands of Hawaii could be explained by such a plume. They found that the volume flow rate of magma and melting spot size resulting from the numerical calculations agree with those deduced from observations. A simpler mathematical model of conductive heat flow from a hot spot by Birch[90] also confirms this possibility.

6.2.5 Self-generating Mechanisms

6.2.5.1 Basic Description

We consider now self-generating plate-tectonic mechanisms. Under this heading we understand those mechanisms in which the moving plate is in itself part of the acting forces. In essence, one assumes that the descending plate slab somehow sets the whole plate-tectonic cycle in motion by the action of gravity[91,92].

It is easy to show that the envisaged mechanism is indeed dynamically possible[93]. If it is simply supposed that the mantle of the Earth is resisting the motion without being in convection, then the velocity must drop from the plate velocity of maybe 30 mm/year (10^{-9} m/s) to zero at, say, 1,000 km depth. This requires a velocity gradient of 10^{-15} s^{-1}. With a viscosity of 10^{20} Pa · s, the drag needed to move the plate is 10^5 Pa. Multiplying with the width of the plate, say 10,000 km, yields a required edge force of 10^{12} N/m. This can easily be supplied if the average push along the mid-Atlantic ridge is calculated. Assuming the width of the flank of the ridge to be 1000 km and the thickness of the plate to be 100 km, the mass of this inclined plate is ($\varrho = 3000$ kg/m^3) 3×10^{14} kg/m per unit width. The slope of the plate is 0.1%, yielding a component of gravity along the plate of 0.01 m/s^2. Hence the force tending to push the plate away from the ridge is 3×10^{12} N/m per unit width of ridge which is three times more than what is needed. This shows that the mechanism is possible.

A somewhat different view was taken by Forsyth and Uyeda[94] who concluded that negative buoyancy in the descending slab could be the primary

89 Crough, S. T., Delrick, R. S.: J. Geophys. Res. 83:1236 (1976)
90 Birch, F. S.: J. Geophys. Res. 80(35):4825 (1975)
91 Hales, A. L.: Earth Planet. Sci. Lett. 6:31 (1969)
92 Isacks, B., Molnar, P.: Nature (London) 223:1121 (1969)
93 Runcorn, S. K.: Tectonophysics 21:197 (1974)
94 Forsyth, D., Uyeda, S.: Geophys. J. R. Astron. Soc. 43:163 (1975)

driving mechanism. At the opposite end of the plate, the spreading center would then become active while, in the mantle, an induced convective motion is set up. There are thus three aspects in the self-generating mechanism: the dynamics of slab descent, the dynamics of the spreading center, and the induced circulation in the mantle. Although, ultimately, the total force balance over an entire plate has to be considered, to give, so to speak, a source-span-sink model[95], it is convenient to discuss first the three aspects, mentioned above, individually.

6.2.5.2 Dynamics of Slab Descent

The descent of the lithospheric slab in the subduction zone by self-gravitational forces is thought to be the actual driving mechanism of the entire plate-tectonic cycle. The general idea of this mechanism was presented, e.g., by Isacks and Molnar[95a], as shown in Fig. 92, who also compared the hypothetical stress patterns with earthquake fault plane mechanisms (they found a reasonable conformity). The reason for the descent of the slab is an increase in density: As material rises at the spreading centers, it is hot; when it moves away it cools so that it attains its coldest stage, implying an increase in density, in the subduction zones.

Actual models of the above self-gravitational slab-pull force were calculated by Chapple and Tullis[96], by Richter[97], and by Jischke[98]. The attempts are based on various types of idealized models (vertical downgoing slab, etc.) and depend, of course, to a very large extent on the choice of the physical parameters involved. Generally, one uses the "observed" plate velocities (cf. Sect. 2.7.2) of a few centimeters per year as input and deduces therefrom the required subduction. In the paper cited above, Jischke introduced a slip zone between the descending slab and the overriding continents and thus explained the observed constancy of the dip angle for most island arc-trench systems. In view of the uncertainly of the nature of so many of the processes involved

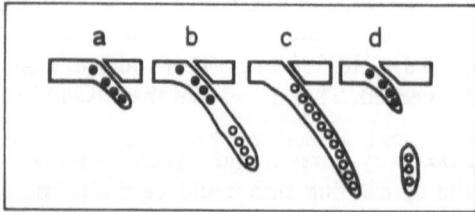

Fig. 92a–d. Sketch (after Isacks and Molnar[95a]) showing qualitatively the distribution of stresses within the downgoing slab as a function of depth extent. The *filled circles* represent down-dip extension, the *unfilled circles* represent down-dip compression; a–d temporal sequence in cycle

95 Lubimova, E. A., Nikitina, V. N.: Tectonophysics 45:341 (1976)
95a Isacks, B., Molnar, P.: Nature (London) 223:1121 (1969)
96 Chapple, W. M., Tullis, E.: J. Geophys. Res. 82(14):1967 (1977)
97 Richter, F. M.: Tectonophysics 38:61 (1977)
98 Jischke, M. C.: J. Geophys. Res. 80(35):4809 (1975)

(hydration, temperature and heat loss, density increase) there is no doubt that the numbers can be made to fit the theory.

Interesting are also estimates of the consequences of the descending slab mechanism on the state of stress that should prevail in such a slab. Figure 92, mentioned already earlier, gives a cartoon-like intuitive view of the expected stress distribution. More serious studies of this problem have been made by Smith and Toksöz[99] who made a static-elastic calculation of the stress in the descending plate, by Sleep[100] who considered fluid interactions, and by Turcotte et al.[101-103] who argued that bending stresses predominate. Again, in view of the uncertainties of the pertinent parameters, it is evidently possible to arrive at various reasonable models.

6.2.5.3 Spreading Centers

The complement to the subduction zones are the spreading centers. The rising of the fluid is thought to be caused somehow as a circulation compensating the active subduction of the slabs. The phenomena occurring at the ensuing "passive" spreading centers have been studied mainly by Lachenbruch[104-106] and his co-workers. The main aspect of these models is the calculation of the passive accretion of the walls. Possible flow patterns within the spreading "wedge" have been calculated.

The above calculations assume a completely passive role of the spreading centers. The possible "pushing" force exerted by the ridges results, in all models[107], as being very small.

6.2.5.4 Countercirculation

Evidently, the descent of material in the descending slabs and the rise of material in the spreading centers has to be compensated somehow by a circulation in the mantle. Some compensation may already take place in the very vicinity of the descending slab. Thus, Ito[108] considered a model involving a slip zone in the vicinity of subduction in which a counterflow occurs. More common are models in which the counterflow extends all the way through the region between subduction and spreading margin of a plate. Generally, it is

99 Smith, A. T., Toksöz, M. N.: Geophys. J. R. Astron. Soc. 29:289 (1972)
100 Sleep, N. H.: Geophys. J. R. Astron. Soc. 42:827 (1975)
101 Turcotte, D. L., Ahren, J. L., Bird, J. M.: Tectonophysics 42:1 (1977)
102 McAdoo, D. C., Caldwell, J. G., Turcotte, D. L.: Geophys. J. R. Astron. Soc. 54:11 (1978)
103 Turcotte, D. L., McAdoo, D. C., Caldwell, J. G.: Tectonophysics 47:193 (1978)
104 Lachenbruch, A. H.: J. Geophys. Res. 78(17):3395 (1973)
105 Lachenbruch, A. H.: J. Geophys. Res. 81(11):1883 (1976)
106 Lachenbruch, A. H.: J. Res. U.S. Geol. Surv. 4(2):181 (1976)
107 Chapple, W. M., Tullis, T. E.: J. Geophys. Res. 82(14):1967 (1977)
108 Ito, K.: J. Geophys. Res. 83:262 (1978)

assumed that the plate motion and the return flow *induce* thermal instability and convection currents (in part by shear heating) in the mantle[109].

Models of this type are generally very complicated, inasmuch as the thermodynamic conditions have to be considered simultaneously with the *external* boundary conditions represented by slab descent, plate motion and counter-circulation. The return circulation, as shear flow, may also "decouple" the plates from the rest of the mantle[110]. A very complete model has been set up by Schubert et al.[111] who investigated the mantle circulation with partial shallow return flow and its effect on the stresses within oceanic plates and on the topography of the sea floor. Inasmuch as there are very many uncertainties in the parameters involved, it comes as no surprise that the model can be made to fit the data.

6.2.5.5 Synthesis

Ultimately, as has been mentioned before, the phenomena occurring at the subduction edge, the spreading edge and below the "intraplate region" have to be taken together to form a single dynamic system. Inasmuch as the instantaneous motion of a plate is kinematically described as a rotation, the dynamic equation determining the motion is to be formulated in terms of *torques*.

The approach to this problem can be entirely phenomenological or model-theoretical. From the phenomenological standpoint, Richardson et al.[112] have tried to balance actually measured (or estimated) intraplate stresses with hypothetical rim stresses. In this fashion, they hoped to arrive at indications of the plate-driving forces. The analysis yielded that the net driving push at spreading centers is at least comparable to the net driving pull at the convergence zones, – in contrast to the prevailing opinion.

Of the theoretical approaches, the already cited study of Chapple and Tullis[113] is worthy of mention. These authors calculated theoretically from models the resulting torques on various plates, balancing intraplate and marginal stresses. A similar theoretical study was made by Richter[114], albeit only for a linear cross-section of a plate. Not surprisingly, reasonable fits with chosen data were obtained in each case. In effect, it is the opinion of the writer that it is rather pointless to construct models as detailed as those that have been attempted. If there are more indetermined parameters than data, evidently any given set of data can be made to fit. Of greater use and significance would be "twiddle calculations" which are only concerned with the orders of magnitude of the phenomena involved.

---

109 Melosh, H. J.: Tectonophysics 35:363 (1976)
110 Yuen, D. A., Schubert, G.: Geophys. Res. Lett. 4(11):503 (1977)
111 Schubert, G., Yuen, D. A., Froidevaux, C., Fleitout, L., Souriau, M.: J. Geophys. Res. 83:745 (1978)
112 Richardson, R. M., Solomon, S. C., Sleep, N. H.: J. Geophys. Res. 81(11):1847 (1976)
113 Chapple, W. M., Tullis, T. E.: J. Geophys. Res. 82(14):1967 (1977)
114 Richter, F. M.: Tectonophysics 38:61 (1977)

6.2.6 Passive Crustal Effects

6.2.6.1 Principles

The active plate-tectonic cycle of rifting, drifting, and subduction-collison will cause the crust to conform to the prevailing dynamical conditions by passive adjustment. It will be convenient to consider the three stages of the cycle separately.

6.2.6.2 Rifting

The passive response of the crust to rifting and accretion has been the subject of some speculative studies. Most of these have been rather qualitative. A typical proposed scheme is shown in Fig. 93, after Voight[115] who suggested a definite sequence of the passive evolution of a spreading ridge. Hayes[116] extended such investigations to asymmetric spreading ridges.

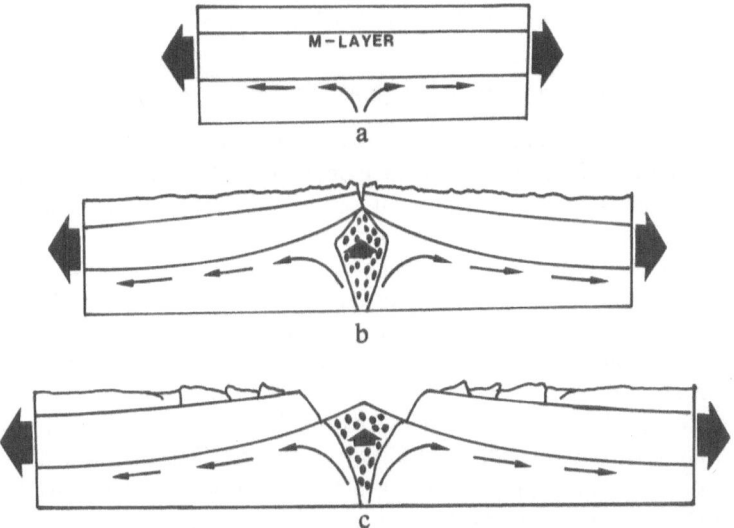

Fig. 93. Schematic cross-sections of the progressive evolution of an oceanic rift. (After Voight[115])

A numerical study, by means of a finite-element technique and assuming non-linear rheology, of the breakup of continents has been reported by Neugebauer and Spohn[117]. This study was carried to the late stage of the development of mature Atlantic-type (cf. Sect. 1.4.3) continental margins.

115 Voight, B.: Bull. Am. Assoc. Pet. Geol. 58(7):1404 (1974)
116 Hayes, D. E.: Bull. Geol. Soc. Am. 87:994 (1976)
117 Neugebauer, H. J., Spohn, T.: Tectonophysics 50:275 (1978)

6.2.6.3 Drifting

In the drifting range, it is the passive effects on intraplate stresses that are of importance. Inasmuch as a circulation below the plate is operative, stress concentrations are caused by it in the upper lithosphere. Numerical (finite-element) studies of such models have been made, e. g., by Kusznir and Bott[118]. In their model, these authors separated the lithosphere into an upper elastic and lower visco-elastic layer. They showed that the stress is concentrated into the upper elastic layer in intraplate regions. The amplification factor is equal to the ratio of thickness of lithosphere to elastic layer. This explains that moderate sources of stress at the plate margins can give rise to substantial, uniform intraplate stresses as they have indeed been observed. The occurrence of intraplate earthquakes is thereby also explained.

Generally, the plates are assumed as essentially elastic in these models. It has been pointed out by Roper[119] that this assumption may be an oversimplification and that *plastic* plates may have to be assumed. Special types of rheological conditions (such as stick-slip friction) have been advocated for the explanation of episodic volcanism and tectonism[120].

6.2.6.4 Converging Plate Margins

At converging plate margins, one of the lithospheric plates is being subducted. The visible effects on the surface depend on the circumstance whether one or both plates bear a continental or oceanic crust at the convergence zone. Depending on conditions, one may encounter a mountain range or a trench in a plate convergence zone. In this connection it may be noted that a study was made by Sorokhtin and Lobkovskiy[121] of the mechanism of drawing sediments into the asthenosphere, based on lubrication theory. They showed that pure subduction occurs without crumpling up if the thickness of the overthrusting plate does not exceed a certain critical value which depends on the viscosity of the material and turns out in practice to be about 500 m. If this thickness is exceeded, crumpling-up occurs. One has here a mechanism to explain the two types of phenomena (mountain building and trench formation) occurring at colliding plates.

Thus, depending on conditions, one will speak of subduction tectonics or of collision tectonics. A sequel to collision tectonics, with an initial pile-up of material, is gravity sliding. We shall discuss the above aspects one by one.

1. Subduction Tectonics. The tectonics of subduction zones is commonly assumed as effected by the sublithospheric convection in the region. The latter, according to one's view, may be induced by the descending slab or it

118 Kusznir, M. J., Bott, M. H. P.: Tectonophysics 43:247 (1977)
119 Roper, P. J.: Geology 1974(5):247 (1974)
120 Ito, K.: Tectonophysics 57:85 (1979)
121 Sorokhtin, O. G., Lobkovskiy, L. I.: Bull. (Izv.) Acad. Sci. USSR Earth Phys. 12(5):3 (1976)

may be primary. Thus, it is possible to calculate various models of subduction tectonics, and so rather detailed models of subduction zones have been proposed[122-125]. In particular, the development of island arcs[126] and of the Pannonian Basin[127] has been studied. "Subduction tectonics" has also been held responsible for the nappe structure in many big mountain ranges (cf. Sect. 1.4.2). Accordingly, the nappes would not be *over*thrusts at all, but passive effects in response to the lithosphere being *under*thrust under the mountains. Other passive effects in subduction tectonics include sediment subduction, subduction accretion and subduction kneading[128].

2. Collision Tectonics. If one of the converging plates consists of a continental crust, one observes in general a pile-up of material (cf. the Andes). If both plates consist of continental crust, one has a continental collision (cf. Himalaya).

A quantitative study of thermal and mechanical models of continent-continent collision zones, using finite-element techniques has been made by Bird et al.[129]. These authors modeled very realistic cases (such as the Zagros orogeny[130]), but, of course, the usual indeterminacy in the parameters of such models makes the results somewhat arbitrary. Nevertheless, realistic temperature and stress profiles were obtained.

3. Gravity Tectonics. A sequel to the crumpling-up of material in its subsequent evolution is its being affected by gravitational effects. One calls this complex of phenomena "gravity tectonics". Thus, there is no question that large-scale gravity sliding has occurred in many instances: North[131] gave a review of possible cases and Ramberg[132] showed that many pertinent forms can be produced by model experiments using a centrifuge. Straightforward scale model experiments of gravity tectonics have also been made by Guterman[133]. Thus, the folded (nappe) structure of many large mountain ranges (cf. Sect. 1.4.2) may well be the result of gravity tectonics, the created features having been uplifted after their genesis. This is a view in opposition to that considering the nappes as a passive response to a lithosphere being underthrust (see 1 above).

122 Toksöz, M. N., Hsui, A. T.: Tectonophysics 50:177 (1978)
123 Hsui, A. T., Toksöz, M. N.: Tectonophysics 60:43 (1943)
124 Davies, G. F.: J. Geophys. Res. 85:6304 (1980)
125 Wang, C. Y.: Geology 8:530 (1980)
126 Marsh, B. D.: J. Geol. 87:687 (1979).
127 Bodri, L., Bodri, B.: Rock Mech. Suppl. 9:233 (1980)
128 Scholl, D. W., Huene, R. v., Vallier, T. L., Howell, D. G.: Geology 8:564 (1980)
129 Bird, P., Toksöz, M. N., Sleep, N. H.: J. Geophys. Res. 80(32):4405 (1975)
130 Bird, P.: Tectonophysics 50:307 (1978)
131 North, F. K.: Bull. Can. Pet. Geol. 12(2):185 (1964)
132 Ramberg, H.: Gravity Deformation and the Earth's Crust. London, New York: Academic Press 1967
133 Guterman, V. G.: Tectonophysics 65:111 (1980)

According to Haarmann[134], gliding seems to be a rather rapid event which is compared with turbidity currents known to occur on inclined slopes of the ocean bottom. Consequently, the front of a gliding tongue might even detach itself from the main part (presumably owing to the momentum inherent in it) to form a detached mountain chain.

The above ideas have been taken up by a variety of authors, notably Bemmelen[135] whose view of the processes occurring is shown in Fig. 94.

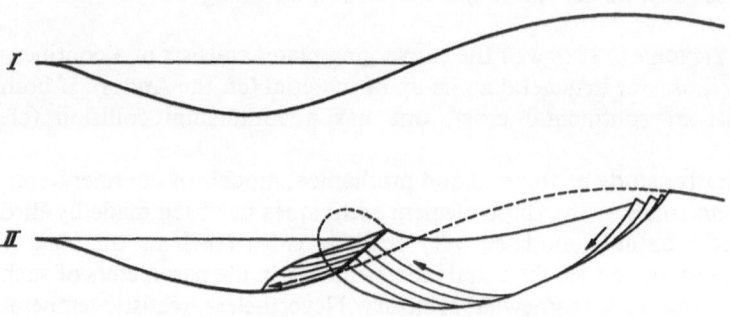

Fig. 94. Scheme of gravitational sliding. (After Bemmelen[135])

6.3 Other Theories of Orogenesis

6.3.1 General Remarks

Although the plate-tectonic theory seems to account for a wide variety of geophysical phenomena, it should be noted that there is some controversy about the actual plate-driving mechanism. Furthermore, although plate tectonics is widely accepted by Earth scientists, a significant number do not follow it to all its speculations and conclusions. Thus, a variety of alternative schemes have been proposed in the course of time. Some of these, in the light of the facts that have come to light during the past decade or so, are now only of historical interest; others, however, still have to be considered at least as possible schemes. Even if none of these "other theories" of orogenesis seem to yield possible schemes for the explanation of orogenesis on as grand and complete a scale as plate tectonics, the effects considered by some of them may still exist on a lesser scale and provide at least some contributing factors in geotectonics.

134 Haarmann, E.: Die Oszillationstheorie. Stuttgart: Enke 1930
135 Bemmelen, R. W. van: Bull. Soc. Belge Geol. 64:No.1, 95 (1955)

6.3.2 The Undation Theory

One theory of orogenesis which is still regarded in some quarters as at least a possible alternative to plate tectonics is the undation theory. This is an attempt at explaining orogenesis in terms of rhythmic oscillations (undations) of the Earth's surface. The theory seems to have been proposed for the first time by Haarmann[136] from purely physiographic reasoning; its mechanical aspects have later been analyzed mathematically by Bemmelen and Berlage[137]; recent accounts of it have been given by Belousov[138] and by Bemmelen[139].

As indicated above, the undation theory assumes that orogenesis is due to oscillations of certain parts of the globe. These oscillations constitute the *primary* cause of orogenesis. During their occurrence, certain parts of the world become elevated and others become lowered with regard to their surroundings. The elevated parts have been called *geotumors*, the lowered ones *geodepressions*. As soon as the limit of stability is exceeded, it is usually assumed that material from the top of the geotumors is supposed to slide down into the depressions. This phase of the process is called *secondary* orogenesis. By it are produced contortions and folds in the strata as the depressions are rapidly being filled in (see Fig. 95). Finally, after the depressions have become geotumors in their own turn during the continuation of the undations, the familiar picture of folded mountain ranges is created. Thus, primary orogenesis is represented by slow irregular oscillations; their periods are of the order of millions of years. Secondary orogenesis is usually assumed to be caused by gliding and sliding of material from the top of the elevations; this produces the detailed patterns of tectonic elements ("gravity

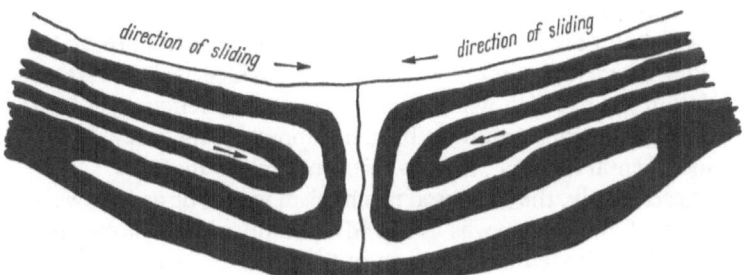

Fig. 95. Fold formation through sliding

136 Haarmann, E.: Die Oszillationstheorie. Stuttgart: Enke 1930
137 Bemmelen, R. W. van, Berlage, H. P.: Gerlands Beitr. Geophys. 43:19 (1935); see also Bemmelen, R. W. van: Proc. 21st Int. Geol. Congr. (Norden) 18:99 (1960)
138 Belousov, V. V.: Trudy Geofiz. In-ta Akad. Nauk SSSR. No. 26 (153), 51 (1955). Endeavour 17:No. 68, 173 (1958). Publ. Bur. Centr. Seismol. Ser. A 20:369 (1959). – Geotektonika. Moscow: Izd. Moscow Univ. (1976)
139 Bemmelen, R. W. van: Acta Geodaet. Geophys. Mont. 11(1 – 2):165 (1976); Tectonophysics 32:145 (1976) and Geol. Mjinb. 56(3):263 (1977)

tectonics"), much in the same fashion as gravity tectonics is also advocated as an important agent in the continental collision zones of plate tectonic theory.

The undation theory was postulated by Haarmann to account for certain physiographic features of the Earth. The uplift of Fennoscandia, the existence of submarine canyons and the fact that various parts of the world seem to have been flooded by the sea at various times, gain indeed a very natural explanation in this way.

Haarmann gives in his book a great number of examples of his interpretation of the physiographic aspects of orogenesis. He maintains, for instance, that the craters on the Moon represent the remnants of primary orogenesis[140], – no secondary orogenesis exists on the Moon. The arcuate strike of most mountain ranges and island chains is explained by the remark that most of the sliding into the depressions would presumably occur in the form of giant tongues (like avalanches); the front of these tongues would then yield the mountain arcs.

Haarmann did not proceed to a detailed investigation into the mechanics of his undation theory. He stated, however, that he expected that the primary undations would be caused by differentiation of magma in the mantle of the Earth into two components, and that the action of gravity would cause the tops of the geotumors to slide down into the depressions. The action of a load (such as ice, mountains, etc.) upon the Earth's surface is by Haarmann entirely discounted.

There have been other speculations regarding the origin of forces which could make the Earth pulsate, such as that of Havemann[141] postulating that this behavior be caused by periodic accumulation and escape of radioactive heat, or by internal pulsations of the Earth's core[142]. Since this is much less specific than the theory mentioned earlier, it is wide open to criticism.

We shall now investigate what the possible causes are that might give rise to the *primary* motions of the Earth's surface which were assumed in the undation theory. Following Haarmann, Bemmelen and Berlage[143] postulated that the undations could be brought about by the uneven differentiation of "fundamental magma" into heavier and lighter material.

Accordingly, the proposed mechanism would be as follows. Originally, the surface of the Earth was flat and a homogeneous layer of "fundamental magma" (envisaged as a mixture of Si-Al and Si-Mg compounds) was resting upon some "basement layer" (possibly identified with the doubtful Birch discontinuity at 900 km depth). In a cross-section through the upper part of the Earth, the situation was therefore as shown in Fig. 96a. Herein, the curvature of the Earth has been neglected, the x-axis represents the bottom boundary of the fundamental magma, the z-axis is the vertical.

---

140 This is, of course, completely at variance with the current view according to which the craters on the Moon have been formed by meteorite impact
141 Havemann, H.: Trans. Am. Geophys. Union 33:749 (1952)
142 Dubourdieu, M. G.: C. R. Acad. Sci. D 277:1109 (1973)
143 Bemmelen, R. W. van, Berlage, H. P.: Gerlands Beitr. Geophys. 43:19 (1935)

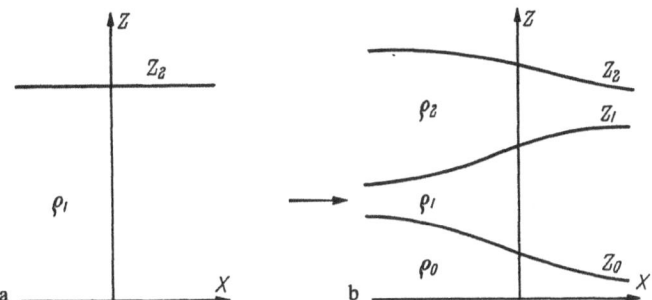

Fig. 96a, b. Magma differentiation from one into 3 layers as envisaged in the undation theory

After the postulated differentiation of the magma into Si-Al and Si-Mg has proceeded for a while, it may be assumed that the picture shown in Fig. 96b results. This picture implies that the speed of differentiation may vary with the geographical location. Furthermore, it is assumed that the density of the fundamental magma is larger than the mean density of the end products, i.e.,

$$\frac{1}{a+b}(a\varrho_0 + b\varrho_2) < \varrho, \tag{6.3.2-1}$$

if the differentiation into material 0 and 2 takes place in the ratio $a:b$. Thus, with continuing differentiation, the surface z_2 is pushed to a higher level than was the case originally. During the differentiation, isostasy is maintained with regard to the material below $z = 0$, but there will of course be a tendency to smooth out the differences in height of the surface z_2 by lateral displacements.

The heights z_0, z_1, z_2 of the corresponding interfaces above the datum-level $z = 0$ are functions of x and t. The speed of differentiation of the fundamental magma depends on the pressure and on the temperature; if this speed is set equal to an undetermined function $F(z_0, z_1, z_2)$, one obtains for the change of thickness of each of the layers under consideration:

$$\left.\begin{aligned}
\varrho_2 \frac{\partial(z_2 - z_1)}{\partial t} &= c_1 F, \\[2ex]
\varrho_1 \frac{\partial(z_1 - z_0)}{\partial t} &= -(c_0 + c_1)F, \\[2ex]
\varrho_0 \frac{\partial z_0}{\partial t} &= c_0 F.
\end{aligned}\right\} \tag{6.3.2-2}$$

Here, c_0 and c_1 are constants. One thus obtains for the vertical component of the change of each interface

$$\frac{\partial z_2}{\partial t} = \left(\frac{c_1}{\varrho_2} - \frac{c_0 + c_1}{\varrho_1} + \frac{c_0}{\varrho_0} \right) F,$$

$$\frac{\partial z_1}{\partial t} = \left(\frac{c_0}{\varrho_0} - \frac{c_0 + c_1}{\varrho_1} \right) F,$$

$$\frac{\partial z_0}{\partial t} = \frac{c_0}{\varrho_0} F.$$

$(6.3.2-3)$

The next task is to calculate the disappearance of the differences in level in the surface z_2 through lateral flow of the rocks. It is quite hopeless to attempt this by recurrence of the fundamental equations of a Maxwell body, and hence Bemmelen and Berlage made the following intuitive assumption

$$\frac{\partial z_2}{\partial t} = \alpha_2 \varrho_2 \frac{\partial^2 z_2}{\partial x^2} + \beta_2 (\varrho_1 - \varrho_2) \frac{\partial^2 z_1}{\partial x^2} + \gamma_2 (\varrho_0 - \varrho_1) \frac{\partial^2 z_0}{\partial x^2},$$

$$\frac{\partial z_1}{\partial t} = \alpha_1 \varrho_2 \frac{\partial^2 z_2}{\partial x^2} + \beta_1 (\varrho_1 - \varrho_2) \frac{\partial^2 z_1}{\partial x^2} + \gamma_1 (\varrho_0 - \varrho_1) \frac{\partial^2 z_0}{\partial x^2},$$

$$\frac{\partial z_0}{\partial t} = \alpha_0 \varrho_2 \frac{\partial^2 z_2}{\partial x^2} + \beta_0 (\varrho_1 - \varrho_2) \frac{\partial^2 z_1}{\partial x^2} + \gamma_0 (\varrho_0 - \varrho_1) \frac{\partial^2 z_0}{\partial x^2}.$$

$(6.3.2-4)$

Here, α, β, γ are all positive constants indicative of the "viscosity" of the substance under consideration. The set of Eq. $(6.3.2-4)$ has been chosen because is has essentially the form of a set of diffusivity equations. It is well known that, if in a system the microscopic resistance to motion is everywhere proportional to the velocity, one ends up, macroscopically, with a diffusivity equation. Proportionality between resistance and flow velocity is, however, characteristic for bodies of a Maxwell type and hence the basic structure of the system $(6.3.2-4)$ of equations is justified. The factors ϱ_2, $(\varrho_1 - \varrho_2)$ etc. are conditioned by the requirement that the corresponding terms must vanish if there is no density difference above and below the interface.

If one combines the motion resulting from the differentiation with that resulting from the smoothing-out of the surface, one obtains finally

$$\frac{\partial z_2}{\partial t} = \left(\frac{c_1}{\varrho_2} - \frac{c_0 + c_1}{\varrho_1} + \frac{c_0}{\varrho_0} \right) F + \alpha_2 \varrho_2 \frac{\partial^2 z_2}{\partial x^2} + \beta_2 (\varrho_1 - \varrho_2) \frac{\partial^2 z_1}{\partial x^2} + \gamma_2 (\varrho_0 - \varrho_1) \frac{\partial^2 z_0}{\partial x^2},$$

$$\frac{\partial z_1}{\partial t} = \left(\frac{c_0}{\varrho_0} - \frac{c_0 + c_1}{\varrho_1} \right) F + \alpha_1 \varrho_2 \frac{\partial^2 z_2}{\partial x^2} + \beta_1 (\varrho_1 - \varrho_2) \frac{\partial^2 z_1}{\partial x^2} + \gamma_1 (\varrho_0 - \varrho_1) \frac{\partial^2 z_0}{\partial x^2},$$

$$\frac{\partial z_0}{\partial t} = \frac{c_0}{\varrho_0}F + \alpha_0\varrho_2\frac{\partial^2 z_2}{\partial x^2} + \beta_0(\varrho_1 - \varrho_2)\frac{\partial^2 z_1}{\partial x^2} + \gamma_0(\varrho_0 - \varrho_1)\frac{\partial^2 z_0}{\partial x^2}.$$

$$(6.3.2-5)$$

It is possible to integrate this for some special cases. First, let us consider the case where the upper layers of the Earth are homogeneous so that $\varrho_0 = \varrho_1 = \varrho_2 = \varrho$, and where there is no differentiation of magma. Then one has

$$\frac{\partial z}{\partial t} = \alpha\varrho\frac{\partial^2 z}{\partial x^2}.$$

$$(6.3.2-6)$$

This is diffusivity equation which describes, for instance, the disappearance of an initial trough. Let the trough be represented by

$$z = -he^{-\mu^2 x^2},$$

$$(6.3.2-7)$$

then the solution of $(6.3.2-6)$ yields

$$z = \frac{-h}{\sqrt{1 + 4\alpha\varrho\mu^2 t}}\exp\left(-\frac{\mu^2 x^2}{1 + 4\alpha\varrho\mu^2 t}\right).$$

$$(6.3.2-8)$$

This confirms the earlier inference that the rheological equation is such that the material behaves like a Maxwell liquid. Thus, if a trough is impressed on the Earth by some mechanism, it will disappear asymptotically with a characteristic time interval determined by the parameter α. If such a trough, i.e., geosyncline, should be filled in (partly or wholly) with light sediments, then we have here a mechanism whereby a mountain range can eventually be created.

The above calculations rest upon the validity of the assumption [used especially in writing down Eq. $(6.3.2-2)$] that, once primary differentiation of the magma into its two components (whatever they be) has started, the latter accumulate *immediately* into their respective layers. This implies that the diffusion of the two components through the original magmatic layer is very rapid. This, however, is very doubtful. The speed of diffusion can be estimated as follows. The viscosity η of the layer in question is of the order of 10^{21} to 10^{22} Pa $\cdot$ s. If we assume a spherical droplet of radius a with a density $\varrho_1 = 1.1\varrho_0$ (i.e., 10% denser than the surrounding fluid), one can calculate the speed v with which it would drop. The latter should be indicative of the order of magnitude of the speed of diffusion of the heavier fluid through the original one. The resistance R of the fluid to the droplet is given by Stokes' law [see Eq. $(3.2.3-6)$]:

$$R = 6\pi a\eta v.$$

$$(6.3.2-9)$$

The force F acting of the droplet effecting its sinking is

$$F = \frac{4}{3}\pi a^3 0.1 \varrho_0 g$$

$$(6.3.2-10)$$

where g is the gravity acceleration. Thus, from $F = R$

$$v = 0.1 \frac{2}{9} \frac{a^2 \varrho_0}{\eta} g ; \qquad (6.3.2 - 11)$$

thus, with $a = 10$ mm (as an upper limit; v increases with a), $\varrho_0 = 3,000$ kg/m^3, $\eta = 10^{22}$ Pa $\cdot$ s, one has

$$v \cong 10^{-23} \text{ m/s} . \qquad (6.3.2 - 12)$$

The time T for the droplet to fall through, say, 100 km (which may be taken as about one half of the thickness of the original layer) is

$$T = 10^{28} \text{ s} = 3 \times 10^{20} \text{ years} . \qquad (6.3.2 - 13)$$

This is about 10^{11} times the total estimated life span of the Earth. It seems therefore extremely doubtful wheter the postulated mechanism could have any real significance, quite apart from any chemical considerations.

The above mechanism is that most commonly quoted in connection with the undation theory. The above theory is somewhat oversimplified; it has been elaborated upon by Shimazu[144], Aslanyan[145], Subbotin[146,147] and others. In all these attempts, some kind of phase or chemical change with subsequent "magma differentiation" is assumed. It is difficult to see, however, how any attempt along these lines could be upheld in view of the slowness of the diffusive processes in the Earth.

In addition to the above investigations, causes completely different from those envisaged above have also been proposed in connection with the "primary" undations. Thus, Bemmelen[148] envisaged some kind of a magmatic diapirism and Lyustikh[149] assumed that magma rises along planetary fractures. Trechmann[150] assumed a cosmic upward pull and Bülow[151] assumed vertical currents in the mantle so that the motion of the crust would be like that of ice on troubled water. Some authors thought to put the energy source for the primary undations entirely into the crust. Thus, Contant[152] attributed it to distilled gases from organic matter in sediments, and Odhner[153,154], followed by Malaise[155,156], to the constriction caused by climatic temperature

144 Shimazu, Y.: J. Earth. Sci. Nagoya Univ. 7:91 (1959)
145 Aslanyan, A. T.: Tr. Upr. Geol. Okhr. Nedr. 2:141 (1959)
146 Subbotin, S. I.: Soobshch. Akad. Litovsk. SSSR. Inst. Geol. Geogr. 5:5 (1957)
147 Subbotin, S. I.: Heol. Zhur. Akad. Nauk Ukr. SSSR. 22:No.5, 3 (1960)
148 Bemmelen, R. W. van: Madj. Ilmu Alam Untuk Indones. 113:1 (1957)
149 Lyustikh, E. N.: Bull. (Izv.) Akad. Nauk SSSR. Ser. Geofiz. 1960:No.3, 402 (1960)
150 Trechmann, C. T.: Geol. Mag. 95:426 (1958)
151 Bülow, K. v.: Geotektonisches Symposium zu Ehren von Hans Stille (ed. Lotze), publ. by Dtsch. Geol. Ges., p. 45, 1956
152 Contant, H.: C. R. Cong. Soc. Sav. Paris, 70^e Congr., Alger, Sect. Sci. 144 (1954)
153 Odhner, N. H.: Geogr. Ann. Stockh. 16:109 (1934)
154 Odhner, N. H.: Ark. Mineral. Geol. 2:No.24, 353 (1958)
155 Malaise, R.: Atlantis. Stockholm: Kalmar 1951
156 Malaise, R.: Geol. Foeren. Foerh. 79:195 (1957)

variations *("constriction theory")*. Finally, as has been mentioned already, Dubourdieu[157] ascribed the cause of the geoundations to some primary pulsations in the Earth's core. Needless to say, these various theories are mostly speculations as a true analysis of their quantitative aspects has yet to be supplied.

Nevertheless, the undation theory yields an alternative to plate tectonics, inasmuch as there, too, some sort of "primary undation" is postulated in the form of continental collisions and oceanic rifts; the subsequent evolution of the crust by gravity tectonics is similar in the two thories. On the other hand, the overwhelming evidence from paleomagnetism indicating large continental displacements is not accounted for in the undation theory: The stand has to be taken that all this evidence is somehow untrustworthy.

In addition to advocating the geo-undations as the primary cause of orogenesis, there is, of course, the possibility that the latter occur in *addition* to other geotectonic phenomena (such as plate motions). It would then be much easier to account for the various types of effects than by the undation theory alone.

6.3.3 The Contraction Hypothesis

One of the oldest and most beloved theories of orogenesis was the contraction hypothcsis. Unfortunately, it can hardly be maintained any longer as a possible scheme, for reasons which will be shown below. However, inasmuch as it has held such a prominent position amongst mountain-bulding theories for a very long time, it may be worth while to review it briefly here.

The contraction hypothesis of orogenesis goes back at least to the time of Descartes. In its modern form, it is usually presented as stated by Jeffreys[158]. Accordingly, it is assumed that the Earth began as a hot, celestial body. Early in its history, it differentiated into an iron core and an essentially silicate mantle. The mantle solidified outward from its base at the liquid iron core and has since been cooling by conduction without convection currents. From the center of the Earth to within about 700 km of the surface there has not been time since the Earth solidified for any appreciable cooling or change in volume to have taken place. Within the region from about 700 to 70 km, cooling by conduction is taking place and hence this layer is contracting and being stretched about an unchanging interior. Hence it is in a state of internal tension.

Near the surface the rocks have already largely cooled so that they are in thermal equilibrium with the heat provided by solar radiation. They are therefore not changing very much in temperature and the cooling and contraction of the layer or shell beneath them puts the outermost shell into a state of internal compression above a level of no strain at 70 km depth.

---

157 Dubourdieu, G.: C. R. Acad. Sci. D277:1109 (1973)
158 Jeffreys, H.: The Earth. London: Cambridge University Press 1929

Thus, the contraction hypothesis divides the Earth upon grounds of thermal and mechanical behavior into three shells: the non-contracting part of the Earth below a depth of about 700 km, the contracting part of the mantle above 700 km and below the level of no strain at about 70 km, and the "exterior" which is crumpling up due to the contraction below. These shells are not dependent upon the Earth's composition and hence should not be confused with such terms as core, mantle, and crust. As the Earth cools, the boundaries between the shells move deeper into the Earth. The stress state assumed in the contraction hypothesis is shown in Fig. 97.

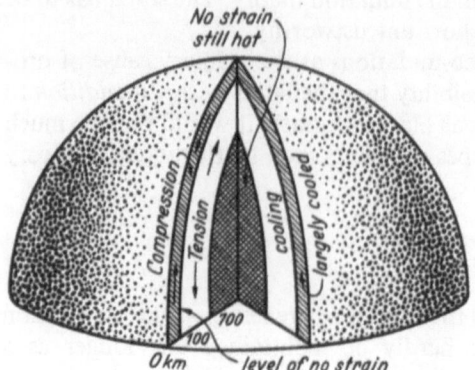

Fig. 97. Stresses in the Earth according to the contraction hypothesis

The level of no strain has been taken at 70 km depth, which is Jeffreys' value. However, there is nothing magic about this number and it may be preferable to put it, say, at 140 km depth which would correspond to Gutenberg's low velocity layer (cf. Sect. 2.1.4).

At first glance, the contraction hypothesis explains many physiographic facts about the Earth. The cross-section of an island arc can be envisaged as the outcome of a deep faulting process in the contracting Earth. Furthermore, the phenomenon of crustal shortening in continental mountain ranges obtains a most natural explanation.

However, an inspection of the exposition given above shows that the contraction hypothesis is essentially a "fixistic" theory; i.e., plate motions or continental displacements are not possibly accounted for. As in the undation theory, all the paleomagnetic evidence indicating continental drift has therefore to be discounted. This, however, is not the only problem. More serious is the fact that the tectonic stress state envisaged in the contraction hypothesis is completely uniform over the whole Earth: A compression above, and a tension below the level of no strain. Such a homogeneous stress state is, however, not what has been observed. The various ramifications of the contraction hypothesis, thus, have to be regarded today as only of historical interest.

The above remarks, however, do not mean to preclude the possibility that there has been *some* contraction of the Earth in the course of geological history: Such contraction, though, would be *in addition* to more significant orogenetic effects. There is, at the present time, no way of telling whether or not such a moderate contraction might have occurred.

6.3.4 Expansion Hypothesis of Orogenesis

It is a most notable feature of most theories of fracture of materials that the predicted failure patterns are independent of the sign of the applied stresses. With the one exception of the possible occurrence of chasms under tension, the expectable failure patterns in an expanding Earth would therefore be the same as those in a contracting Earth.

In an expansion theory, one would assume that the Earth consists of several layers, the topmost representing the crust, and the next one down representing an expanding layer. The orogenetic effects would originate in this expanding layer; the crust, in turn, would yield under the stresses which are created. According to the above remarks all that has been said regarding plastic and brittle failure patterns in the contraction theory of orogenesis, can also be said with regard to an expansion theory of orogenesis. The formation of island arcs, their position on the Earth's surface, the junctions between orogenetic elements, etc. could equally be explained by assuming a slight expansion in a layer beneath the crust. In addition, there is the possibility of deep chasms occurring in an expansion theory which could be thought of as an explanation of deep ocean trenches.

In this connection, it may be interesting to note that most oceanic features can indeed be explained by the assumption of tensional forces. Thus, if only oceanic orogenesis be considered, some support might be gained for an expansion hypothesis orogenesis. Because much knowledge about oceanic features has been collected rather recently, this has contributed to a revival of the expansion theory [157-161].

However, there is one fundamental difficulty. This is that in an expansion theory, it is no longer easy to account for the observed crustal shortening as there is no reason for the "skin" of an expanding sphere to become crumpled up. It would therefore appear that all the expansion could create is a pattern of fissures through which the liquid "magma" below could rise to cause mountains. There seems to be no possibility of explaining nappes and similar phenomena.

In spite of these difficulties, Egyed[160] followed the hypothesis of an expanding Earth into some detail, claiming that it could account for many

---

159 Egyed, L. wrote many papers on this subject, some of which have already been cited in Sect. 5.4.3. For other authors, also cf. Sect. 5.4.3

160 Egyed, L.: Acta Geolog. Magyar Tudom. Akad. Föld. Közl. 4:43 (1956)

161 Steiner, J.: Geology 5:313 (1977)

observed facts. According to Egyed, the tension in the crust would affect the modulus of rigidity which, in turn, would produce warping of parts of the crust. In this fashion, one again arrives basically at an undation theory with all its drawbacks. In an expansion-undation theory, however, it seems doubly impossible to obtain the crustal shortening necessary for mountain formation, because of the tensional character of the stresses involved. If a slight expansion of the Earth's interior did occur, it may be responsible for such features as rift valleys, the mid-Atlantic rift, etc., but certainly not for the folding-up of mountains.

Thus, the main problem of the expansion theory, besides its being a fixistic theory unable to account for continental displacements, is again, as in the contraction theory, that it requires a uniform geotectonic stress state over the entire Earth. Such a postulated uniformity of the stress state runs counter to experience.

Nevertheless, the expansion theory has found recent advocates who base their arguments mainly on geological grounds, claiming that many features (particularly on the Southern Hemisphere) cannot be explained by plate tectonics[162-164]. However, it is not clear at all whether the physiographic evidence indeed requires expansion. Thus, McElhinny et al.[165] found that paleomagnetic data limit the expansion to less than 0.8%. The principal mechanical and physical difficulties of a theory postulating substantial expansion of the Earth have already been discussed in Sect. 5.4.3.

6.3.5 Membrane Tectonics

As has been outlined in the earlier sections of this book, there is overwhelming evidence that there have been substantial changes in the latitude positions of parts of the lithosphere during geologic time. This, when one is referring to one "observational" area only, may also be expressed by saying that "the pole has wandered". For each relative position of the pole with regard to the area under consideration, the latter has to adjust itself to the equilibrium figure of the Earth. Inasmuch as the Earth is not a sphere, this latitude change causes inner strains; the study of these strains and their geodynamic effects has been called "membrane tectonics".

Thus, if the polar axis moves through an angle Θ, i.e., from the position represented by PQ to the position $P'Q'$ in Fig. 98, the crust of the Earth must adjust itself to the new shape. It is a reasonable conjecture that this adjustment will cause stresses that might have orogenetic significance. In the most

162 Steiner, J.: Geology 5:313 (1977)

163 Smith, P. J.: Nature (London) 268:200 (1977)

164 Carey, S. W.: Expanding Earth Symposium, Sydney, Feb. 11 – 13, 1981. Proceedings to be published in Tectonophysics

165 McElhinny, M. W., Taylor, S. R., Stevenson, D. J.: Nature (London) 271:316 (1978)

extreme case the pole would move from its original position on the former equator through 90° normal to the latter. That would mean that the former equator becomes a meridional circle which would imply a shortening due to the ellipticity of the Earth of

$$\frac{10018758 \text{ m}}{10001962 \text{ m}} = 1.00168 \qquad\qquad (6.3.5-1)$$

or of 17 km for a quadrant. Since an orogenetic system consists of worldwide mountain ranges of the shape of roughly two great circles, it will presumably be intersected four times by any other great circle so that the above-mentioned shortening (17 km) would be that available for each mountain range. From various sources of evidence, it must be assumed that two orogenetic cycles took place in the time interval since the Devonian in which the pole moved, maybe, through 80°. This leaves a shortening of about 8 km per mountain range. However, it should be noted that the above value is for *maximum* shortening, valid only for that meridian which goes through the new pole at right angles to the direction of the polar shift. If we consider *any* quadrant, the available shortening is on the average again about halved which leaves approximately 4 km for a mountain range:

$$s_T = 4 \text{ km} . \qquad\qquad (6.3.5-2)$$

This seems very little and it is seen that latitude changes cannot be the cause of mountain building. They cannot be the cause of plate motions, either. Nevertheless, although it has thus been shown that latitude changes are inadequate to cause orogenesis, they might still give rise to shearing of the crust on a global scale. This question has been analyzed by various people [166–170].

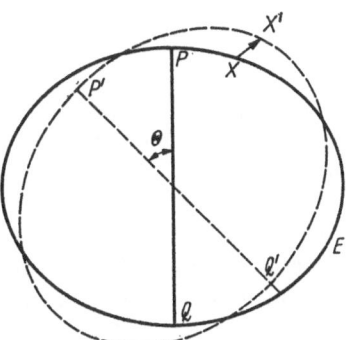

Fig. 98. Geometry of the shift of the Earth's axis of rotation

166 Vening Meinesz, F. A.: Trans. Am. Geophys. Union 28:1 (1947)
167 Milankovitch, M.: Kanon der Erdbestrahlung und seine Anwendung auf das Eiszeitenproblem. Belgrade: Éd. spéc. Acad. R. Serbe, Tome 133, 1941
168 Schmidt, E. R.: Földtani Közlöny 18:94 (1948)
169 Scheidegger, A. E.: J. Alberta Soc. Pet. Geol. 6:266 (1958)
170 Turcotte, D. L.: Geophys. J. R. Astron. Soc. 36:33 (1974)

Thus, referring to Fig. 98, let the original state of the Earth be represented by the ellipse drawn in *solid lines*, and the state corresponding to that after a shift of the polar axis through the angle Θ, by *dotted lines*. This picture implies that the axis of rotation and the geoidal axis always coincide corresponding to an infinitely great adaptability of the Earth to the prevalent dynamic forces. During the displacement, each point of the Earth shifts to a new position; such a shift is indicated for a point on the surface by an *arrow* pointing from X to X'.

What are the conditions governing the displacement? It is evident that there are two types of conditions necessary which can be designated as "kinematic" and as "dynamic". The kinematic condition requires that the change of shape of the total Earth be such that the original ellipsoid changes into another ellipsoid, the dynamic condition requires that the state before and after the displacement be an equilibrium state.

The kinematic condition can be taken care of as follows. In the original state we describe each point on the Earth's surface by its (geocentric) longitude α and corresponding latitude φ (or polar distance δ instead). It is to be understood that these coordinates represent an ellipsoid of revolution of a given ellipticity. The position of the pole is given by $\delta = 0°$. A similar set of coordinates α', φ' (or δ'), then, corresponds to the deformed state with the new polar axis $P'Q'$ instead of PQ (cf. Fig. 98). Again, these coordinates are presumed to describe an ellipsoid of revolution of the same ellipticity as that above. The new position of the pole is given by $\delta' = 0°$. In terms of the present scheme, the deformation is thus given by the following transformation equations

$$\alpha' = \alpha'(\alpha, \delta)\,, \tag{6.3.5 – 3a}$$

$$\delta' = \delta'(\alpha, \delta)\,. \tag{6.3.5 – 3b}$$

Thus, if the coordinates α, δ are taken as variables over the surface of a unit sphere, the shift of the polar axis represents a mapping of that sphere upon itself.

The dynamic condition requires that the transformation expressed by Eq. (6.3.5 – 3a/b) leads from an equilibrium state to another equilibrium state. In general, any possible set of equations of the above type will prescribe a set of *strains* in the Earth. However, only such strains are permissible whose associated *stresses* satisfy equilibrium conditions.

Upon inspection of the above discussion, it is evident that one is faced with (a) a set of constraints for the strains, and (b) with a set of constraints for the stresses. It is therefore obvious that the problem of determining the deformation of the Earth's crust due to polar wandering cannot possibly be solved unless some stress-strain relations are assumed. These, however, constitute the "rheological condition" for the material in question and the difficulties for estimating the latter for the various layers of the Earth are well known.

From the above considerations it is clear that the orogenetic significance of a shift of the polar axis can be determined only if a certain rheological condition is assumed for the crust of the Earth. The two extreme conditions that have been investigated are (a) the assumption that the crust of the Earth is ideally elastic (Vening Meinesz), and (b) the assumption that the crust of the Earth has no strength at all (Milankovitch). These are, of course, the two logical extremes which one might attempt to investigate.

Interestingly enough, it turns out that, if only the lines of maximum shear for an infinitesimal polar shift are wanted, the rheology of the material is of no import. The picture of the shear lines[171] is always as shown in Fig. 99. (Conformal mapping; the infinitesimal shift is from the top point in the picture in the plane of projection.) The maximum shear occurs on that great circle which contains the direction of polar shift at a distance of 45° from the pole.

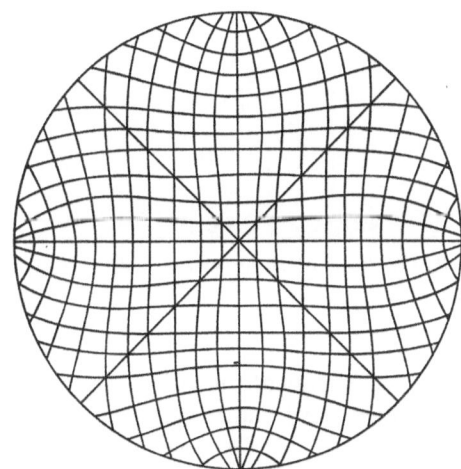

Fig. 99. Lines of maximum shear induced by an infinitesimal shift of the pole in the plane of projection. (After Scheidegger[171])

It remains to correlate tectonic features on the Earth's surface with the possible shear patterns. The position of the shear pattern with regard to the Earth's surface is given firstly by the present position of the pole, and secondly by the tangent to the polar path at the present time. Whereas the present position of the pole is quite definite, the direction of the tangent to the path is not. However, assuming this to be along the 180° meridian, a comparison can be made with Menard's fracture zones (Sect. 1.5.4). This comparison is shown in Fig. 100. It is really quite amazing how well Menard's fracture zones fit into this net. In spite of this success, the explanation of the zones as being caused by polar wandering should not be regarded as more than a reasonable possibility. One of the chief drawbacks of such an explana-

171 Scheidegger, A. E.: J. Alberta Soc. Pet. Geol. 6:266 (1958)

tion is that any shear net produced by a polar shift would be perfectly symmetrical. Fracture zones similar to those found by Menard should therefore be found in other parts of the world. Somes zones found by Hess[172] near China fit the pattern very well. It is not known whether others exist.

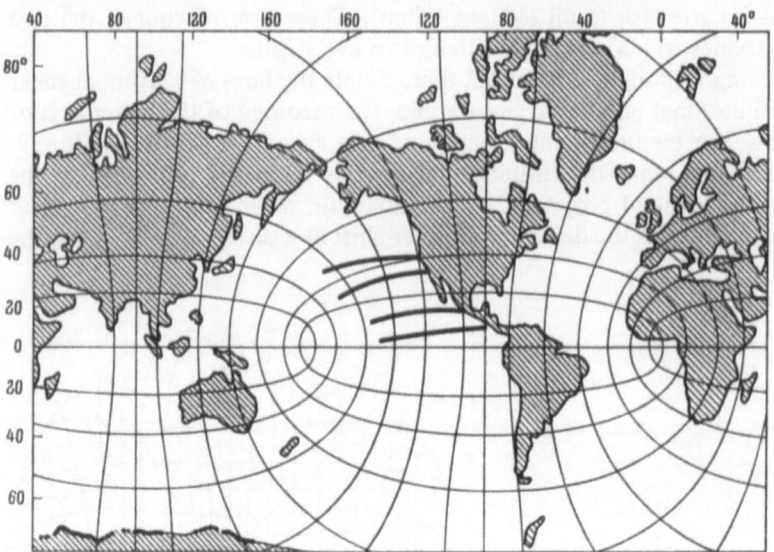

Fig. 100. Shear net in a weak Earth, assuming the polar path to be tangent to the 180° meridian. Comparison with Menard's fracture zones. (After Scheidegger[172a])

Refinements of the above theory have been made by Turcotte[173], who investigated the fracture patterns induced by a pole shift in a shell of finite thickness. In this fashion, an explanation of the origin of the East African Rifts was proposed[174].

6.3.6 Effects of Despinning and Tidal Flexure

We have already discussed some indirect effects of the rotation of the Earth in the section on polar wandering (Sect. 6.3.5). However, there may also be a much more direct effect: The ellipticity of the Earth is essentially conditioned by the angular velocity of the rotation. Thus, if the rotational speed were to change, the ellipticity must change and this, in fact, could cause the crust to adjust itself to the changed shape beneath, thereby initiating orogenesis.

172 Hess, H. H.: J. Mar. Res. 14:423 (1955)
172a Scheidegger, A. E.: J. Alberta Soc. Pet. Geol. 6:266 (1958)
173 Turcotte, D. L.: Geophys. J. R. Astron. Soc. 36:33 (1974)
174 Oxburgh, E. R., Turcotte, D. L.: Earth Planet. Sci. Lett. 22:133 (1974)

The above mechanism has been advocated not so much for the Earth[175] as for Mercury[176] whose surface exhibits large systems of lineaments which could be explained as the result of "despinning".

A modification of the above ideas has been proposed by Schmidt[177]. Accordingly, the length of the day was not only increasing during geologic history, but was subject to a pulsation; during the accelerating and decelerating phases of the rotation, the stresses induced in the Earth are then as shown in Fig. 101. In this fashion, Schmidt attempts to account for many aspects of the orogenetic cycles. Unfortunately, it is not quite clear how the pulsation should come about; Schmidt[177] assumes that in addition to the well-known Newtonian attraction there is also a repulsion in cosmic bodies. Needless to say, this is higly speculative.

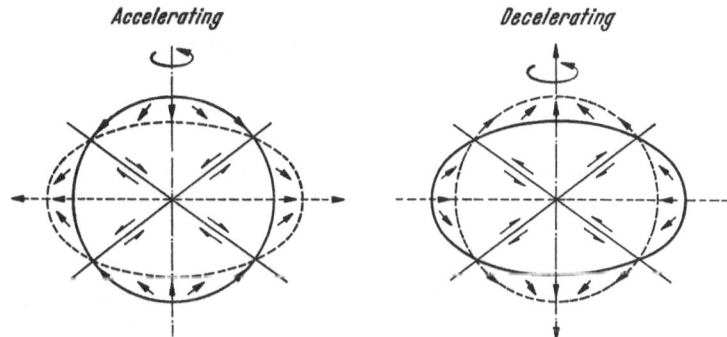

Fig. 101. Stresses in Earth due to accelerating and decelerating rotation. (After Schmidt[177])

Finally, it may be mentioned that (daily) tidal flexure has also been advocated as a cause of orogenetic phenomena[178]. Thus, it has been suggested that the crust is expanding owing to tidal flexing, which would lead to a buildup of horizontal stresses. Estimates of the latter seem to yield reasonable orders of magnitude. However, the variability of the stress state (compressional, tensional) cannot be accounted for by such a theory.

6.3.7 Cosmological Speculations

Finally, we may mention some cosmological speculations regarding the origin of orogenesis. Thus Tamrazyan[179] correlated geological diastrophisms with the passage of the Earth through the plane of the galaxy. It is thought that the

175 Melosh, H. J.: Icarus 31:221 (1977)
176 Melosh, H. J., Dzurisin, D.: Icarus 35:227 (1978)
177 Schmidt, E. R.: Rel. Ann. Inst. Geol. Publ. Hung. B10:157 (1948)
178 Woodriff, R., Eliezer, I.: Tectonophysics 62:T1 (1980)
179 Tamrazyan, G. P.: Bull. (Izv.) Akad. Az. SSSR. 12:85 (1957)

latter occurrence, taking place every 200 million years, would trigger orogenesis. It is difficult, however, to envisage the suggested connection between the galaxy and mountain building.

Another interesting attempt along these lines is due to Haites[180] who noted that there are approximately equal perspective ratios between the times when orogenesis occurred and the distance of the planets from the Sun. Again, this observation is offered only as a correlation, but perhaps one might be tempted to speculate that orogenesis was connected with the creation of the planets.

6.4 Conclusion

The last word on geotectonic hypotheses has certainly not yet been spoken. At the present time, some version of plate tectonism is the most favored theory. Unfortunately, a satisfactory mechanism causing the motion of the plates has not yet been proposed. Most favored is a self-exciting process initiated by downgoing slabs in the subduction zones, but the difficulties encountered in these models have been shown above. An alternative possibility is the presence of thermal convection currents, but the very mechanical possibility of the existence of such currents is by no means assured.

Therefore, it is not entirely possible to rule out completely "other" theories of orogenesis. Only the contraction hypothesis can safely be discarded, at least as a *sole* or *significant* cause of tectonic activity. However, the other schemes that were mentioned could some day again receive more credibility as further facts are accumulated about our Planet.

---

180 Haites, T. B.: J. Alberta Soc. Pet. Geol. 8:345 (1960)

7. Geotectonics

7.1 General Remarks

We are now turning our attention to some features of the Earth's surface that are smaller than those considered heretofore. Such features include faults and joints, folds and petrofabric structures. They are relatively easily amenable to an investigation of their physical causes because of the very smallness of the phenomena.

The study of faults leads directly to the consideration of the earthquake source, which is by many considered as a faulting mechanism of some sort.

All these features are direct manifestations of the action of the geotectonic stress field. This field has also very peculiar surface manifestations caused by the presence of the latter as a boundary condition. Finally, we consider actual numerical geomechanical models.

The present chapter of this book, therefore, will be found to be nothing but the discussion of the theory of the various manifestations of the geotectonic stress field.

7.2 Theory of Tectonic Features

7.2.1 Introduction

We first turn to those manifestations of the geotectonic stress field which we have called and described as "tectonic features" in Sect. 1.6. Such features include faults and related matters such as joints and valley structures in plan, folds and petrofabric structures. We shall consider these features in their turn below.

7.2.2 Theory of Geotectonic Failure Phenomena

7.2.2.1 Introductory Remarks

From the physiographic appearance of the faults (cf. Sect. 1.6.2) it seems likely that they are simply the expression of localized mechanical failure of the material of the Earth's crust. Using one of the theories of such mechanical failure, it should therefore be possible to reconstruct the field of stresses which must have been in existence when a fault was created. If the prevalence of such a field of stresses can be rendered plausible from other considerations about the Earth's crust, the phenomenon of faulting will have been "explained".

Phenomena related to faults are *joints*. Inasmuch as the latter are small cracks in rocks (cf. Sect. 1.6.3), it is likely that they, too, are some sort of failure phenomenon. They will, therefore, be discussed in the present section in conjunction with the faults.

Finally, we have noted (in Sect. 1.6.4) that the orientation trends of valleys in plan also show superficially the phenomenology of fracture patterns. They will therefore also be discussed here.

The theories for the mentioned features are based on the various theories of failure in Earth materials. In addition, the question can be asked regarding the provenance of the *energy* initiating such failure. It must be reasoned that this energy is derived from plate-tectonic processes.

7.2.2.2 Phenomenological Theory

The subject of mechanical failure has been discussed in Sect. 3.3 of this book. Accordingly, Mohr's "engineering" theory gives at least a qualitative description of how fractures occur. Based upon it, Anderson has evolved a theory of fault formation which will now be discussed. Anderson[1,2] observed that an undisturbed or "standard" state of stress in the Earth's crust cannot be entirely arbitrary. First, it must be near the breaking point as is evidenced by the frequency of earthquakes. Second, there can be no pressure or tension perpendicular to the surface and no shearing force parallel to it. The latter condition implies that the normal to the surface is one of the principal directions of stress at or near the surface. Thus, except in strongly folded areas, one principal direction of stress is nearly vertical, the other two are nearly horizontal.

Now, assuming comparatively small disturbances of this standard stress state in Earth's crust, Anderson arrived, based on Mohr's engineering theory of fracture (cf. Sect. 3.3.2), at the following explanation of the three types of faults observed by geologists (see also Figs. 11 and 71):

(1) *Normal faults*. Assuming that there is relief of pressure in all horizontal directions, the greatest pressure is the vertical pressure which is due to gravity. In general, the horizontal stresses will not be equal so that the greatest tension will prevail in a certain direction. Thus, the intermediate stress will also be horizontal, but at right angles to the direction of greatest tension (whatever

1 Anderson, E. M.: Trans. Edinb. Geol. Soc. 8(3):387 (1905)
2 Anderson, E. M.: The Dynamics of Faulting and Dyke Formation, 2nd edn. London: Oliver & Boyd 1951

this may be) and, if fracture occurs, this will happen along a plane containing the intermediate stress and inclined at an angle $\varphi \leq 45°$ toward the vertical, which is the direction of largest principal stress. One thus obtains the characteristics of a normal fault. From the geometrical pattern of the stresses it is obvious that the motion of the two parts must be such that the horizontal extent is increased.

(2) *Transcurrent faults.* Assume that there is an increase of pressure in one horizontal direction and a relief of pressure in a horizontal direction at right angles to it. The smallest principal stress is then horizontal and the intermediate one is vertical. Now, if fracture occurs, this must happen according to Mohr's theory in a vertical plane inclined at an angle $\varphi \leq 45°$ toward the greatest pressure. One obtains thus a fault with a vertical dip, the motion of the two parts being essentially horizontal. This is the characteristic pattern of a transcurrent fault.

(3) *Reversed faults.* Assume that there is an increase of pressure in all horizontal directions. In general, one horizontal direction will be characterized by the fact that along it the pressure will be greatest. Thus the minimum pressure will be vertical and the intermediate principal stress will be horizontal, at right angles to the greatest pressure. If conditions are such that fracture occurs, this will happen according to Mohr's theory along a plane inclined at an angle $\varphi \leq 45°$ toward one horizontal direction, the motion of the two parts being toward each other. One thus obtains the characteristics of a reversed fault: the dip is shallow, and the motion is such that the horizontal extent is shortened.

It may be noted that Anderson's theory has been confirmed in the field, e. g., by Lensen[3].

It may be well to add to the presentation of Anderson's theory a few remarks on energy implications. The theory is based essentially on a frictional model. Thus, the energy released in faulting, as was shown by Dahlen[4], is simply

$$E = \int_S (\sigma'_{ij} + \sigma_{ij}/2) u_i dS_j \qquad (7.2.2-1)$$

where σ'_{ij} is the ambient stress before faulting, σ_{ij} and u_i are the stress change and the displacement associated with the faulting, respectively. The integral is to be taken over the fault surface S. Savage and Walsh[5] have modified the above equation to include graviational effects. Estimates of the frictional resistance[6] and consequent frictional heating[7-9] have also been published.

3 Lensen, G. J.: N. Z. J. Geol. Geophys. 1:533 (1958)
4 Dahlen, F. A.: Geophys. J. R. Astron. Soc. 48:239 (1977)
5 Savage, J. C., Walsh, J. B.: Bull. Seismol. Soc. Am. 68:1613 (1978)
6 Olsson, W. A.: Int. J. Rock Mech. Min. Sci. 11:267 (1974)
7 Cardwell, R. K., Chinn, D. S., Moore, G. F., Turcotte, D. L.: Geophys. J. R. Astron. Soc. 52:525 (1978)
8 Lachenbruch, A. H., Sass, J. H.: J. Geophys. Res. 85:6185 (1980)
9 Lachenbruch, A. H.: J. Geophys. Res. 85:6097 (1980)

The slippage along a fault surface may, in effect, become unstable leading to a stick-slip process[10]. Faulting has also been considered phenomenologically as a Kolmogorov-type stochastic process upon whose basis observed length distributions in fault populations have been explained[11].

Apart from the faulting phenomena discussed above, Anderson's theory also provides for an explanation of *dikes*. Dikes are in the main, nearly vertical fissures between 3 and 30 m wide that have been infilled with some intrusive material. The two sides of a dike appear to have moved apart in a direction normal to the fissure such that there is neither a lateral nor a vertical dislocation. Anderson explained dikes by the remark that Mohr's theory of fracture does not apply in the case of a tensile stress state. In the latter case, fracture is, according to engineering theories, normal to the "tensile stress". Thus, dikes may be explained by considering them as evidence of "tension fracture".

7.2.2.3 Analytical Attempts

Anderson's theory, as outlined in the previous paragraphs, seems to account for the types of faults and related phenomena that have been observed by geologists. It might be desirable, however, to seek a more analytical description of its, after all very qualitative statements. In this connection, we note that Hafner[12] gave an analytical representation of Anderson's standard states. He then proceeded to calculate analytically such deviations from this standard state as would seem reasonable and which could produce faulting. The faulting patterns to be expected were then also calculated upon the assumption of Mohr's criterion of fracture. The results were compared with geologically observed facts.

Thus, Anderson's standard state can be expressed as a two-dimensional stress state. Let x and y be two Cartesian coordinates, x horizontal, y downward, in the direction of gravity ($y = 0$ surface), and let the components of the stress tensor be σ_x, σ_y, τ_{xy}. It is then convenient to express the stresses by means of Airy's stress function Φ so that [cf. Eq. (3.2.1 – 8)]

$$\sigma_x = \partial^2 \Phi/\partial y^2, \quad \sigma_y = \partial^2 \Phi/\partial x^2 - \varrho g y, \quad \tau_{xy} = -\partial^2 \Phi/\partial x \partial y \quad (7.2.2-2)$$

if gravity is the only body force and ϱ = density. The stress function, furthermore, has to satisfy the following equation

$$\frac{\partial^4 \Phi}{\partial x^4} + 2 \frac{\partial^2 \Phi}{\partial x^2 \partial y^2} + \frac{\partial^4 \Phi}{\partial y^4} = 0 . \qquad (7.2.2-3)$$

10 Weertman, J.: J. Geophys. Res. 85:1455 (1980)
11 Ranalli, G.: Math. Geol. 12:399 (1980)
12 Hafner, W.: Bull. Geol. Soc. Am. 62:373 (1951)

Then, Anderson's standard state can be expressed by the following choice of the stress function:

$$\Phi = -\tfrac{1}{6}\varrho g y^3 \tag{7.2.2-4}$$

which yields

$$\left.\begin{array}{l} \sigma_x = \sigma_y = -\varrho g y \\ \tau_{xy} = 0 . \end{array}\right\} \tag{7.2.2-5}$$

As an example of a practical stress state, one can assume the presence of an "additional" horizontal component in addition to the standard stress state, but the absence of an associated additional vertical component. This is expressed by: (no body force for additional stress)

$$\sigma_y = \frac{\partial^2 \Phi}{\partial x^2} = 0 \quad \text{for all values of } y . \tag{7.2.2-6}$$

Integrating, one obtains the stress function:

$$\Phi = cf_1(y)x + ax + bf_2(y) + d . \tag{7.2.2-7}$$

To satisfy Eq. (7.2.2-3), the fourth order derivatives of f_1 and f_2 must vanish. Hence the second order derivatives must be either linear functions of y, constants, or zero. The stress components then are

$$\sigma_x = cf_1''(y)x + bf_2''(y); \quad \sigma_y = 0; \quad \tau_{xy} = -cf_1'(y) . \tag{7.2.2-8}$$

The boundary conditions at the surface require that $f_1' = 0$ for $y = 0$. Keeping within the limits of the above restrictions, one can set up the following subgroups:

(a) $f_1''(y) = 0; \quad f_2''(y) = y + d ,$ (7.2.2-9a)

$\quad\;\; \sigma_x = b(y + d); \quad \sigma_y = 0; \quad \tau_{xy} = 0 .$ (7.2.2-9b)

(b) $f_1' = y; \quad f_2''(y) = 0 .$ (7.2.2-10a)

$\quad\;\; \sigma_x = cx; \quad \sigma_y = 0; \quad \tau_{xy} = -cy .$ (7.2.2-10b)

(c) $f_1'(y) = \dfrac{1}{2}y^2; \quad f_2''(y) = 0 ,$ (7.2.2-11a)

$\quad\;\; \sigma_x = cxy; \quad \sigma_y = 0; \quad \tau_{xy} = -\dfrac{c}{2}y^2 .$ (7.2.2-11b)

The most general expression for the stress systems satisfying the assumption of absence of a vertical stress component is given by the superposition of Eq. (7.2.2-9) to Eq. (7.2.2-11). It is seen that the stipulation $\sigma_y = 0$ is associated with two additional general properties of the internal stress system: (1) that the shearing stress is a function of y only, i.e., constant in all horizontal planes, and (2) that σ_x has linear gradients in both the horizontal and vertical directions.

Of practical importance are the stress systems of the first two subgroups. The combination of Eqs. (7.2.2 – 9), (7.2.2 – 10) with the standard stress state yields (with $d = 0$)

$$\sigma_x = cx + by - ay; \quad \sigma_y = -ay; \quad \tau_{xy} = -cy \qquad (7.2.2-12)$$

where $a = \varrho g$. An analysis of these expressions yields that the trajectories of maximum principal pressure are curved lines dipping downward away from the area of maximum compression. The curvature is strong if the vertical gradient of σ_x is small. From these trajectories, the potential fault surfaces can be calculated according to Mohr's criterion. The potential faults obviously belong into the class of thrust faults. The set dipping toward the area of maximum pressure is slightly concave upward, the complementary set concave downward. Thrust faults of the former type are very common in nature and the theoretically deduced curvature is frequently observed. The latter type appears to occur only rarely.

The above theory, which is essentialy that of Hafner, has been elaborated upon by Sanford[13]. Calculating the elastic response of a rock layer to a two-dimensional stress distribution, Sanford assumed a Mohr-type fracture criterion for the actual occurrence of the faults. His results were compared with scale-model experiments and a reasonable substantiation of the theory was obtained.

The above theories of fault formation are all based on the simple fracture theory of Mohr. A more exact description would be obtained if the dynamic equation were solved in an elastic medium in which a discontinuity (fault) is in the process of forming. For the development of the discontinuity, some failure criterion has to be used; the evolution with time of the stress and strain state in the elastic medium adjacent to the fault (which in turn affects the stress state by representing a boundary condition) ought to be calculated. Unfortunately, this is a very difficult problem; therefore, one usually had to make rather severe simplifications. Weertman[14] made a quasi-static approximation, King[15] calculated a kinematic model of a shallow fault, Haskell and Thomson[16] determined the *near-field* of a finite propagating tensile fault, and Turcotte and Spence[17] investigated a two-dimensional crack model. Analysis of faulting in three-dimensional strain fields were made by Reches[18], and by

13 Sanford, A. R.: Bull. Geol. Soc. Am. 70:19 (1959)
14 Weertman, J.: Bull. Seismol. Soc. Am. 55:945 (1965)
15 King, C. Y.: Bull. Seismol. Soc. Am. 62:551 (1972)
16 Haskell, N. A., Thomson, K. C.: Bull. Seismol. Soc. Am. 62:675 (1972)
17 Turcotte, D. L., Spence, D. A.: J. Geophys. Res. 79:4407 (1974)
18 Reches, Z.: Tectonophysics 47:109 (1978)

Archuleta and Frazier[19]. The *inception* of faulting was studied by Rudnicki[20], the stress field near the *end* of a fault by Hildebrand[21]. Chinnery and Petrak[22] used a quasi-static dislocation model[23] to allow for variable discontinuities (cf. Sect. 3.2.1). Using the elastic dislocation model, Alpan and Teisseyre[24] calculated the energy balance in faults, and Madariaga[25] the wave emission from faults.

The mathematical difficulties become even more formidable if not a simple elastic model containing a discontinuity is considered, but if additional physical factors are taken into account. The main efforts concern the inclusion of friction. Thus, Weertman considered dislocations[26] with finite friction[27] and Walsh[28] investigated the mechanics of long shallow strike-slip faults with friction. Furthermore, Turcotte[29] investigated the effects of thermal conditions, suggesting that transform faults might be contraction cracks. Other specific geological conditions (viz. rift fault patterns) were studied by Withjack[30].

7.2.2.4 Second-order Faults

Anderson's theory as outlined above is, naturally, somewhat of an oversimplification. The stresses in the vicinity of a fault are affected by the presence of the fault itself. The modified stress state gives rise to faults that are not immediately explainable by Anderson's theory if only the large-scale stress state is taken into account. One calls faulting which is due to the stresses modified by the presence of the original faults, "second-order" faulting.

Thus, Anderson calculated the change of stresses due to the development of a transcurrent fault by using the solution of Inglis (cf. Sect. 3.2.1) of stress around an elliptic crack in a plate. Thus, he assumed that there exists a vertical transcurrent fault of length $2c$ with the coordinate x being taken along the strike and y normal to the strike of the fault in a horizontal plane. The "additional" stress system (i.e., the stresses "additional" to Anderson's standard state), which produced this fault, must have had its principal axes inclined at 45° and 125° to the strike of the fault. Introducing elliptic coordinates α, β with

19 Archuleta, R. J., Frazier, G. A.: Bull. Seismol. Soc. Am. 68:573 (1978)
20 Rudnicki, J. W.: J. Geophys. Res. 82:844 (1977)
21 Hildebrand, N.: Tectonophysics 64:211 (1980)
22 Chinnery, M. A., Petrak, J. A.: Tectonophysics 5:513 (1967)
23 Steketee, J. A.: Can. J. Phys. 36:1168 (1958)
24 Alpan, I., Teisseyre, R.: Bull. Int. Inst. Seismol. Earthquake Eng. 3:53 (1966)
25 Madariaga, R.: Bull. Seismol. Soc. Am. 68:869 (1978)
26 Weertman, J.: Bull. Seismol. Soc. Am. 54:1035 (1964)
27 Weertman, J.: J. Geophys. Res. 85:1455 (1980)
28 Walsh, J. B.: J. Geophys. Res. 73:761 (1968)
29 Turcotte, D. L.: Geophys. Res. 79:2573 (1974)
30 Withjack, M.: Tectonophysics 59:59 (1979)

$$x = c \cosh \alpha \cos \beta,$$
$$y = c \sinh \alpha \sin \beta \qquad \qquad \qquad \qquad (7.2.2-13)$$

permits one to express the additional stresses at inifinity after the formation of the fault as follows:

$$\tau_{\alpha\alpha} = K \sin 2\beta,$$
$$\tau_{\alpha\beta} = K \cos 2\beta, \qquad \qquad \qquad \qquad (7.2.2-14)$$
$$\tau_{\beta\beta} = -K \sin 2\beta$$

where K is some constant related to the strength of the rock. After the fault has occurred, one has the further condition that all the stresses must vanish at the fault surface. Inglis has given the solution for this case; one obtains

$$\tau_{\alpha\alpha} = K \sin 2\beta \, (\cosh 2\alpha - 1) \left(\frac{1}{F} - \frac{1}{F^2} \right),$$

$$\tau_{\alpha\beta} = K \sinh 2\alpha \left(\frac{\cos 2\beta}{F} - \frac{1 - \cos 2\beta}{F^2} \right), \qquad (7.2.2-15)$$

$$\tau_{\beta\beta} = -K \sin 2\beta \left(\frac{\cosh 2\alpha}{F} + \frac{1 - \cos 2\beta}{F^2} \right)$$

with

$$F = \cosh 2\alpha - \cos 2\beta. \qquad \qquad \qquad (7.2.2-16)$$

In the stress system before the formation of the fault, the additional stresses are given by Eq. (7.2.2 – 14); the stress system after the formation of the fault is given by Eq. (7.2.2 – 15). The difference between the two stress systems is due to the creation of the fault.

The system of Eq. (7.2.2 – 15) shows that the fault causes a stress concentration near the tips of the original crack. Furthermore, the stress trajectories intersect the fault near its tip at roughly right angles. Thus, additional (transcurrent) faults caused by the stress concentration after the formation of the "main" fault, branch off from the latter at acute angles since, according to Anderson's theory, their strikes must bisect the stress trajectories. This explains the often observed occurrence of "splay faulting" in fault systems, i.e., of faults that branch off at an acute angle from the main faults in an otherwise more or less parallel system.

In a more sophisticated analysis, Chinnery[31] used Steketee's[32] quasi-static dislocation model (cf. Sect. 3.2.1) to discuss the stress changes around

31 Chinnery, M. A.: Can. J. Earth Sci. 3:163 (1965)
32 Steketee, J. A.: Can. J. Phys. 36:1168 (1958)

primary faults that lead to secondary faulting. Finally, Segall and Pollard[33] gave an analytical solution of cracks en echelon and attempted to calculate the elastic interaction between such cracks.

The difficulty of obtaining mathematical solutions of the faulting problem induced various researchers to make model experiments or inferences from field studies. Thus, Rangers and Müller-Salzburg[34] studied the general kinematics of the faulting process in this fashion, Lajtai[35] studied en echelon tension gashes, Friedman and others[36,37] microscopic feather fractures, Berger[38] oblique internal foliations in dikes, and Oldenburg and Brune[39] transform faults. Real progress, however, was achieved in following up fracture propagation only after the introduction of finite-element methods in computer simulation. By this means, the study of the formation of primary[40] and secondary faults[41] became possible.

7.2.2.5 Non-brittle Fracture

The theories of faulting discussed thus far are based on the idea that the mechanical process involved is basically that of *brittle* fracture. However, one may expect that, on occasion, other modes of fracture may also be of importance in connection with the formation of faults. This is particularly relevant with regard to the development of faulting with time. For the latter, one normally assumes some frictional stick-slip phenomenon. If it is assumed that the friction stress at a fault decreases with increasing slip velocity (in the limit case this is represented by the difference between sticking and sliding friction), then there arises an inherent instability in the quasi-static creep slippage on a fault[42]. Similar effects are observed if a strain-softening model is assumed: The result is episodic faulting[43].

The problem has also been studied experimentally[44-46]. Thus, Mandl[46] et al. have made an experimental study of this type of phenomenon with a view to explaining tectonic faulting. In it, they experimented with granular materials and demonstrated the origination of shear zones. The experiments clearly highlighted basic differences between shear zones in densely packed

33 Segall, P., Pollard, D. D.: J. Geophys. Res. 85:4337 (1980)
34 Rangers, N., Müller-Salzburg, L.: Rock Mech. Suppl. 1:20 (1969)
35 Lajtai, E. Z.: Bull. Geol. Soc. Am. 80:2253 (1969)
36 Friedman, M., Logan, J. M.: Bull. Geol. Soc. Am. 83:3073 (1970)
37 Conrad, R. E., Friedman, M.: Tectonophysics 33:187 (1976)
38 Berger, A. R.: Bull. Geol. Soc. Am. 82:781 (1971)
39 Oldenburg, D. W., Brune, J. N.: J. Geophys. Res. 80(17):2575 (1975)
40 Malina, H.: Rock Mech. 2:1 (1970)
41 Bock, H.: Rock Mech. 3:225 (1970)
42 Weertman, J.: J. Geophys. Res. 84:2146 (1979)
43 Stuart, W. D.: Tectonophysics 52:613 (1979)
44 Schneider, H.: Bull. Int. Assoc. Eng. Geol. 16:235 (1977)
45 Hildebrand-Mittlefehlt, N.: Tectonophysics 57:131 (1979)
46 Mandl, G., De Jong, L. N. J., Maltha, A.: Rock Mech. 9:95 (1977)

granular materials such as elastic sediments under stress conditions that either allow or suppress dilatation of the sheared material. In such materials shearing always starts in a narrow band only a few grains thick. The "interlocking resistance" of the particles inside this band is overcome by two mechanisms: dilatation and grain breakage. The former occurs when the compressive stresses in the pack are low enough to allow particles to override each other. If the stress level is too high, shearing will be accomplished by severe crushing of the grains themselves inside the shear zone (no dilatation). As shearing proceeds, the thickness of the shear zone increases, which effect is caused by the accumulation of grain fragments in the interstices increasing the local resistance to shearing. In some of the experiments shear zones were produced which showed a striking similarity with natural shear zones observed in clay. The stress measurements indicated that shear bands produced between rigid plates are bounded by planes of maximum shear stress rather than by Coulomb-type slip planes [cf. Eq. (3.2.2 – 12a)]. On the other hand, freely developing sets of minor shear indicate a tendency of the material to deform in accordance with Coulomb's slip concept. This observation explains the fact that certain types of tectonic faults in "plastic" sediments are found to develop parallel to competent rock boundaries, whereas others behave on a large scale as Coulomb slips. Comparisons of the predicted shear patterns with those observed in the field have been reported by Cummings[47] and by Dailly[48].

7.2.2.6 The Role of Pore Pressure

A further modification of the simple theories of faulting is based on the consideration of the role of pore pressure, first introduced by Hubbert and Rubey[49]. As noted in Sect. 3.1.3, the presence of a pore fluid has the effect that the mechanical behavior of the material is determined by the effective rather than by the total stress.

Without the influence of pore pressure, overthrust faulting would probably be impossible because the friction betwen the fault planes would be much too great (in fact, greater than the breaking strength of the material!)[50]. A discussion of overthrust faulting was, indeed, the starting point for Hubbert's and Rubey's[49] suggestion of the importance of taking pore pressure into account. The original theory was further refined by Hsü[51] by taking the cohesive strength of the material into account. Nur and Byerlee[52] showed that

47 Cummings, D.: Bull. Geol. Soc. Am. 87:720 (1976)
48 Dailly, G. C.: Bull. Can. Pet. Geol. 24:92 (1976)
49 Hubbert, M. K., Rubey, W. W.: Bull. Geol. Soc. Am. 70:115 (1959)
50 Ramberg, H.: Geol. Foeren. Foerh. 99(2):111 (1977)
51 Hsü, K. J.: Bull. Geol. Soc. Am. 80:927 (1969)
52 Nur, A., Byerlee, J. D.: J. Geophys. Res. 76:6414 (1971)

the case of Hubbert and Rubey[49] constitutes a limit of a more general one and Forristall[53] considered the full state of stress in the overthrusting block. Rambach and Deramond[54] made a phenomenological calculation of overthrusting by considering the waterlogged medium as a thin visco-plastic layer upon which a thick rigid-plastic block slides.

The geological implications of the general theory in overthrust faulting were investigated with regard to the toe of thrust sheets by Raleigh and Griggs[55] and with regard to layered sequences by Gretener[56]. Müller and coworkers[57,58] made a specific application of overthrusting theory, using a finite-element computer model, to the explanation of the Jura mountains in Switzerland. Finally, Martin[59] investigated (by means of simple Mohr diagrams) the effect of fluid pressure on induced faulting, particularly on the triggering of earthquakes, and Chapman[60] analyzed the modifications necessary for thrust sliding in an aqueous medium.

Nevertheless, in spite of the above theoretical attempts, the phenomenology of large thrusts has not yet been mechanically explained in a satisfactory manner. In particular, the tongue-type structure of some thrusts (cf. Sect. 1.6.2) has not yet been accounted for[61]. One of the difficulties in explaining faulting at depth is that the pore pressure must be of the order of the lithostatic pressure in order to have required effect of lowering the friction. Goguel[62] discussed this problem in general terms and showed that recrystallization and compaction could produce high pore pressures. Bredehoeft and Hanshaw[63] considered a series of physical processes which could create high pore pressure and determined a series of geneal solutions for the corresponding hydrodynamic models. They showed that already a sedimentation rate of 500 m/10^6 years in a sedimentary column whose permeability is $\sim 10^{-10}$ m/s as well as source layers at depth (dehydrating gypsum or montmorillonite) in low-permeability surroundings could produce the required pressures. In this, the existence of horizons of material of low permeability was found to be most important, inasmuch as otherwise the creation and continuation of high fluid pressures seems unlikely. Lachenbruch[64] suggested that the pore fluid might be heated by the fault friction, thereby

53 Forristall, G. Z.: Bull. Geol. Soc. Am. 83:3073 (1972)
54 Rambach, J. M., Deramond, J.: Tectonophysics 60:T7 (1979)
55 Raleigh, C. B., Griggs, D. T.: Bull. Geol. Soc. Am. 74:819 (1963)
56 Gretener, P. E.: Bull. Can. Pet. Geol. 20:583 (1972)
57 Müller, W. H., Briegel, U.: Eclogae Geol. Helv. 73(1):239 (1980)
58 Müller, W. H., Hsü, K. J.: Rock Mech. Suppl. 9:219 (1980)
59 Martin, J. C.: J. Geophys. Res. 80(26):3783 (1975)
60 Chapman, R. E.: Bull. Geol. Soc. Am. Part I, 90:19 (1979)
61 Gretener, P. E.: Bull. Can. Pet. Geol. 25(1):110 (1977)
62 Goguel, J.: Rev. Geogr. Phys. Geol. Dyn. 9:153 (1969)
63 Bredehoeft, J. D., Hanshaw, B. B.: Bull. Geol. Soc. Am. 79:1097 (1968)
64 Lachenbruch, A. H.: J. Geophys. Res. 85:6097 (1980)

enhancing mobility. A related idea assumes that the heating is produced by the concentration of isothermal surfaces above basement culminations[65].

The pore pressure in a geological formation does not only affect the mechanism of overthrust faulting. Its general significance has been reviewed by Gretener[66]. Accordingly, the sonic velocity, the electrical resistivity, and the mechanical behavior of such layers are all affected. Apart from the faulting mechanism which has already been discussed, the pore pressure also affects the formation of open joints and the compaction (as an example, compaction around a reef was discussed by Labute and Gretener[67]) of sedimentary sequences.

7.2.2.7 Joints[68]

1. Introductory Remarks.   Joints are ubiquitous structures, as has already been remarked in connection with their physiographic description. Although they have been intensively studied, the mechanics of their formation is still somewhat of a mystery[68,69]. In fact, there are a number of different theories as to their origin which will be reviewed below. In this connection it should be noted that it is quite clear that not all joints have been formed in the same fashion; to some types, e. g., columnar jointing in lava, an obvious special type of origin (cooling!) can immediately be assigned. However, the bulk of the "ordinary" joints in rock outcrops do not fall into such an obviously "special" category, and thus their origin is subject to some controversy.

2. Tension Cracks.   A brittle material under tension becomes cracked. A tensional stress state in the Earth occurs primarily during the cooling of hot masses, for instance in a lava flow, whose density increases during cooling so that they shrink. During the shrinkage, typical columnar joints are formed which are pentagonal or hexagonal in cross-section.

The mechanics of cracks due to thermal concentration has been studied by Lachenbruch[70] in quite general terms. Although the latter author had mainly applications to permafrost in mind, he stressed explicitly that his argument is also applicable to cooling cracks in rock.

Lachenbruch[70], in fact, made an analysis of the formation of tension cracks. He noted that a straight, vertical crack, when it is being formed, causes a stress relief in the medium which ranges from a maximum, say, σ_m in the direction perpendicular to the crack to a minimum $m\sigma_m$ (where m is Poisson's ratio) parallel to it. Thus, a crack induces an appreciable stress an-

65 Ayrton, S.: Geology 8:172 (1980)

66 Gretener, P. E.: Bull. Can. Pet. Geol. 17:255 (1969)

67 Labute, G. J., Gretener, P. E.: Bull. Can. Pet. Geol. 17:304 (1969)

68 This section after Scheidegger, A. E.: Riv. Ital. Geofis. Sci. Aff. 5:1 (1979), by permission

69 Price, N. J.: Fault and Joint Development in Brittle and Semi-Brittle Rock. Oxford: Pergamon Press 1966

70 Lachenbruch, A. H.: Geol. Soc. Am. Spec. Pap. 70:1 (1962)

isotropy within the medium in which it propagates. Since a tension crack tends to propagate in the direction perpendicular to maximum tension, a crack advancing in an oblique direction will tend toward the normal to a pre-existing crack after entering its zone of stress relief. Therefore, the intersection of a crack with a preexisting one tends to be orthogonal. Conversely, an orthogonal intersection between tension cracks suggested that one of the cracks involved predates the other.

As noted above, simultaneously arising cracks in cooling substances do not intersect each other orthogonally, but at some obtuse angle. One can, in effect, show that a system of regular hexagonal columns releases the maximum elastic strain energy per unit crack area in a medium under tension. However, regarding the actual mechanics of formation of the triple junction, there are only conjectures. One of the more satisfactory theories assumes that obtuse junctions form from bifurcation or branching of growing cracks. If each of the branches diverges from the original direction by an angle of about 60°, and each branch again bifurcates when it attains a prescribed length, patterns of the type actually observed in nature could be generated.

In this instance, Bankwitz[71] claimed that columnar tension jointing in basalt starts at the interior and is wavy. Ryan and Sammis[72] suggested that the fracture mechanism in cooling basalt columns is essentially cyclic, which may be the cause of the horizontal striations that are occasionally observed.

Tension fracture phenomena have also been invoked as the origin of the "ordinary" ubiquitous joints[73], although it is difficult to see how the phenomenology of the latter can be reconciled with a tensional origin. A possible mechanism for the formation of tension joints has been based on the assumption of the occurrence of natural hydraulic fracturing in the Earth's crust. This approach has mainly been taken by Secor[74,75]. If the formation pressure is high enough, fractures would form normal to the least principal stress direction (cf. Sect. 3.3.2).

Again, the problem arises as to how these fractures are to remain open to form the ubiquitously visible surface joints: At the surface, the formation pressure is certainly zero.

3. *Pressure Cracks*. A elongated sample under axial pressure tends to buckle. In this connection, cracks may open parallel to the maximum pressure direction as this is well known from the phenomenon of exfoliation. The case of a package of parallel laminae being subject to a pressure acting parallel to the layering was discussed by Brunner and Scheidegger[76] based on the well-known analysis of forces acting on penny-shaped cracks that had been made long ago

71 Bankwitz, P.: Z. Geol. Wiss. Berlin 6:285 (1978)
72 Ryan, M. P., Sammis, C. G.: Bull. Geol. Soc. Am. 89:1295 (1978)
73 Bankwitz, P.: Z. Geol. Wiss. Berlin 6:301 (1978)
74 Secor, D. T.: Am. J. Sci. 263:633 (1965)
75 Secor, D. T.: Proc. Geol. Surv. Can. Pap. 68 – 52:3 (1969)
76 Brunner, F. K., Scheidegger, A. E.: Rock Mech. 5:43 (1973)

by Griffith. The result of all such calculations is that the crack is most likely to be parallel to the largest compression and will open up in the direction of the smallest compression. This, however, can happen only if this smallest compression is very small indeed. The tension σ_t induced at the tip of an elliptical crack in a two-dimensional stress state (σ_1, σ_3) is

$$\sigma_t = -\sigma_1 + \sigma_3 + 2\frac{a}{b}\sigma_3 \qquad (7.2.2-17)$$

where σ_1 is the largest, σ_3 is the smallest principal stress, and a is the axis of the ellipse parallel to the σ_1, b the axis parallel to the σ_3-direction. If σ_t is larger than the tensile strength s_t of the material, then the crack will extend itself. If one takes the overburden pressure for σ_3

$$\sigma_3 = \varrho g H \qquad (7.2.2-18)$$

where ϱ is the density, g the gravity acceleration, and H the depth, then one finds for the limiting depth at which a crack can extend itself

$$H_{\text{limit}} = \frac{\sigma_1 + s_t}{\varrho g(1 + 2a/b)}. \qquad (7.2.2-19)$$

For reasonable values of the material constants ($s_t = 6$ MPa, $\varrho = 2{,}500$ kg/m^3) and a representative value of $\sigma_1 = 10$ MPa, one finds for H approximately 50 m which shows that the envisaged mechanism can reach at most to this depth. This bears out the fact that the envisaged mechanism can be invoked for an explanation of exfoliation: The force normal to the surface is evidently zero, and thus "leafing" can be induced by the compression. It is difficult to see, however, how this mechanism could be invoked for an explanation of joints other than those parallel to a free surface.

4. *Slaty Cleavage.* "Joints" can also be in the form of so-called slaty cleavage[77]. This is a form of cleavage which occurs normal to the maximum principal pressure and involves rotation and deformation of the constituent grains, as well as pressure solution transfer and recrystallization of the components of the rock. The literature on the subject is very large; it has recently been summarized and reviewed by Siddans[78], by Wood[79], and by Williams[80]. Inasmuch as a reorientation and reformation of the fabric of the rock is involved, a discussion of the mechanics of the formation of slaty cleavage goes beyond an analysis of fracture mechanics, but involves the mechanism of the actual genesis of the rocks in question during the metamorphosis of earlier ones. Whereas "joints" as considered heretofore are the result of tectonic

77 A discussion of the nomenclature of cleavage has recently been given by Powell, C. M.: Tectonophysics 58:21 (1979)
78 Siddans, A. W. B.: Earth Sci. Rev. 8:205 (1972)
79 Wood, D. S.: Annu. Rev. Earth Planet. Sci. 2:369 (1974)
80 Williams, P. F.: Tectonophysics 39:305 (1977)

forces acting on the "finished" rock, the formation of slaty cleavage always involves some degree of actual metamorphism and never occurs in rock that is not being changed in the process. The pattern of slaty cleavage, therefore, depends on the forces that were acting during the genesis of the rocks and not on the tectonic forces acting at later stages of their history.

A similar "genetic" mechanism of "joint" formation is that the formation of "spaced cleavage" which consists of numerous discontinuity surfaces bounding undeformed slats of water-soluble rocks oriented subparallel to the axial planes of associated folds. These "joints" are thought to be caused by dissolution phenomena related to shortening in a direction roughly normal to the axial plane of the folds[81].

A type of pressure-solution phenomenon is possibly also involved in "crenulation cleavage". Inasmuch as the scale of this phenomenon is near-microscopic, it will be discussed in connection with the theory of petrofabrics (Sect. 7.2.4).

5. Shear Fractures. It has been seen that the ubiquitous joints visible in fresh outcrops can be interpreted reasonably, at least phenomenologically, as Mohr-type fractures produced by the present-day tectonic stress system.

In such an interpretation there is, however, an immediate paradoxon: The vertical stress at the surface of the Earth, where most joints are observed, is evidently zero, and is, thus, the *smallest* compression. The two other principal tectonic stress directions are horizontal, and according to Mohr's theory, the potential fracture surfaces should contain the *intermediate* principal stress directions; the joints, if they are interpreted as Mohr-fractures, should there-fore be of the type of (miniature) normal or reversed faults. However, the overwhelming phenomenological evidence is that the tectonic joints are near-vertical and geometrically arranged as *if* the intermediate principal stress direction *were* vertical, the largest and smallest *were* horizontal. In fact, inter-preting joints as Mohr-type fracture surfaces in a two-dimensional (parallel to the Earth's surface) tectonic stress field yields results on a worldwide basis which are consistent with data form in-situ stress measurements and earth-quake source mechanisms.

A second paradoxon in the above interpretation of joints as Mohr-type fractures is that the observed angles of intersection between distinct joint sets at any one outcrop are too large, i.e., there are too close to 90°. Mohr's theory states that the potential fracture surfaces are inclined at about 30° toward the maximum compression, which yields an angle of around 60° be-tween conjugate fractures. In nature, conjugate joint sets are overwhelmingly found to enclose angles close to 90°, indicating that joints tend to form in the planes of maximum shear of the horizontal tectonic stress field.

An obvious attempt at a solution of the paradoxons is that of invoking the possibility that the joint pattern was imprinted on the rocks at depth (where

---

81 Alvarez, W., Engelder, T., Lowrie, W.: Geology 4:698 (1976)

the weight produces enough vertical pressure for the latter to become the intermediate stress) prior to their being exposed in an outcrop. For the horizontal tectonic stresses at the surface one finds commonly values of some 10 MPa. This value is reached by the overburden pressure at some 300 m depth; — a depth at which it is certainly not unreasonable to assume the rocks to have been buried prior to their exposure. The unloading hypothesis does not, however, explain the large (near-orthogonal) fracture angles. Barton[82, 83] (see Sect. 3.3.5) has shown that these angles come close to 90° if the rock is in a critical state. Assuming reasonable values, e.g., for limestone, yields a critical state for an overburden pressure of upward from 250 MPa which would require a depth of burial of minimally 7,500 m. In view of the fact that jointing is observed ubiquitously, even in post-glacial holocene deposits, such necessary depths of burial evidently obviate the theory.

A further problem is how the "fracture design" can be "preserved" in the rock so that it becomes evident when the outcrop is exposed after unloading. One possibility is to assume the stress to be "locked in" during the unloading process because of lateral constraints so that it is preserved right to the point of fracture. Voight[84] suggested a mechanism by which stress could be "locked in" in this fashion, but Holzhausen and Johnson[85] pointed out the difficulties occurring in this connection so that it appears as most doubtful that a stress state at depth could be "preserved" until the rock reaches the surface.

6. Shear Flow. The large angles between conjugate joint sets may, in fact, indicate that the fracture is not of the Mohr type at all. The joint sets align themselves very closely with the planes of maximum shear in the tectonic stress field which may indicate that they are the result of some ductile or plastic slippage process. Thus, an obvious phenomenological explanation of joints would be that they are the response to an instantaneous creep process induced in the horizontal plane by the momentarily acting tectonic stresses. In fact, one knows that rocks respond to stresses by logarithmic creep so that yielding phenomena in the planes of maximum shear would at least be qualitatively expected. The joints, then, would be simply the surface expression of the maximum-shear phenomena. This type of explanation entirely circumvents all the problems posed by the "paradoxons", inasmuch as the flow pattern would naturally reach up to the Earth's surface.

The interpretation of joints in outcrops would therefore be that they represent planes of maximum shear in a creep-flow state of the lithosphere. Local deviations simply represent irregularities in this flow whose average properties can be ascertained by corresponding averaging. The trajectories following the

82 Barton, N.: Int. J. Rock Mech. Min. Sci. 13:255 (1976)
83 Barton, N., Choubey, V.: Rock Mech. 10:1 (1977)
84 Voight, B.: Am. J. Sci. 272:662 (1974); Voight, B.: Proc. 3d Congr. Int. Soc. Rock Mech. Denver A 2:580 (1974)
85 Holzhausen, G., Johnson, A. M.: Tectonophysics 58:237 (1979)

(mean) directions (strikes) of the joints on a map represent the slip lines of the "tectonic flow" present in the area.

7.2.2.8 Theory of Valley Orientations in Plan

Finally, we may mention the problem of vallery orientations in plan. We have seen (Sect. 2.6.4) that the general trends of valleys seem to fit well together with the directions of hypothetical shear lines in a neotectonic stress system. Naturally, valleys are not "fissures" in the ground, but caused by the fact that the erosion due to exogenic agents is enhanced in their direction.

It may come somewhat as a surprise that valleys are of recent origin. However, if it is kept in mind that rates of rising in mountainous areas are commonly of the order of millimeters per year, i.e., kilometers per million years (cf. Sect. 1.7.6), and hence erosion rates of the same order of magnitude for there to be equilibrium, it should no longer come as a surprise that the surface of the Earth, as we see it, is of recent making. The neotectonic stress field can therefore be assumed to be the governing factor in forming the orientation structure of river courses; the valleys, in the mean, corresponding to the lines of maximum shear.

7.2.3 Folding

7.2.3.1 Introduction

Folds have evidently been produced by the geophysical stress field, since they represent the expression of tectonic displacements. However, the actual mechanical processes that are involved in fold formation are by no means fully understood. Ramsay[86] and Johnson[87] have written books reviewing the subject.

Qualitatively, various physical processes have been invoked as occurring in fold formation.

The first is elastic buckling which has been proposed by Smoluchowski[88]. The opposite extreme to assuming elastic buckling is the conjecture that the appearance of folding is due to differential rheid flow in the folded rock[89,90]. In between the two extreme assumptions are those postulating some kind of intermediate rheological behavior (such as plastic buckling) for the rock strata in question.

86 Ramsay, J. C.: Folding and Fracturing of Rocks. New York: McGraw-Hill 1967
87 Johnson, A. M.: Styles of Folding. Amsterdam: Elsevier 1977
88 Smoluchowski, M.: Anz. Akad. Wiss. Krakau Math. Naturwiss. Kl. 1909:3 (1909)
89 Carey, S. W.: J. Geol. Assoc. Aust. 1:67 (1954)
90 Wynne-Edwards, H. R.: Am. J. Sci. 261:793 (1963)

Because of many difficulties with the theoretical approach, folds have also been investigated by mechanical model experiments. We shall discuss all these possibilities below.

7.2.3.2 Elastic Buckling

As noted, elastic buckling as a theory of fold formation was originally proposed by Smoluchowski[91]. The theory was later extended to selective buckling of composite layers by Ramberg[92]. A theory of folding based on monoclinal flexuring of elastic multilayers was proposed by Reches and Johnson[93] and a theory based upon the model of an elastic-buckling layer floating upon a viscous substratum by Golecki[94]. These attempts at explaining buckling were based upon straightforward analysis. Because the problems can become formidable, Anthony and Wickham[95] used finite-element simulations by means of a computer. The possibility of stress concentrations in buckling folds giving rise to slip cleavage was also investigated[96,97].

In order to demonstrate how elastic buckling could give rise to folds, let us consider the following idealized case. A rectangular piece of the crust of the Earth is represented by a thin plate of the same shape (but thought of as plane). A force F is acting in the original plane of this plate normal to one of its edges. Then, for an elastic material, the condition of equilibrium is

$$-Fy(x) = M d^2 y/dx^2 \qquad (7.2.3-1)$$

where F is the force, y the deflection of the plate from its original plane, x the coordinate in the direction of the acting force, and M a constant indicative of the resistance of the plate to bending depending on its elastic parameters. In Eq. (7.2.3 – 1) the left-hand side represents the moment of the force exercised upon the plate at the point x, the right-hand side is the resistance to bending of the plate, being assumed as proportional to the local curvature which is equal to $d^2 y/dx^2$ for small deflections.

The solution of Eq. (7.2.3 – 1) is

$$y = A \sin\left(\sqrt{\frac{F}{M}}\, x\right) + B \cos\left(\sqrt{\frac{F}{M}}\, x\right) \qquad (7.2.3-2)$$

where A, B are constants of integration to be determined from the boundary conditions. If we assume for the latter, say,

91 Smoluchowski, M.: Anz. Akad. Wiss. Krakau Math. Naturwiss. Kl. 1909:3 (1909)
92 Ramberg, H.: Tectonophysics 1:307 (1964)
93 Reches, Z., Johnson, A. M.: Tectonophysics 35:295 (1976)
94 Golecki, J. J.: Int. Rock Mech. Min. Sci. Geomech. Abstr. 16:93 (1979)
95 Anthony, M., Wickham, J.: Tectonophysics 47:1 (1978)
96 Groshong, R. H.: Geology 3:411 (1975)
97 Alvarez, W., Engelder, T., Lowrie, W.: Geology 4:698 (1976)

$$y(0) = y(X) = 0 \tag{7.2.3-3}$$

it becomes at once obvious that, in general, no solutions of the assumed type (i. e., bulging) exist; that is, one obtains

$$A = B = 0 . \tag{7.2.3-4}$$

The plate, if it be subjected to the indicated force, will simply contract a little under the load and will not bulge. However if F has a certain particular value given by (i. e., if it is equal to an eigenvalue of the system)

$$F = M \frac{n^2 \pi^2}{4X^2} \tag{7.2.3-5}$$

with n denoting any even integer, then the solution becomes ($B = 0$)

$$y = A \sin \left(\frac{n \pi}{2} \frac{x}{X} \right), \tag{7.2.3-6}$$

and this satisfies the boundary conditions for *any* A. Thus, if F reaches an eigenvalue of the equation, the plate will buckle. The lowest eigenvalue of F produces a sinusoidal half wave with arbitrarily great amplitude. For F above the first eigenvalue, the deformation becomes *unstable*. The first mode of buckling, being in the shape of a sinusoidal half wave, has some resemblance to a fold. This resemblance has been used for an explanation of folding.

7.2.3.3 Theories Assuming Infinitely Flexible Strata

The purely kinematical aspects of folding can be investigated without recurrence to the rheology of the material involved. De Sitter[98] proposed several models, notably one of concentric folding which was later extended by Hara[99]. Based on a purely kinematical argument, Esz[100] was able to show that the supposed "drag folds" between closely spaced parallel shear surfaces are almost impossible to arise under natural geological conditions. Based on essentially kinematical arguments, Kuenen[101] showed that the origin of ptygmatic folds must be sought in crumpling of an originally planar vein by subsequent compression. Similarly, Reches[102] investigated the kinematics of fold formation in a foliated medium consisting of many parallel plane layers. Finally, Caron[103] used general analytical strain theory for the investigation of the deformation of lineations.

98 de Sitter, L. U.: Proc. K. Ned. Akad. Wet. 52:No.5 (1939). See also de Sitter, L. U.: Structural Geology. New York: McGraw-Hill Book Co. 1956
99 Hara, I.: Hiroshima Univ. J. Sci. Ser. C 5:217 (1967)
100 Esz, V. V.: Geotektonika (USSR) 1969(3):52 (1969)
101 Kuenen, P. H.: Tectonophysics 6:143 (1968)
102 Reches, Z.: Tectonophysics 57:119 (1979)
103 Caron, J. M.: Eclogae Geol. Helv. 72:485 (1979)

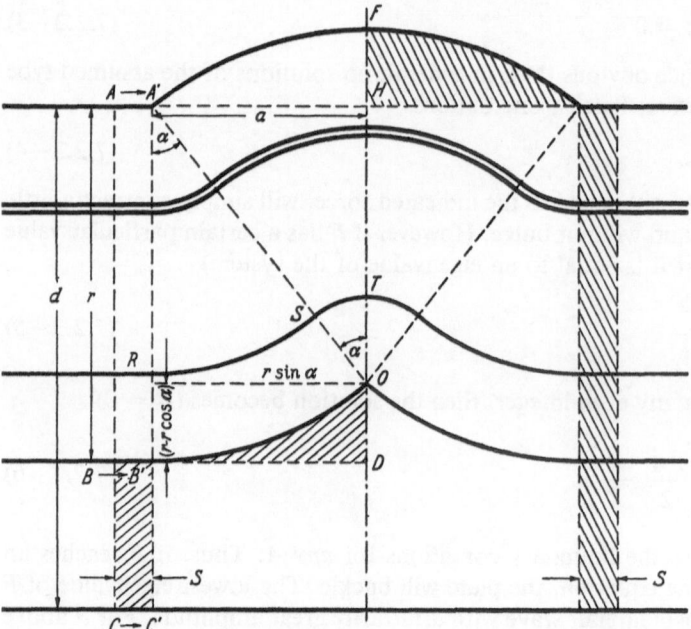

Fig. 102. De Sitter's[104] model of folding

We present here the approach of de Sitter to the problem. According to de Sitter, the mechanism of folding is the result of the following conditions: (a) during folding, the volume of the strata is conserved, and (b) each infinitesimal layer undergoes only bending. Furthermore, de Sitter assumed the "principle of concentric folding" which is expressed by the assumption that the surface of a folded layer is formed by three circles (cf. Fig. 102). De Sitter thus arrived at the picture of folding illustrated in Fig. 102; in this figure, *unprimed letters* refer to the situation before folding, whereas *primed letters* refer to the situation after folding. The folding has been caused by the compression by the amounts 2*S* of the original strata. It is easy to see that above the line through *B*, the two laws stipulated by de Sitter are indeed satisfied, below that line this is, however, not the case. Thus, the equivalent of the material from the first *shaded area* must be transferred into that of the second ony by plastic flow or some such phenomenon. Moreover, although de Sitter is certainly satisfying his two assumptions above the line through *B*, it is quite obvious that this is not the *only* solution satisfying those principles. It may thus the observed that the solution of de Sitter is not a real "explanation" of folds as the cause of the latter is not reduced to a field of forces, nor is any attempt made at a rationalization of what forces could produce the particular type of bending assumed by de Sitter. That the two laws are satisfied, is not

104 de Sitter, L. U.: Proc. K. Ned. Akad. Wet. 52:No.5 (1939)

sufficient to account for this, as they are only an expression of the conservation of area and matter.

It is obvious that de Sitter's theory can be used to estimate the basal shearing plane (located at depth d beneath the surface in Fig. 102) from a knowledge of the volume of rocks pushed above the original level by the folding and the crustal shortening involved. Duska[105] has shown that d can also be estimated from the dip of the strata should the folds be eroded.

The theory of de Sitter has been modified by Tiedemann[106]. The latter author replaced the circles which make up the form of a concentric fold, by sine curves. As in de Sitter's scheme, the two fundamental assumptions are adhered to. It is fairly easy to calculate the shapes of a series of sine curves that make up the strata in a layer of the Earth, and one thus obtains a picture as shown in Fig. 103. It will be observed that the lower boundary of possible folding is now not a surface below which one has to assume plastic deformation or such like, but rather a "shearing plane" above which a displacement takes place, but below which everything remains fixed.

Referring to Fig. 103 one has for a point on the curve:

$$y = \tfrac{1}{2}h \sin \pi x/\varphi . \qquad\qquad (7.2.3-7)$$

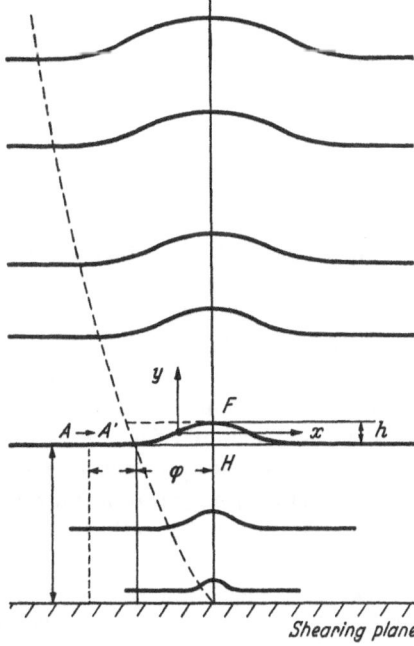

Fig. 103. Tiedemann's[106] model of folding

105 Duska, L.: J. Alberta Soc. Pet. Geol. 9:20 (1961)
106 Tiedemann, A. W.: Geol. Mijnb. 3:199 (1941)

From the geometry apparent in Fig. 103, one can form the following equations which are based upon the fundamental assumptions:

$$\text{area}\, A'FH = \tfrac{1}{2}\varphi h = sh,$$ (7.2.3 – 8)

$$L \equiv \text{length}\, A' \text{ to } F = AF \equiv \varphi + s.$$ (7.2.3 – 9)

However, the length of the sine curve can be calculated; one has:

$$dL = (1 + y'^2)^{1/2}dx = (1 + n^2)^{1/2}(1 - [n^2/(1 + n^2)]\sin^2 \pi x/\varphi)^{1/2}dx$$ (7.2.3 – 10)

with $n^2 = \pi^2 h^2/4\,\varphi^2$, and hence

$$L = 2(1 + n^2)^{1/2}E(k) \cdot \varphi/\pi$$ (7.2.3 – 11)

where $E(k)$ is a standard elliptic integral:

$$E(k) = \int_0^{\pi/2} (1 - k^2 \sin^2 x)^{1/2}dx$$ (7.2.3 – 12)

with

$$k = n/(1 + n^2)^{1/2}.$$ (7.2.3 – 13)

The integral $E(k)$ has been tabulated; using its values, one can calculate the values for h for various assumptions for s and d. It can be readily seen that the process of folding can be explained by a continuous movement, simply by adjusting the parameter n for the neighboring strata accordingly. One thus arrives at a series of folds as depicted in Fig. 103.

7.2.3.4 Viscous Layers

The usual analytical treatment of folding in recent years has been based on the assumption of a sequence of viscous layers of different viscosities. Particulary Biot and co-workers[107–111] helped to advance the corresponding mathematical theory. The solutions to their equations were based on mathematical approximations. Fletcher[112,113] later produced an exact infinitesimal amplitude solution of the equations governing folding of a single viscous layer embedded in a medium of lower viscosity.

The analytical approach is evidently not satisfactory, inasmuch as fold development for *large* (not infinitesimal) amplitudes is required. This problem can only be solved numerically by means of a computer, usually employing

107 Biot, M. A.: Bull. Geol. Soc. Am. 75:563 (1964)
108 Biot, M. A.: Bull. Geol. Soc. Am. 76:251 (1965)
109 Biot, M. A.: Bull. Geol. Soc. Am. 76:371 (1965)
110 Biot, M. A.: Bull. Geol. Soc. Am. 76:833 (1965)
111 Biot, M. A., Odé, H.: Geophysics 30:213 (1965)
112 Fletcher, R. C.: Tectonophysics 39:593 (1977)
113 Fletcher, R. C.: Tectonophysics 60:77 (1979)

finite-element techniques. There are a large number of investigations of this subject matter in existence [114–118].

Amongst these investigations, those of de Bremaecker and Becker [117] are particularly complete, inasmuch as many characteristic cases were investigated. As with most computer solutions, these are generally "one-shot" affairs; i. e., one cannot recognize the dependence of the solutions generally on the parameters involved, but has to produce a variety of cases on the computer. As an example, we show in Fig. 104 the results for the compression of a high-viscosity central layer with low-viscosity outer layers in a medium-viscosity matrix.

The model of fold formation by a viscous mechanism has been particularly successful in connection with the explanation of folds in rhyolite flows [120].

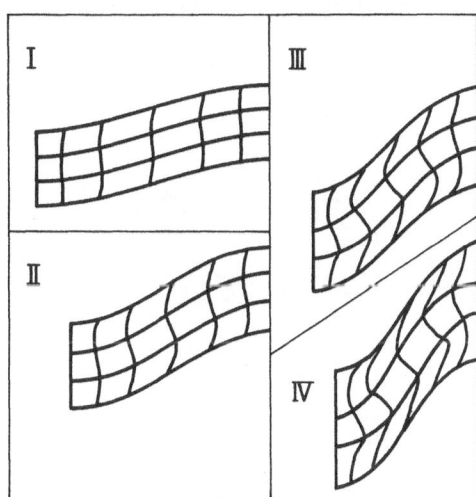

Fig. 104. Four stages of deformation of a high-viscosity central layer with low-viscosity outer layers in a medium-viscosity matrix (not shown). (After de Bremaecker and Becker [119])

7.2.3.5 Plastic Materials

The folded strata may be regraded as having undergone plastic buckling. For this purpose the material is regarded as being essentially elastic, but that, through the yield limit being exceeded, plastic bands develop in certain regions. The latter can be represented as a continuous distribution of (infinitesimal) dislocations; – an approach which has been taken by Teisseyre [121,122].

114 Dietrich, J. H.: Can. J. Earth Sci. 7:467 (1970)
115 Shimamoto, T., Hara, I.: Tectonophysics 30:1 (1976)
116 Cobbold, P. R.: Tectonophysics 38:339 (1977)
117 Williams, J. R., Lewis, R. W., Zienkiewicz, O. C.: Tectonophysics 45:187 (1978)
118 Lewis, R. W., Williams, J. R.: Tectonophysics 44:263 (1978)
119 De Bremaecker, J., Becker, E. B.: Tectonophysics 50:349 (1978)
120 Fink, J.: Geology 8:250 (1980)
121 Teisseyre, R.: Bull. Seismol. Soc. Am. 54:1059 (1964)
122 Teisseyre, R.: Bull. Earthquake Res. Inst. 44:153 (1966)

A related model assumes visco-plastic (rather than elastic-plastic) folding, for which Chapple[123] made computer calculations. The instability of bands, incidentally, seems to be general in any finite-amplitude theory of folds[124].

7.2.3.6 General Rheology

Systems of folds can also be created by an unspecified rheological behavior of the materials in question. By this we mean that the materials are capable of flow with well-defined flow lines, assuming that the stresses and their durations are of the right order of magnitude. The actual type of rheological behavior (plastic, Maxwell-, Bingham-type) need not be specified in detail. Such materials have simply been called[125] "rheid" and it stands to reason that, as long as flow *does* occur, it makes little difference what the actual equations

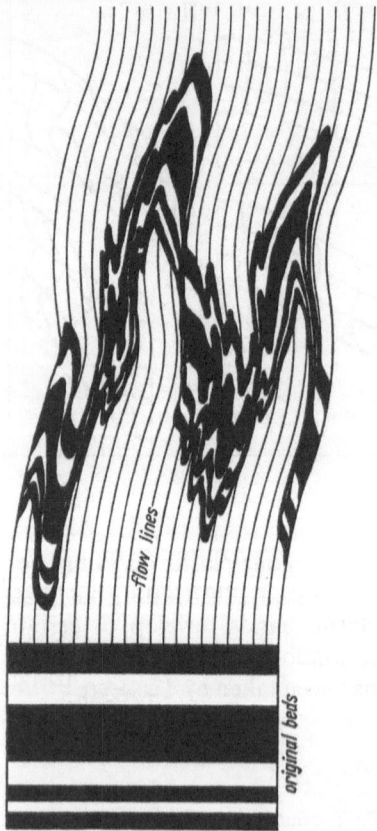

flow lines

original beds

Fig. 105. Formation of folds owing to differential flow. (After Carey[125])

123 Chapple, W. M.: Tectonophysics 7:97 (1969)
124 Chapple, W. M.: Bull. Geol. Soc. Am. 79:47 (1968)
125 Carey, S. W.: J. Geol. Soc. Aust. 1:67 (1953)

of motion are. In effect, differential flow phenomena can simulate tectonic fold phenomena, although they are not tectonic at all (Fig. 105).

Thus, Cobbold[126] has shown that the folding of the components of a rock system can be viewed as a general unstable strain-dependent process. Naturally, the individual fold shapes depend on the initial morphology, the mechanical behavior of the rock, and the overall finite strain. Whether the folded components develop periodic wave forms depends on the relative rates of propagation versus amplification of the folds and the boundary conditions of the rock system.

7.2.3.7 Model Experiments of Folding

The analytical treatment of the problem of fold formation is fraught with considerable difficulties. Hence, recourse was taken to some sort of experiments when it was realized that the connection between the geophysical stress field and its geological manifestations should be established.

The basis of such experimental investigations is the mechanical theory of scaling. The general principles of scaling have been known in physics for a long time, but it is to the credit of Hubbert[127,128] to have pointed out their significance in connection with geodynamical problems.

The general principles of mechanical scaling are based upon the fact that the scaling factors between the model and nature must be chosen in such a manner as to cause all the relevant dynamical equations to become identities if the scaling factors are inserted in place of the quantities themselves. This can be exemplified as follows. Assume a dynamical system whose behavior is completely described by Newton's law of motion

$$F = M d^2 X / d T^2 \qquad (7.2.3-14)$$

where F signifies the force, M the mass, X the displacement, and T time. If we denote the corresponding scaling factors by f, m, x, t respectively and insert them into the equation of motion, we obtain the following scaling condition:

$$f = mx/t^2 . \qquad (7.2.3-15)$$

The relationship between the various scaling factors is therefore the same as that between the dimensions of the various quantities involved. Since, in a mechanical system, there are *three* independent dimensional units (usually chosen as *length*, *mass*, and *time*) it follows that there are, in general, three independent scaling factors that one is able to choose at will. All other scaling factors are then prescribed.

After Hubbert's fundamental investigations, many people have made model studies of geololgical structures. The most comprehensive set of experi-

126 Cobbold, P. R.: Philos. Trans. R. Soc. London Ser. A 283:129 (1976)
127 Hubbert, M. K.: Bull. Geol. Soc. Am. 48:1459 (1937); 62:355 (1951)
128 Hubbert, M. K.: Bull. Am. Assoc. Pet. Geol. 29:1630 (1945)

ments has been made by Ramberg and co-workers[129-135], who investigated not only straightforward folds, but also ptygmatic structures, intersecting folds, drag folds, gravity folding, etc. In this, they used various materials (rubber sheets, putty, viscous fluids) to produce patterns that looked very much like the folds observed in the field. Another series of experiments was performed by O'Driscoll[136-141] who investigated the deformations occurring in layered models of plastic substances and even in packs of cardboards (such as packs of playing or computer cards). Decks of cards to model geomechanic structures were also used by Müller's group in Karlsruhe[142].

Models using viscous fluids were also studied by Lebedeva[143,144], Sycheva-Mikhaylova[145], Hudleston[146], and Shimamoto[147]. Plastic substances were used by Vikhert and Kurbatova[148] (clay), by Agostino[149] (plasticene), by Ramsay[150] (plasticene), by Latham[151], and by Gairola[152]. Dennis and Häll[153] duplicated the conditions obtaining in Jura-type folding (a stiff sedimentary overlying a highly ductile evaporitic layer) by using a wood block and plasticene in a centrifuge experiment, and Horsfield[154] modeled the conditions obtaining in the North Sea by using a specially built sandbox. Finally, folding experiments were also made in actual rock specimens[155-158].

129 Ramberg, H.: Geol. Rundsch. 51:405 (1961)
130 Ramberg, H.: Bull. Am. Assoc. Pet. Geol. 47:484 (1963)
131 Ramberg, H.: Geol. Mag. 100:97 (1963)
132 Ramberg, H.: Bull. Geol. Inst. Univ. Uppsala 42:1 (1963)
133 Ramberg, H., Stephansson, O.: Tectonophysics 1(1):101 (1964)
134 Ghosh, S. K.: Tectonophysics 6(3):207 (1968)
135 Gosh, S. K., Ramberg, H.: Tectonophysics 5(2):89 (1968)
136 O'Driscoll, E. S.: J. Alberta Soc. Pet. Geol. 10:145 (1962)
137 O'Driscoll, E. S.: Nature (London) 1962:1146 (1962)
138 O'Driscoll, E. S.: Nature (London) 201:672 (1964)
139 O'Driscoll, E. S.: Nature (London) 203:832 (1964)
140 O'Driscoll, E. S.: Econ. Geol. 59:1061 (1964)
141 O'Driscoll, E. S.: Bull. Can. Pet. Geol. 12:279 (1964)
142 Rengers, N., Müller, L.: Rock Mech. Suppl. 1:20 (1970)
143 Lebedeva, N. B.: Sov. Geol. 2:70 (1966)
144 Lebedeva, N. B.: Tectonophysics 7:339 (1969)
145 Sycheva-Mikhaylova, A. M.: Geotektonika (USSR) 1969(4):47 (1969)
146 Hudleston, P. J.: Tectonophysics 16:189 (1973)
147 Shimamoto, T.: Tectonophysics 22:253 (1974)
148 Vikhert, A. V., Kurbatova, N. S.: Geotektonika (USSR) 1968(2):136 (1968)
149 Agostino, P. N.: Bull. Geol. Soc. Am. 82:2651 (1971)
150 Ramsay, J. G.: Bull. Geol. Soc. Am. 85:1741 (1974)
151 Latham, J. P.: Tectonophysics 57:T1 (1979)
152 Gairola, V. K.: Tectonophysics 41:291 (1977)
153 Dennis, J. G., Häll, R.: Tectonophysics 45:T15 (1978)
154 Horsfield, W. T.: Geol. Mijnb. 56:363 (1977)
155 Paterson, M. S., Weiss, L. E.: Nature (London) 195:1046 (1962)
156 Paterson, M. S., Weiss, L. E.: Bull. Geol. Soc. Am. 79:795 (1968)
157 Handin, J. et al.: Bull. Geol. Soc. Am. 87:1035 (1976)
158 Friedman, M., Hugman, R. H. H., Handin, J.: Bull. Geol. Soc. Am. (I) 91:307 (1980)

7.2.4 Theory of Petrofabrics

7.2.4.1 Introduction

"Petrofabrics" (structural petrology) concerns the small-scale (of the order of millimeters) features of rocks. Such features include the shape and orientation of pebbles, the strain effects on crystals, the orientation of grains in polycrystalline rocks, the formations stylolites as well as the orientation of microjoints and -folds. The original impetus for the study of such features was undoubtedly given by Sander[159-161]. A recent review of the subject as a whole was given by Friedman and Sowers[162] which contains many references.

Many petrofabric features are caused by sedimentation processes. These are of no direct concern in connection with a discussion of geodynamics because they were not produced by the geophysical stress field. It is, however, the *changes* in the original configurations that are of interest; – in this context, a knowledge of the original features is, of course, important. Nevertheless, the mechanics of deposition itself belongs to another field of study.

We shall now discuss the various theories of petrofabric features and their genesis by the geophysical stress field one by one.

7.2.4.2 Pebble Shapes

A pebble may give an indication of the strain it has undergone, and therewith of the stresses that have been acting upon it, during its history. A pebble forming part of a conglomerate may, however, owe its shape partly to previous abrasion, not only to the subsequent straining. If the pebbles are assumed to be spherical before being strained, then the strains undergone by them can be inferred from their final shapes in a straightforward manner. However, if the pebbles had an unknown elliptical shape before deformation, this is not so easy. Nevertheless, it is still possible to do this if all the pebbles are assumed to have had the *same* shape but various orientations before deformation; then the original shape as well as the strain tensor can be determined. The theory and useful templates for applying it have been reviewed by Elliott[163] and by Siddans[164]. Evidently, the method can be put onto a statistical basis so that the *average* strain having acted on *average* pebbles can be determined from a pebble-shape analysis in a conglomerate.

Elliott's method was further expanded by Shimamoto and Ikeda[165] who devised a simple algebraic method for strain estimation from deformed

159 Sander, B.: Gefügekunde der Gesteine. Wien: Springer 1930
160 Sander, B.: Einführung in die Gefügekunde der geologischen Körper I. Wien: Springer 1948
161 Sander, B.: Einführung in die Gefügekunde der geologischen Körper II. Wien: Springer 1950
162 Friedman, M., Sowers, G. M.: Can. J. Earth Sci. 7:477 (1970)
163 Elliott, D.: Bull. Geol. Soc. Am. 81:2221 (1970)
164 Siddans, A. W. B.: Tectonophysics 64:1 (1980)
165 Shimamoto, T., Ikeda, Y.: Tectonophysics 36:315 (1976)

ellipsoidal objects with random orientation in the undeformed state, and by Lisle[166] who gave approximate estimates of the tectonic strain ratio from the mean shape of deformed elliptical markers. Specific cases referring to conditions of simple shear and combined simple shear and pure shear, were discussed by Le Theoff[167]. Finally, the deformation of arbitrary polygonal strain markers[168,169] and of general bodies[170] in a strain field has also been discussed.

7.2.4.3 Mechanical Deformation of Crystals

The situation with crystals in a homogeneous substance is very much akin to that which occurs with pebbles, except that the original shape of the crystal is usually known.

As possible deformation mechanisms the following have been suggested: (1) the crystals are passively deformed by the mechanical movement of the matrix in which they are embedded, − the rheology of the grains and matrix may be assumed as viscous, plastic or even more general; (2) the crystals *align* themselves, without being actually deformed, according to the movement in the matrix in which they are embedded. Let us discuss briefly these two possibilities.

1. Deformation of Crystals in a Matrix. The first model in this category is one in which the deformation of a viscous spherical particle embedded in a viscous medium was studied. Based on a set of velocity equations originally derived by Taylor[171], Gay[172] found that the particle changes its shape according to the equation

$$\ln \left(\frac{a}{b} \right) = \ln \left(\frac{a_i}{b_i} \right) + \frac{5}{2R + 3} \ln \sqrt{\frac{\lambda_1}{\lambda_2}} \qquad (7.2.4-1)$$

where a_i, b_i are the original half-axes of the particle, λ_1, λ_2 are the major and minor principal elongations of the system, and R is the viscosity ratio (particle/medium). The above solution was verified by finite-element calculations on a computer[173].

The model of viscous grains is probably not very widely applicable. Generally, crystals (particularly halite, calcite) may be regarded as *plastic*. Based on simple relations between maximum shear in plastic flow and principal stresses, Carter and Raleigh[174] calculated the direction of the

166 Lisle, R. J.: Geol. Mijn. 56(2):140 (1977)
167 Le Theoff, B.: Tectonophysics 53:T7 (1979)
168 Robin, P. F.: Tectonophysics 42:T7 (1977)
169 Roder, G. H.: Tectonophysics 43:T1 (1977)
170 Willis, D. G.: Bull. Geol. Soc. Am. 88:883 (1977)
171 Taylor, G. I.: Proc. R. Soc. London 138:41 (1932)
172 Gay, N. C.: Tectonophysics 5:211 (1968)
173 Shimamoto, T.: J. Geol. Soc. Jpn. 81(4):255 (1975)
174 Carter, N. L., Raleigh, C. B.: Bull. Geol. Soc. Am. 80:1231 (1969)

principal stress axes of the geotectonic stress tensor from typical fabrics. The general aspects of the problem were discussed by Nicolas[175] and a completely theoretical approach to the problem was made by Etchecopar[176] by setting up a purely geometrical model of dislocation gliding treated numerically. The type of model is illustrated in Fig. 106. Evidently, it is possible to calculate the resulting shapes for many original configurations.

Most basic investigations into the resulting deformation patterns were made experimentally. These refer generally to specific materials, such as quartz[177], quartzites[178,179], olivine[180], and other silicates[181]. These studies are usually carried out by the application of electron microscopy.

A particular type of mechnical deformation in a crystal is that of mechanical twinning. From the investigations of such twins, the stress system producing them can often be inferred. The basic relationships are determined from experiments. Corresponding studies have been reported by Burger[182] for calcite, by Kirby and Christie[183] for diopside, and by Livshits and Prishchepov[184] for zinc.

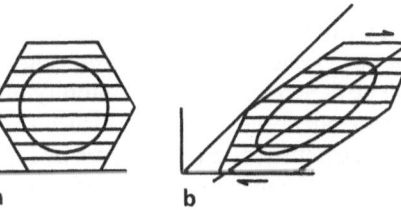

Fig. 106. a Initial cell, **b** Cell deformed by homogeneous slip (horizontal lines are slip lines). (After Etchecopar[176])

a   b

2. Alignment of Particles in a Matrix. Petrofabric effects may be induced in a medium subject to strain without an actual deformation of the particles occurring. The latter are simply passively lined up within the matrix. Generally, one finds that the long axes of the particles align themselves parallel to the shear planes of the matrix in the direction of the flow. This is the case whether the matrix is assumed as viscous or not. Two possible mechanisms have been suggested; one relies on the varying angular velocity of the grains as they rotate in the sheared matrix, the other on grains tending to

175 Nicolas, A.: Eclogae Geol. Helv. 72(2):509 (1979)
176 Etchecopar, A.: Tectonophysics 39:121 (1977)
177 Tullis, I.: Tectonophysics 39:87 (1977)
178 Mitra, S.: Contrib. Mineral. Petrol. 59:203 (1976)
179 Mitra, S.: J. Geol. 86:129 (1978)
180 Gueguin, Y.: Les dislocations dans l'olivine des peridotites. Ph. D. Thesis, Univ. Nantes (1979)
181 Christie, J. M., Ardell, A. J.: In: Electron Microscopy in Mineralogy (eds. H. R. Wenk et al.). Berlin-Heidelberg-New York: Springer, p. 374, 1978
182 Burger, H. R.: Bull. Geol. Soc. Am. 83:2439 (1972)
183 **Kirby**, S. H., Christie, J. M.: Phys. Chem. Miner. 1:137 (1977)
184 Livshits, L. D., Prishchepov, V. F.: Bull. (Izv.) Acad. Sci. USSR Earth Phys. 1977(6):19 (1977)

take attitudes such that minimum angular momentum is transferred during collisions[185,186].

7.2.4.4 Theory of Orientation of Grains in Polycrystalline Rocks

The orientation of the grains in a polycrystalline material may depend in a large measure on the stresses present during the formation of the rock. This can be due to *nucleation* or to *growth* in a preferential direction.

Referring to nucleation, Goguel[187-189] noted that the solubility of various minerals in a melt is a function of the prevailing stress such that the orientation of the grains in metamorphic rocks gives an indication of the stresses acting at the moment of crystallization. For feldspars, Goguel[189] predicted a grain orientation preferentially normal to the acting maximum pressure. For pyrite, galena, and rock salt the faces of the crystal cubes are preferentially normal to the principal stress directions, in other cubic minerals (magnetite, diamond) the diagonals of the cubes are parallel to the principal stress directions. In hexagonal crystals (β-quartz, ice, beryl) the crystal axis is preferentially parallel or perpendicular to the maximum compression. For other types of crystals, the relationships are more complicated.

The dependence of the solubility of a component on stress also leads to the general phenomenon of "pressure solution" whereby a change in the petrofabric structure of a rock occurs by internal diffusion[190]. This mechanism was investigated with regard to various specific materials, such as quartz[191], quartzite[192,193], and rutile[194]. It has been advocated as the reason for the existence of "tabular" (flattened) mineral grains (normal to the maximum stress) often observed in deformed rocks, although shear displacement of segments of initially equant grains has also been invoked in the latter case[195].

A nucleation process can also occur in the origination of kink bands in a perfectly plastic material with one slip surface[196]. Whilst this has no physical relation to the nucleation of crystals, there is a superficial analogy between the two processes, inasmuch as in both cases an energetic instability in present.

The above remarks refer to the mechanics of *nucleation* of crystals. The stresses and strains acting during the *growth* phase of crystals (visible as inclusions in metamorphic fabrics) can also be inferred from petrofabric

185 Rees, A. I.: J. Geol. 76:457 (1968)
186 Rees, A. I.: Tectonophysics 55:275 (1979)
187 Goguel, J.: C. R. Acad. Sci. 260:6145 (1965)
188 Goguel, J.: Bull. Soc. Geol. Fr. (7) 7:747 (1965)
189 Goguel, J.: Bull. Soc. Geol. Fr. (7) 9:481 (1967)
190 Elliott, D.: Bull. Geol. Soc. Am. 84:2645 (1973)
191 Le Corre, C.: Bull. Soc. Geol. Fr. (7) 19:1109 (1977)
192 Wilson, C. J. L.: Tectonophysics 57:T19 (1979)
193 Gapais, D.: Bull. Mineral. 102:249 (1979)
194 Mitra, S.: Bull. Geol. Soc. Am. 90:227 (1979)
195 Gregg, W.: Tectonophysics 49:T19 (1978)
196 Weiss, L. E.: Tectonophysics 65:1 (1980)

analyses[197]. If a crystal is considered when it is growing during progressive deformation, the relative rotation between crystal and matrix occurs in increments whose axis is dependent on the shape of the incremental strain ellipsoid. Wilson[197] applied this fact to the determination of the strains undergone by porphyroblasts of Norway.

7.2.4.5 Stylolites

Stylolites are another feature which can be ascribed to pressure solution; they occur mainly in limestones. The solution proceeds mainly from layering and jointing surfaces in the rock. Because of stress-dependent differences in solubility the stylolite plugs grow in the direction of the largest compression.

There is strong evidence that stylolites are indeed sutures formed by the removal of soluble rock material owing to stress-dependent differences in solubility[198]. The stylolite plugs grow in the direction of the largest compression[199-203], as this produces a consolidation in that direction (the stylolite plug is denser than the porous limestone). The envisaged process is, hence, in conformity with the principle that a state tends to change in such a fashion that it corresponds to a minimum potential energy content.

7.2.4.6 Microjoints, Microfolds, and Similar Features

If minute cracks, small folds, and such "petrofabric" features are analyzed on a microscopic scale, similar types of inferences regarding the stress and strain history as from the corresponding macroscopic features can be drawn.

The minute cracks in rocks are commonly referred to as "cleavage". Some of the implications regarding cleavage were already discussed in Sect. 7.2.2. Let us simply recapitulate that cleavage is developed in relation to lithology as a function of regional strain variation. The various possibilities have been summarized by Le Corre[204] who distinguished five types of cleavage. The terminology, however, is entirely qualitative. An interesting type of cleavage is crenulation cleavage; these are thin, sharply defined quasi-planar discontinuities which truncate a crenulated preexisting fabric (Fig. 107). While mostly (see a review by Gray[205]) a purely mechanical (shear) origin of such features has been preferred, there has recently been a shift toward aspects of mineralogical differentiation. Thus, an explanation of the phenomenon has been

197 Wilson, M. R.: J. Geol. 80:421 (1972)
198 Stockdale, P. B.: J. Sediment. Petrol. 13:3 (1943)
199 Arthaud, F., Mattauer, M.: Bull. Soc. Geol. Fr. 11:738 (1969)
200 Arthaud, F., Mattauer, M.: Bull. Soc. Geol. Fr. 14:12 (1972)
201 Plessmann, W.: Geol. Rundsch. 61:332 (1972)
202 Schäfer, K.: Fridericiana 23:30 (1979)
203 Sellier, E., Morlier, P.: C. R. Acad. D 282(10):953 (1976)
204 Le Corre, C.: Bull. Mineral. 102:273 (1979)
205 Gray, D. R.: Lithos 10:89 (1977)

advanced which assumes that solution and removal of quartz along the cleavages occurs ("solution transfer process"), resulting in a passive concentration of phyllosilicates[206-209]. One would, thus, again be faced with some process akin to "pressure solution". The fact that the cleavage planes are connected with the crenulation and are more or less orthogonal to the direction of maximum shortening (indicated by σ_{max} in Fig. 107) would appear to obviate the interpretation of crenulation cleavage as a shearing phenomenon which is independent of the fold formation and indeed to support the pressure-solution hypothesis[210]. However, the question has evidently not yet been entirely resolved.

Microcracks may also occur in the constituents of conglomerates, such as in pebbles and clasts. Normally, these cracks correspond entirely to the neotectonic stress field (cf. Sect. 7.2.2), but, inasmuch as pebbles and clasts may be well cemented into a matrix, some of the ancient strains may be preserved by them. Upon this basis, Eisbacher[212] tried to deduce Paleozoic stress conditions from fractures in pebbles in a carboniferous conglomerate in Nova Scotia.

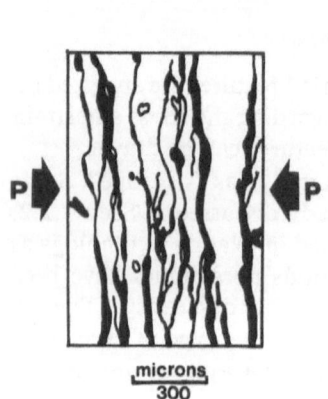

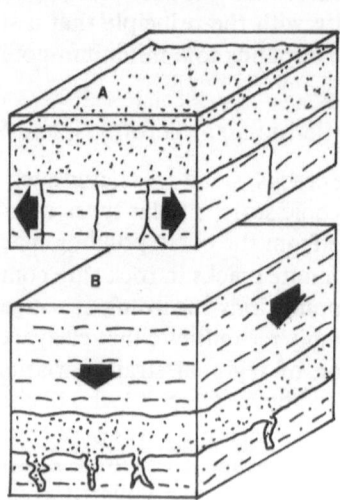

Fig. 107

Fig. 108

Fig. 107. Crenulation cleavage (schematic). (After Gray[211])

Fig. 108. Genetic model of synsedimentary crack formation. (Modified after Mojica and Herrera[213])

206 Gray, D. R.: Bull. Geol. Soc. Am. 89:577 (1978)
207 Gray, D. R.: Am. J. Sci. 279:97 (1979)
208 Gray, D. R., Durney, D. W.: J. Struct. Geol. 1:73 (1979)
209 Granath, J. W.: J. Geol. 88:589 (1980)
210 Gray, D. R., Durney, D. W.: Tectonophysics 58:35 (1979)
211 Gray, D. R.: J. Geol. 85:763 (1977)
212 Eisbacher, G. H.: Can. J. Earth Sci. 6:1095 (1969)

For many, evidently extension-type, small cracks a synsedimentary origin has been postulated: The sediments being deposited cause a stress state in the underlying stratum in which the vertical is the direction of maximum, one of the horizontal directions that of minimum compression. An extension crack would form parallel to the maximum compression, i.e., parallel to the vertical[213]. Thus, the genetic model of the process would be as shown in Fig. 108. Many instances of cracks of this type have been found in the field[213].

A corresponding mechanism in crystalline rocks is that of a sequence of cracking and sealing: The rock fails by brittle fracture and is then sealed by crystalline material derived from the rock matrix by pressure solution[214]. However, no stress estimates have been made for either of the two extension-crack mechanisms discussed last.

The grains in polycrystalline materials may undergo deformations after their nucleation and growth. Thus microcracks occur in the constituents of granite[215], mylonitic rocks[216], and tectonites[217].

Finally, small-scale folds can have a petrofabric aspect which, in the usual manner described earlier, can be interpreted in terms of stresses and strains. Thus, Tchalenko[218] investigated the evolution of kink bands in sheared clays and Hancock[219] and Ramsay[220] analyzed en-echelon veins cutting small chevron folds in terms of strains undergone by the corresponding matrix rocks. Similar analyses have also been reported for mylonites[221,222]. Crenulation folds in argillaceous sediments have been investigated by Maltman[223], in polycrystalline rocks by Vialon[224].

7.3 Theory of the Seismic Source

7.3.1 General Remarks

As noted, earthquakes occur when stresses that had been built up in the ground are suddenly released by the formation (or increase) of a discontinuity in the medium involved. A valid theory of the phenomenon of earthquakes must account for the various aspects of its characteristic features. First of all,

213 Mojica, J., Herrera, A.: Geol. Norandina 1:19 (1980)
214 Ramsay, J. G.: Nature (London) 284:135 (1980)
215 Douglass, P. M., Voight, B.: Geotechnique 19(3):376 (1969)
216 Eisbacher, G. H.: Bull. Soc. Am. 81:2009 (1970)
217 Stauffer, M. R.: Can. J. Earth Sci. 7:498 (1970)
218 Tchalenko, J. S.: Tectonophysics 6(2):159 (1968)
219 Hancock, P. L.: Geol. Mag. 109(3):269 (1972)
220 Ramsay, J. G.: Philos. Trans. R. Soc. London Ser. A 283:3 (1976)
221 Themistocleous, S. G., Schwerdtner, W. M.: Can. J. Earth Sci. 14(8):1708 (1977)
222 Carreras, J., Estrada, A., White, S.: Tectonophysics 39:3 (1977)
223 Maltman, A. J.: Tectonophysics 39:417 (1977)
224 Vialon, P.: Eclogae Geol. Helv. 72(2):531 (1979)

one must find a mechanism which causes stresses to be built up in those layers of the Earth which are prone to contain earthquake foci. Second, the particular cyclic pattern in which the energy is released in earthquake sequences must be explained. Third, one has to account for the characteristic way in which energy is radiated from the focus during an earthquake. Finally, the actual process at the focus during the occurrence of an earthquake is of much interest. Needless to say, these are all very difficult problems and no complete explanation can be hoped for to date.

7.3.2 Models of Earthquake Preparation

In principle, the preparation of an earthquake occurs by the buildup of tectonic strain to such a point where the material "breaks". In various regions of the world, it has been observed that time intervals of high seismic activity alternate with time intervals of low activity, the characteristic period being about 10 years. It is therefore necessary to seek a mechanism which could produce patterns of alternating activity.

The oldest idea to solve this problem was to consider earthquakes as some "stick-slip" type of brittle fracture process. This would imply the validity of Mohr's fracture criterion, i. e., the postulate that the stresses producing the fractures are such that the fracture surface is inclined by about 30° toward the maximum principal pressure and contains the intermediate principal stress. Equally, turning to the microscopics of brittle fracture, one would assume the Griffith mechanism and a crack propagation velocity derived from Eq. (3.3.3 – 10). In this interpretation, an earthquake would correspond to a sudden extension at the margin of an old or to a creation of a new fracture, the sweep of the edge at high speed providing the shock; the energy for the fracture would be provided by the release of elastic energy in the strained region. This theory has been called "strain rebound theory" and seems to be due to Reid[225].

Straightforward as such a theory might appear, it has some unsatisfactory aspects. Brittle fracture represents essentially the opening up of a crack, and it is very doubtful whether at the depths at which earthquakes occur, cleavage could occur at all. Furthermore, it is most doubtful whether the material can be assumed as brittle.

The next possibility is therefore to think of some high-velocity ductile fracture. However, this also represents essentially the opening up of a crack with the extension of the crack at its edges causing the shock. This seems improbable.

It appears therefore indicated that one should look for a mechanism where the shock is produced by a sudden slippage along a preexisting fault surface. Analytically, the intermittent sliding along a preexisting fracture surface can

---

225 Reid, H. F.: Phys. Earth 6:100 (1933)

be described by the mechanism of snapping dislocations. The general idea of dislocations has been explained in Sect. 3.2.1.3. However, it is now necessary to envisage physical conditions for the start and continuation of the snapping process along an earthquake fault.

It has already been stated that it is quite impossible to try to give a proper analytical description of a fracture process, and therefore it is also impossible to give an exact analysis of the snapping of dislocations. The best that can be hoped for are, therefore, statistical considerations. In order to do this, Housner[226] split the whole earthquake fault into slip areas A in which the slip is constant, i.e., areas which are active in any one earthquake. He then assumed that the expected number of slips having areas between A and $A + dA$ would be proportional to dA/A, i.e., he assumed a statistical frequency distribution f of slip areas A which may be written

$$f = C\frac{a_0}{A} \equiv C\frac{1}{x} \tag{7.3.2-1}$$

where a_0 is the minimum value of A. Normalization yields

$$C = \frac{1}{\log x_1 - \log x_0} \tag{7.3.2-2}$$

where x_1 is the largest and x_0 the smallest possible value of x. Hence the frequency distribution can be expressed as follows

$$f = \frac{1}{\log x_1 - \log x_0} \cdot \frac{1}{x}. \tag{7.3.2-3}$$

As a measure of an earthquake, Housner took the average slip S occurring along the fault. This he connected by a logarithmic measure with the Richter magnitude M which led to the following formula:

$$dM = dS/S. \tag{7.3.2-4}$$

Integrating, Housner obtained

$$S = S_0 e^M. \tag{7.3.2-5}$$

If it be further assumed that all dislocations are geometrically similar, the slips must be proportional to the square root of the area A. Hence:

$$A = A_0 e^{2M} \tag{7.3.2-6}$$

and finally with Eq. (7.3.2-1):

$$f = \text{const } e^{-2M}. \tag{7.3.2-7}$$

The last Eq. (7.3.2-7) may be compared with the frequency distribution of earthquakes in any one area, M being taken as the Richter magnitude.

226 Housner, G. W.: Bull. Seismol. Soc. Am. 45:197 (1955)

According to Housner, the agreement is good, at least for the earthquakes that occur in the Imperial Valley of California.

A similar test can be made with Eq. (7.3.2 – 6) regarding the areas affected by earthquakes of any one magnitude. From an analysis of observational data Housner obtained

$$A_0 = 3.1 \times 10^3 \, \text{m}^2 . \tag{7.3.2 – 8}$$

Similarly, he obtained for the constant S_0 in Eq. (7.3.2 – 5):

$$S_0 = 6.4 \, \text{mm} . \tag{7.3.2 – 9}$$

These are values which appear to be reasonable. They permit one to calculate the average slip and area of slip for earthquakes of any one magnitude.

In spite of the apparent success of the above theory, there remain, in fact, very grave difficulties of a fundamental nature. These are connected with the observation that it appears as impossible that any form of fracture could exist at the depths at which earthquakes occur. The difficulty can be circumvented by increasing the complexity of the model. The original stick-slip model of Reid (elastic rebound) was further investigated on several occasions, putting it upon a more sophisticated theoretical basis [227,228]. A review of the possibilities has been given by Stuart [229].

Thus, attempts at introducing different types of rheology (not only elasticity and brittle fracture) into various types of models have been reported. Savage and Prescott [230] proposed a mechanism based on the behavior of a screw dislocation in an elastic plate (lithosphere) overlying a visco-elastic substratum (asthenosphere). The relaxation of the asthenosphere in conjunction with the mechanics of the dislocation process in the lithosphere produces a cyclic strain release. Similarly, Andrews [231] envisaged a multilayered model of a seismic area with various properties of the individual layers.

By assuming the presence of a series of circular faults in an earthquake region, which are activated if the stress buildup reaches a critical value on each individual fault, Caputo [232] set up a model for explaining the observed magnitude-frequency laws, i. e., the statistics of recurrence times. This model again produces intermittently occurring earthquakes although the stress accumulates linearly.

A visco-elastic model of the earthquake buildup-and-release process was also investigated by Burridge [233], whilst Stuart [234] assumed strain softening.

227 Ohnaka, M.: Bull. Seismol. Soc. Am. 66(2):433 (1976)
228 Bufe, C. G., Harsh, P. W., Burford, R. O.: Geophys. Res. Lett. 4(2):91 (1977)
229 Stuart, W. D.: Rev. Geophys. Space Phys. 17(6):1115 (1979)
230 Savage, J. C., Prescott, W. H.: J. Geophys. Res. 83:3369 (1978)
231 Andrews, D. J.: J. Geophys. Res. 83:2259 (1978)
232 Caputo, M.: Ann. Geofis. (Roma) 29(4):277 (1976), also Bull. Seismol. Soc. Am. 67(3):849 (1977)
233 Burridge, R.: J. Geophys. Res. 82(11):1663 (1977)
234 Stuart, W. D.: J. Geophys. Res. 84:1063 (1979), Science 203:907 (1979), also Stuart, W. D., Mavko, G. M.: J. Geophys. Res. 84:2153 (1979)

Finally, the stick-slip mechanism has been physically modeled by an experiment using foam rubber as medium[235].

A periodic strain release can also be obtained by assuming that the strain buildup itself is intermittent. Thus, Matuzawa[236] proposed that earthquakes arise in "fields" which are heat engines that function intermittently. The driving force of these engines is the constantly supplied heat flow from the interior of the Earth where it is assumed that in seismic areas the heat flow is somewhat larger than in seismically inactive zones. This would cause the temperature in seismic areas to rise which in turn would produce stresses which eventually could cause an earthquake swarm as soon as the breaking strength of the rocks is exceeded. Once an earthquake swarm is started, it may itself be capable of dissipating the excess heat beneath it rather rapidly so that the process then could start all over again. This would produce the cyclic appearance of earthquake sequences.

Matuzawa[236] considered two thermal mechanisms which might be capable of producing stresses. The first is the obvious one of producing the stresses by thermal expansion. Matuzawa[236] found (with reasonable assumptions for the constants involved) that the seismic zones in the upper part of the mantle would have to be heated to a temperature by 100°C higher than the surrounding aseismic zones to reach stresses equal to the breaking strength of rocks. To expect this to occur every 10 years or so is certainly somewhat unreasonable. Matuzawa therefore considered a second possibility, viz. the assumption that a solid-liquid phase transformation would occur at the bottom of the crust. Such phase transitions are governed by the well-known Clausius-Clapeyron equation

$$\Delta p = \frac{L}{T(v_1 - v_2)} \Delta T \tag{7.3.2 - 10}$$

where Δp is the increase in pressure necessary in order to maintain equilibrium if the temperature is raised by the amount $\Delta T(\text{K})$; L is the heat of melting (per unit mass) and v_1, v_2 are the specific volumes (per unit mass) of the liquid and solid phases, respectively. Using suitable constants ($v_1 = 0.385 \times 10^{-6}\,\text{m}^3/\text{g}$, $v_2 = 0.346 \times 10^{-6}\,\text{m}^3/\text{g}$, $T = 1,500\,\text{K}$, $L = 360\,\text{joule/gram}$) one obtains the result that a temperature increase of 5 K requires an increase of pressure of 30 MPa to maintain equilibrium. If the mantle is assumed to be in a critical equilibrium state, it follows that a greater pressure is required to maintain this equilibrium in a region where the temperature is higher than in the surrounding region; for a 5° temperature increase, the required pressure increase is, as outlined above, 30 MPa. Now, if a temperature increase of 5° is assumed as a reasonable one to occur in the region of an earthquake field, this means that, under the above assumption of a critical equilibrium state, the pressure beneath the crust in an earthquake area must be higher by the calcu-

235 Hartzell, S. H., Archuleta, R. J.: J. Geophys. Res. 84:3623 (1979)
236 Matuzawa, T.: Bull. Earthquake Res. Inst. Tokyo 31:179, 249 (1953), 32:231, 341 (1954)

lated amount as compared with the surrounding areas. Matuzawa shows that such an increase in pressure is indeed capable of producing stresses in the crust which are of the order of the breaking strength of rocks. In order to do this he took as models of earthquake fields plates of various shapes (circular, elliptic) with fixed rim, subject to a pressure on one side[237]. In the case of a circular model, Matuzawa chose the following constants and dimensions: thickness $h = 25$ km, radius $a = 2h$, modulus of elasticity $E = 1.25 \times 10^{11}$ Pa, Poisson's ratio $m = 1/4$. The result was that the maximum shearing stress is of the same order of magnitude as the pressure increase (30 MPa) and hence of the order of the breaking strength of rocks. The corresponding volume increase ΔV, owing to the upbulging of the plate, turns out to be $\Delta V = 5.1 \times 10^{10}$ m^3. It must be assumed, of course, that an equal volume ΔV below the plate is being filled with molten material during the buildup of stresses. As soon as the breaking occurs, the pressure below the considered part of the crust will collapse, heat will escape and great seismic activity will follow. This process may be assumed to repeat itself at regular intervals.

The possibility of the above mechanism hinges on whether enough heat can be supplied to cause the required differential heating. Taking the heat of melting as equal to 9.4×10^7J/m^3 yields that a total amount of heat of $H = 9 \times 10^{20}$ Joules is required to melt the above-calculated volume ΔV. This amount of heat must be conducted into and out of the area of the earthquake field during the time interval of one cycle (about 10 years). Thus, the total heat flow in an earthquake area consistent with the above mechanism would have to be

$$\frac{H}{\pi a^2} \times \frac{1}{10 \text{ years}} \cong 4 \times 10^5 \, \text{mW/m}^2 . \tag{7.3.2 – 11}$$

This is by about the factor 10^4 higher than any normally observed heat flows.

The explanation of the required high heat flow, thus, constitutes one of the main difficulties in the field theory of earthquakes although some investigations regarding this point have been published by Aki[238]. In addition, Lomnitz[239] has also investigated the thermodynamics of planets from a more general standpoint.

Further geological models intending to explain the occurrence of earthquake swarms have been proposed on various occasions. Thus, Hill[240] suggested that a rather specific geological process occurs involving the existence of clusters of magma-filled dikes in which a sequence of progressive failure occurs along a system of conjugate fault planes joining the dike tips en

237 These plates are assumed to be *elastic*. Kasahara [Kasahara, K.: Bull. Earthquake Res. Inst. 37:39 (1959)] amplified the above theory by assuming *plastic* plates
238 Aki, K.: J. Phys. Earth 4:53 (1956)
239 Lomnitz, C.: Geophys. J. R. Astron. Soc. 5:157 (1961)
240 Hill, D. P.: J. Geophys. Res. 82(8):1347 (1977)

echelon at oblique angles. Similarly, Budiansky and Amazigo[241] assumed a direct interaction of fault slip and lithospheric creep. Thatcher and Rundle[242] proposed a specific model of earthquake buildup based upon the subduction phase of the plate-tectonic cycle and Chapple and Forsyth[243] connected earthquakes with the bending of tectonic plates at trenches. A very extreme view has been taken by Gold and Soter[244] who ascribed earthquake buildup to the periodic release of gases (mostly methane) from deep in the Earth's mantle.

Phenomena precursory to a fracture process, such as pore dilatancy (cf. Sect. 3.3.3.3) have been used in attempts at predicting earthquakes. The whole subject matter of forecasting of seismic events, however, is beyond the scope of a treatise on geodynamics; it has been recently reviewed in a monograph on catastrophes[245].

7.3.3 Models of Earthquake Foci

7.3.3.1 General Remarks

Models of earthquake foci have been referred to already in this book in Sect. 2.2.4. We shall investigate here somewhat more closely some of the assumptions adopted there.

The whole issue in the construction of models is to explain the pattern of displacements observed at seismological observatories in terms of a focal mechanism. The observed displacements are partly conditioned by the properties of the medium transmitting the waves; it is therefore first necessary to reduce them to the abstract case of seismic rays emanating in all directions from the focus into a homogeneous medium. This can be done by the device (also mentioned in Sect. 2.2.4) of introducing a focal sphere; i. e., sphere of homogeneous material enclosing the focal region, the sphere being large compared with the focal region. In the focal sphere, all seismic rays are straight. As was outlined earlier, each seismic station corresponds to a point on the surface of the focal sphere; the equivalent displacement on that surface can be inferred from seismic observations. This can be done with the aid of travel-time curves and with the aid of results of wave-transmission theory which is beyond the scope of the present book. Then, the task is to construct such models of the focal mechanism so that the induced motions on the surface of the focal sphere tally with those which are inferred from actual observations.

The types of models that have been proposed for this purpose fall into three categories: (1) a fault with rigid motion of the two halves of the focal sphere; (2) a point singularity embodying a singularity in the stress field; (3) a

241 Budiansky, B., Amazigo, J. C.: J. Geophys. Res. 81(26):4897 (1976)
242 Thatcher, W., Rundle, J. B.: J. Geophys. Res. 84:5540 (1979)
243 Chapple, W. M., Forsyth, D. W.: J. Geophys. Res. 84:6729 (1979)
244 Gold, T., Soter, S.: Sci. Am. 242(6):154 (1980)
245 Scheidegger, A. E.: Physical Aspects of Natural Catastrophes. Amsterdam: Elsevier 1975

point singularity embodying a singularity in the strain field (a dislocation), and (4) a complicated fracture-formation process.

We shall discuss these possibilities in detail below.

7.3.3.2 Rigid Fault-plane Model

A rigid fault-plane model has already been discussed in Sect. 2.2.4. It was shown that it corresponds essentially to the picture of a fault, i.e., of "an orange being sliced down the middle" (cf. Fig. 44). With regard to the distribution of first onsets of seismic waves, this model has been very successful. It has led to the determination of the directions of the principal tectonic stresses causing the earthquakes.

In view of the success of the rigid fault-plane model with regard to the accounting of observed facts, it appears that the assumption of a simple fault has at least a great likelihood of being correct. However, in order to get a proper understanding of the earthquake process, it is necessary to have a detailed picture of the mechanism of slip along the fault.

At first glance it would appear that earthquakes are not even mechanically possible[246]. The surface roughness of broken rocks causes, under ordinary (laboratory) circumstances, the latter to have a coefficient of friction of $f = 2$. At 100 km depth the pressure is roughly

$$p = 3 \times 10^9 \, \text{Pa} \, . \tag{7.3.3 – 1}$$

This means that, with a coefficient of friction of $f = 2$, one needs a tangential stress of

$$\tau = fp = 6 \times 10^9 \, \text{Pa} \tag{7.3.3 – 2}$$

to slide one rock face over another. However, this required tangential stress is about ten times larger than the yield stress of rocks. Hence earthquakes appear to be impossible.

The only possible way to resolve the above dilemma is by assuming that the coefficient of friction must, in effect, be much smaller than $f = 2$ suggested by laboratory experiments. Indications that this is so, come from experiments by Jaeger[247] who found friction coefficients of the order of 0.6 under pressure. Such values, however, are still too high for earthquakes to be possible. Further indications that the coefficients of friction might be much lower at high pressures come also from the work of Bowden[248] on the microscopics of friction. Accordingly, under ordinary circumstances, the actual area of contact (not to be confused with the apparent area of contact) between

---

246 This has been noted, for instance, by Orowan [Orowan, E.: Geol. Soc. Am. Mem. 79:323 (1960)]

247 Jaeger, J. C.: Geofis. Pura Appl. 43:148 (1959)

248 Bowden, F. P., Tabor, D.: The Friction and Lubrication of Solids. Oxford: Clarendon Press 1954

two sliding surfaces is in fact very small, due to minute surface irregularities. During sliding, pressures across the actual area of contact are always very great; in fact so great as to cause plastic yielding and flow. If the pressure between the two sliding bodies is increased, the actual area of contact is simply increased proportionally which yields the customary linear law of dry friction implied in Eq. (7.3.3 – 2).

Extrapolating the above picture to very high pressures, it can be argued that one should expect that, at a certain stage, complete contact (i. e., a saturation point) would be reached between the sliding surfaces and hence that at that stage the linear law of dry friction should break down. It is easy to calculate the pressure beyond which the law of dry friction is *certain* to break down. Bowden's[248] (p. 31) data show that in the case of steel, the area of contact is a fraction of $1/(9.5 \times 10^3)$ of the total area per pressure of 100 kPa. Thus, if the law of dry friction were valid to very high pressures, i. e., if the area of contact would remain proportional to the load, the saturation point would be reached at a pressure of 9.5×10^8 Pa. Beyond this pressure, the mechanism of dry friction as envisaged by Bowden could certainly no longer be valid. This "saturation" pressure, however, is lower than the pressure at 100 km depth.

The above arguments account for *some* lowering of the coefficient of friction. However, for a *substantial* lowering, the presence of high-pressure pore fluid is required[249,250]. Such pore fluid could also be responsible for local melting of the material [cf. Sect. 7.3.3.3 below].

7.3.3.3 Stress Singularities

The most obvious procedure to introduce a physically realistic earthquake-source model is by assuming a singularity at the center of an otherwise homogeneous and completely elastic focal sphere.

Considering first singularities in the stress field, we can assume various types of multipole forces which are suddenly applied at the center of the focal sphere. The displacements at the surface of the sphere can then be calculated by methods of the theory of elasticity[251–253]. If this is done, it turns out that a dipole force with a moment, applied at the center of the focal sphere, produces exactly the same nodal lines as those obtained by the model of rigid faulting. Other types of forces have also been tried, and it turns out that various types of sources produce exactly the same P nodal lines (see Fig. 109). The attendant questions have been analyzed, for instance, by Byerly and Stauder[254]. Thus, a distinction between the various types of sources can be

249 Barley, B. J.: Tectonophysics 51:T9 (1978)
250 Rudnicki, J. W.: Bull. Seismol. Soc. Am. 69:1011 (1979)
251 Malinovskaya, L. N.: Tr. Geofiz. Inst. Akad. Nauk SSSR. No. 22(149):143 (1954)
252 Nakano, H.: Seismol. Bull. Centr. Met. Obs. Jpn. 1:92 (1923)
253 Gassmann, F.: Geofis. Pura Appl. 40:55 (1958)
254 Byerly, P., Stauder, W.: Earthquake Not. 29(3): 17 (1958)

made only if either S nodes or else amplitudes are taken into account. Attempts at such calculations and comparisons with observational data have been made by various people. Thus Knopoff and Gilbert[255] as well as Herrmann[256] attempted to calculate radiation patterns, Keylis-Borok[257] compared some calculations with earthquake records and found that a dipole source with a moment is the most likely type (at least in Russia), and Honda and Masatuka[258], in a similar investigation, found that quadrupole sources fit Japanese data best. It appears thus that the problem of the source type has not yet been finally resolved.

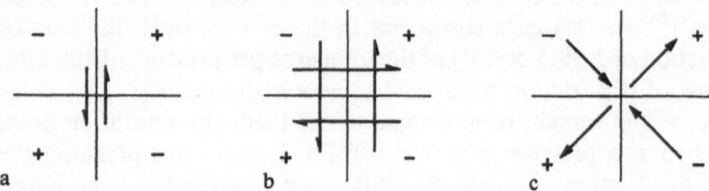

Fig. 109a–c. Identical quadrants of compression and dilatation produced (a) by a dipole force, (b) by a quadrupole force, (c) by a orthogonal regional compression and tension

In addition to the above investigations, one might mention that Kasahara[259] assumed an ellipsoid instead of a spherical focal region so as to obtain an asymmetry. Furthermore, Knopoff and Randall[260] used a compensated linear vector dipole. Brunelle[261] investigated Rayleigh-type surface discontinuities, rather than singularities in a plane stress state.

It may be interesting to compare the above theory with model experiments. Such experiments have been performed by Healy and Press[262–264] with the express purpose of elucidating the position of the P and S nodes in relation to the seismic source. Whereas for the P nodes the result was as expected, this was not the case for the S nodes[264], if the classical theory of a dipole was taken as a basis of comparison. However, the validity of the models has admittedly not been established beyond doubt. In fact, some investigations based on observations of near earthquakes gave the result that something akin to a

255 Knopoff, L., Gilbert, F.: Bull. Seismol. Soc. Am. 49:163 (1959)
256 Herrmann, R. B.: Bull. Seismol. Soc. Am. 69:1 (1979)
257 Keylis-Borok, V. I.: Bull. (Izv.) Akad. Nauk SSSR. Ser. Geofiz. 1957, No. 4:440 (1957).
    Ann. Geofis (Roma) 12:205 (1959) and references given there
258 Honda, H., Masatuka, A.: Sci. Rep. Tohoku Univ. (5) 8:30 (1952)
259 Kasahara, K.: Bull. Earthquake Res. Inst. 35:532 (1957), 36:21 (1958)
260 Knopoff, L., Randall, M. J.: J. Geophys. Res. 75:4957 (1970)
261 Brunelle, E. J.: Bull. Seismol. Soc. Am. 63(6):1885 (1973)
262 Healy, J. H., Press, F.: Bull. Seismol. Soc. Am. 49:193 (1959)
263 Healy, J. H., Press, F.: Geophysics 25:987 (1960)
264 Press, F.: Publ. Dom. Obs. Ottawa 20:271 (1958)

conical source should be assumed[265]. Incidentally, such conical models have also been studied theoretically[266-268].

Finally, we present some calculations of the size of the strained region V prior to an extreme earthquake. If S and μ be the strength and the rigidity of the medium, respectively, then Bullen[269,270] showed that the following relationship is valid

$$12 q \mu E = S^2 V \qquad (7.3.3-3)$$

where E is the energy released in seismic waves and qE the distortional strain energy in the focal region. The energy E can be obtained from the various magnitude relationships given in Sect. 2.2.3, and hence an estimate of V can be arrived at. For an earthquake of the maximum magnitude known ($M = 8.6$), Bullen[270] obtained

$$V = 6 \times 10^3 \, \text{m}^3 . \qquad (7.3.3-4)$$

This is equivalent to a sphere of radius 25 km.

Analogous calculations have also been made by Teisseyre[271] and by Bath and Duda[272].

7.3.3.4 Strain Singularities

Instead of considering singularities in the stresses in the focal sphere, one can consider singularities in the strains. This leads to the theory of dislocations in an elastic medium. The dislocation models advocated in this connection have been shown in Fig. 65 (Sect. 3.2.1). A dislocation theory of earthquake faulting has been proposed particularly by Saito[273], by Vvedenskaya[274], by Droste and Teisseyre[275], and by Andrews[276]. Reviews of this theory have been given by Constantinescu[277], by Press[278], and by Jobert[279].

265 Mikumo, T.: Mem. Coll. Sci. Univ. Kyoto A 29(2):221 (1959)
266 Inouye, W.: Bull. Earthquake Res. Inst. 14:582 (1936)
267 Takagi, S.: Q. J. Seismol. 14(3):1 (1950), 17(3):53 (1953), 17(4):1 (1953)
268 Hirono, T., Usami, T.: Pap. Met. Geophys. 7:287 (1956)
269 Bullen, K. E.: Trans. Am. Geophys. Union 34:107 (1953)
270 Bullen, K. E.: Bull. Seismol. Soc. Am. 45:43 (1955)
271 Teisseyre, R. K.: Acta Geophys. Pol. 6:260 (1958)
272 Bath, M., Duda, S.: Ann. Geofis. (Roma) 17:353 (1964)
273 Saito, Y.: Geophys. Mag. 28 No. 3:329 (1958)
274 Vvedenskaya, A. V.: Bull. (Izv.) Akad. Nauk SSSR. Ser. Geofiz. 1958:175 (1958), 1959:516 (1959), 1960:513 (1960), 1961:261 (1961)
275 Droste, S., Teisseyre, R.: Ann. Geofis. (Roma) 12:179 (1959). Sci. Rep. Tohoku Univ. Ser. V Geophys. 11(1):55 (1959). Bull. Seismol. Am. 50:57 (1960), see also Teisseyre, R.: Acta Geophys. Pol. 8(2):107 (1960)
276 Andrews, D. J.: Bull. Seismol. Soc. Am. 65(1):163 (1975)
277 Constantinescu, L.: Rev. Roum. Geol. Geophys. Geogr. Ser. Geophys. 9(1):3 (1965)
278 Press, F.: Vesiac Rep. 7885-1-X:269 (1977)
279 Jobert, G.: J. Geophys. 43:329 (1977)

Steketee[280] has analyzed such models rather closely mathematically and showed that a dislocation sheet can be built up from nuclei which correspond to those shown in Fig. 109 where the *arrows* are now interpreted as displacements. He noted that the type (*a*) and (*b*) nuclei are dynamically significantly different: while the (*b*) type is in equilibrium, the (*a*) type is not; i. e., if the resulting force and torque are calculated for a part of the medium which contains the nucleus, these vanish for the (*b*) nuclei but not for the (*a*) nuclei. Thus, for an (*a*) nucleus, at least in the static case, an external torque is required to maintain equilibrium (the force vanishes automatically). It should be noted, however, that the above remarks refer to *static* nuclei only; for dynamic nuclei, the conditions applying are still somewhat uncertain. Steketee's calculations have been further elaborated upon by Chinnery[281].

7.3.3.5 Fracture Theory of Earthquakes

A somewhat realistic theory of the earthquake focal mechanism can evidently only arrived at if the dynamic faulting process occurring therein is studied more closely, This includes dynamical studies.

One series of attempts[282-287] is based on studying the elastodynamics (usually in a "far-field" solution) of a discontinuity moving with a certain velocity on a given surface (the fault plane of the earthquake). The theoretically calculated far-field radiation is then compared with seismograph recordings. The most successful model of this type is that of a dynamic shear crack which is produced by a mechanical instability of the rocks unable to withstand the preexisting shear stress. The slippage at the crack is given by the difference between existing and frictional stress.

The calculations are evidently quite complex, and hence, as noted, usually a far-field approximation was made for comparison with seismograph records. However, for engineering purposes, a "near-field" approximation valid for the epicentral region could be more significant. Such a study has been made by Anderson[288].

Because of the difficulty of obtaining analytical solutions to the elastodynamic earthquakes, finite difference calculations using computers[289,290] have also been undertaken, as well as scale-model experiments[291].

280 Steketee, J. A.: Can. J. Phys. 36, 192:1168 (1958)
281 Chinnery, M. A.: Bull. Seismol. Soc. Am. 51:355 (1961). J. Geophys. Res. 3852 (1960)
282 Fossum, A. F., Freund, L. B.: J. Geophys. Res. 80(23):3343 (1975)
283 Andrews, D. J.: J. Geophys. Res. 81(32):5679 (1976)
284 Husseini, M. I.: Bull. Seismol. Soc. Am. 66(5):1427 (1976)
285 Ben-Menahem, A.: Bull. Seismol. Soc. Am. 66(6):1787 (1976)
286 Madariaga, R.: SIAM-AMS Proc. 12:59 (1979)
287 Freund, L. B.: J. Geophys. Res. 84:2199 (1979)
288 Anderson, J. G.: Geophys. J. R. Astron. Soc. 46:575 (1976)
289 Cohen, S. C.: J. Geophys. Res. 82(86):3781 (1977)
290 Stöckl, H.: J. Geophys. 43:311 (1977)
291 Liu, H. P., Hagman, R. L., Scott, R. F.: Geophys. Res. Lett. 5(5):333 (1978)

From the results of such studies obtained and their comparison with seismograph records one may mention the observation that rupture speeds are likely to lie between the Rayleigh-wave and shear-wave velocities[292]. For a California earthquake, a rupture speed of between 2.4 and 2.5 km/s was found[293].

The above investigations are based upon the assumption of a homogeneous fracture process in an elastic medium. Modifications have been attempted by allowing for inhomogeneities in the fracture process, i.e., by proposing a "fault plane with barriers"[294,295]. This possibility has been studied analytically as well as experimentally. Inasmuch as a host of new parameters is introduced in such a model, it is very versatile. It can explain such features as otherwise inexplicable ripples in seismograms and departures of the scaling law of the seismic amplitude from the pattern expected corresponding to a simple similarity assumption. An application to the near-field radiation in a California earthquake (San Fernando, 1971) was reported by Bouchon[296]. It should be noted, however, that the introduction of the new parameters (relative to the barriers) simply throws ununderstandable facts upon hypothetical ones.

Modifications to the idealized elastodynamic procedure have also been made by introducing different types of rheology, such as creep[297] and deformation-diffusion effects in fluid-infiltrated porous media[298,299].

7.3.4 Mechanical Triggering of Earthquakes

7.3.4.1 Introduction

Once we have studied the problem of the focal mechanism of an earthquake, the next problem is that of triggering of the mechanism. If the material in the focal region is close to the breaking point, then it may be expected that relatively minor causes could "trigger" earthquakes. We have seen in Sect. 2.2.6.5 that the time sequence of energy release of earthquakes does not appear to contain any periodicities; thus periodically acting effects, such as the tides of the solid Earth, must be excluded as possible earthquake triggers, and the search for possible agents must be confined to aperiodic phenomena.

292 Richards, P. G.: Bull. Seismol. Soc. Am. 66(1):1 (1966)
293 Trifunac, M. D., Uwadia, F. E.: Bull. Seismol. Soc. Am. 64(3):511 (1974)
294 Scholz, C. H., Molnar, P., Johnson, T.: J. Geophys. Res. 77:6392 (1972)
295 Moore, H., Sibson, R. H.: Tectonophysics 49:T9 (1978)
296 Bouchon, M.: Bull. Seismol. Soc. Am. 68(6):1555 (1978)
297 Ranalli, G.: Rheol. Acta 13:586 (1974)
298 Rice, J. R., Simons, D. A.: J. Geophys. Res. 81(29):5322 (1978)
299 Bonafede, M., Mulargia, F., Boschi, E.: Nuovo Cim. (I) 3C:180 (1980)

7.3.4.2 Solar Activity

In this instance, Sytinskiy[300] and Simpson[301] claim to have found that solar activity plays a significant part in the release of earthquakes. However, more convincing are correlations between earthquakes and hydrological phenomena. Such hydrological phenomena can act in two ways: by increasing the surface load owing to the accumulation of water, and by increasing the pore-water pressure in the formation.

7.3.4.3 Increase of Surface Load

An increase of the load at the surface is, naturally, first of all presented by the filling up of an artificial lake. The problem of accidentally inducing earthquakes by the filling up of artificial reservoirs has been studied on many occasions. Simpson[302] and Gough[303] have given a review of cases where earthquakes were induced in this fashion. Accordingly, earthquakes with magnitudes from 5.0 to 6.5 on the Richter scale were induced by five artificial reservoirs. In 12 cases the induced earthquakes had magnitudes between 3.5 and 5. However, not all artificial reservoirs have induced earthquakes, of the five very largest reservoirs on Earth, only one (Kariba Lake in Africa) created earthquakes. Thus, it is a minority (10%) of reservoirs that cause seismicity. Whether this happens or not evidently depends on the tectonic stress state that is present[303a]. The most direct triggering mechanism is caused by the additional load of the water. The stresses induced in a layered-elastic half-space by a non-homogeneous load from above (corresponding to a lake) have been calculated by Lee[304]; simpler models had been calculated earlier by Steinhauser[305] and Caputo[306]. It was found that the stresses induced by a water volume of the order of 10^{10} m^3 are noticeable as far down as the Mohorovičić discontinuity.

The above triggering mechanism, based on the addition of small incremental stresses to the tectonic stresses which are already present, acts without delay. However, delay times of up to 5 years between the filling of the reservoir and the occurrence of the induced earthquakes have been found; similarly, Beck[307] has shown that at least in one instance (Lake Oroville, California), the weight-induced stresses are too small to trigger the observed earthquakes. Thus, another mechanism must, on occasion, be involved: This is the action *via* the pore pressure (see below).

300 Sytinskiy, A. P.: Geomagn. Aeron. 3:120 (1963)
301 Simpson, J. F.: Earth Planet. Sci. Lett. 3:417 (1968)
302 Simpson, D. W.: Eng. Geol. 10:123 (1976)
303 Gough, D. I.: In: The Assessment and Mitigation of Earthquake Risk. UNESCO, 91, 1978
303a See also Gupta, H. K., Rastogi, B. K.: Dams and Earthquakes. Amsterdam: Elsevier 1976
304 Lee, T.: Bull. Seismol. Soc. Am. 62:1597 (1972)
305 Steinhauser, F.: Gerlands Beitr. Geophys. 6:466 (1934)
306 Caputo, M.: J. Geophys. Res. 66:1479 (1961)
307 Beck, J. L.: Bull. Seismol. Soc. Am. 66:1121 (1976)

The load due to surface water is mostly caused by the filling up of artificial lakes. However, corresponding instances can also occur under natural conditions. Thus, numerous small earthquakes have been triggered by the variations in the water level of the Mississippi River near New Madrid[308]. It may be that the correlation between solar activity and earthquakes is hydrological in nature: Varying solar activity[309] causes varying hydrological loads (water and ice) to be present on the Earth's surface which could trigger earthquakes.

7.3.4.4 Increase of Pore Pressure

The most likely mechanism of triggering earthquakes appears to be by an increase in formation pore pressure[310]. According to the well-known Terzaghi relation (cf. Sect. 3.3.2) an increase in pore-water pressure may cause a material to fail. This is, in effect, the mechanism by which many reservoirs trigger earthquakes. The primary cause is not the additional load represented by the water, but the increase in pore pressure caused by the additional column of water present. This explains the long time delays of up to 5 years between the filling of a reservoir and the occurrence of earthquakes which are occasionally observed: The pore pressure increase is not communicated immediately to the strata which are prone to earthquake activity, but only after a long time delay.

The mechanism of inducing earthquakes by a buildup in formation pore pressure can be observed directly in fluid injection procedure. A famous instance of this type occurred near Rangely, Colorado[311] where earthquake activity was directly related to the rate of injection of fluids into the formation.

7.4 Boundary Effects

7.4.1 Introduction

The geophysical stress field existing within the Earth is greatly affected by boundary surfaces. Such a boundary surface is, of course, represented in the first place by the surface of the Earth itself. We have already seen how the general horizontality of the latter gives rise to the standard stress states of Anderson (cf. Sect. 7.2.2.2). Now we are concerned with the effect of small scale boundary irregularities upon the stress field. In principle, there are two types of effects: The effect of topographic irregularities on the Earth's *surface*, and the effect of the presence of holes and inclusions *in the ground*. We shall discuss these subjects in their turn.

308 McGinnis, L. D.: Circ. Ill. State Geol. Surv. 344:1 (1964)
309 Lursmanashvilli, O. V.: Bull. (Izv.) Acad. Sci. USSR Earth Phys. 2:80 (1973)
310 Kisslinger, C.: Eng. Geol. 10:85 (1976)
311 Raleigh, C. B., Healy, J. J., Bredehoeft, J. D.: Science 191:1230 (1976)

7.4.2 Stresses on Surface Irregularities

7.4.2.1 Theory

The topographic irregularities on the Earth's surface can generally be represented by models of notches or of protrusions. A further possibility is the existence of a single step (such as represented by a rock wall); however, the equilibrium conditions can be satisfied in this case only if a symmetrical step is assumed at infinity. For a calculation of the effects of such surface features, the ground is usually assumed as elastic and the stresses far away from the irregularity as given.

Thus, it is evidently the theory of notches in elastostatics which is of primary importance. A classic treatise on this subject has been written by Neuber[312]. Some further basic problems of the theory of elasticity bearing upon notches were reviewed by Musschelischwili[313]. A complex variable method suitable for geodynamic problems was used by Sturgul[314]. Finite-element techniques were used by Voight and Samuelson[315], and by Ladanyi and Archambault[316]. Special calculations for the stress behavior near sharp corners were made by Karal and Karp[317], and for the stress distribution near a vertical wall by Sturgul and Scheidegger[318]. The results of these investigations will be presented in connection with the discussion of typical features.

7.4.2.2 Valleys

A valley can immediately be regarded as a "notch" in a plane. Hence, the general theory of notch stresses can be applied. This has been done by Scheidegger[319], and by Sturgul and Scheidegger[320]. Under certain circumstances, a stress reversal may occur at the "shoulders" of the valley. As an illustration, Fig. 110 shows the stresses along a notched boundary line. Along this boundary line, the normal stress and the shearing stress vanish; the remaining stress component (normal stress parallel to the boundary line) varies from a maximum value of $3.333\,p$ (p being the value at infinity, i.e., the tectonic stress) at the base of the notch to a minimum of around $0.829\,p$, and then goes asymptotically to $1.000\,p$. This shows that, at the shoulder of the valley, there is a stress relief (the compression becomes less than the tectonic stress).

312 Neuber, H.: Theory of Notch Stresses. Ann Arbor: Edwards 1946
313 Musschelischwili, N. I.: Einige Grundaufgaben zur mathematischen Elastizitätstheorie. München: Carl Hanser Verlag 1971
314 Sturgul, J. R.: Pure Appl. Geophys. 68:66 (1976)
315 Voight, B., Samuelson, A. C.: Pure Appl. Geophys. 76:40 (1969)
316 Ladanyi, B., Archambault, G.: Proc. 24th Int. Geol. Congr. Montreal 13:249 (1972)
317 Karal, F. C., Karp, S. N.: Geophysics 29:360 (1964)
318 Sturgul, J. R., Scheidegger, A. E.: Rock Mech. Eng. Geol. 5:137 (1967)
319 Scheidegger, A. E.: Proc. R. Soc. Victoria 76:141 (1963)
320 Sturgul, J. R., Scheidegger, A. E.: Pure Appl. Geophys. 68:49 (1967)

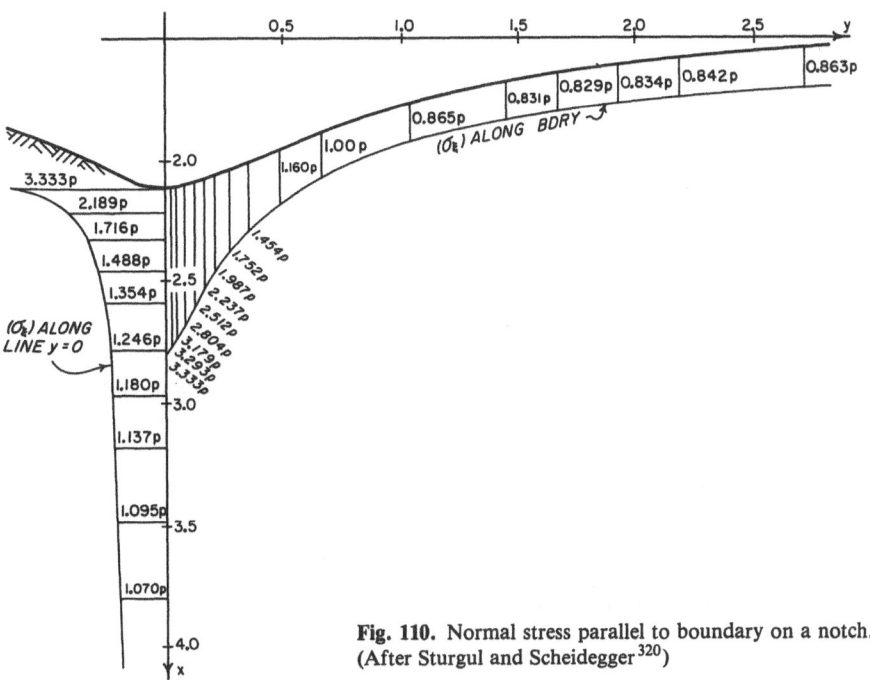

Fig. 110. Normal stress parallel to boundary on a notch. (After Sturgul and Scheidegger[320])

7.4.2.3 Protrusions

The reverse case to a notch is a protrusion. The mathematics is the same for both cases; thus the investigations presented under Sect. 7.4.2.2 also refer to mountain ranges. Again, a stress reversal may occur in a mountain range, this time at the top. Thus, Gerber and Scheidegger[321] considered an elevation in a folded mountain range (Fig. 111) in which basically a standard state of incipient thrust faulting is assumed in the sense of Anderson's theory. Elasticity theory shows that, at the crest of such an elevation subject to lateral compression, a stress reversal occurs at its top. Thus, the weight acting vertically downward is the largest principal pressure P at the summit of such a mountain range. The pressure parallel to the surface becomes the smallest principal stress (it is, in fact, a tension) and the intermediate principal stress is in a horizontal direction normal to an exposed wall. The two potential fracture surfaces will then correspond to the directions denoted by M in Fig. 111; they lead to a wedge-shaped breakout. Such breakouts are commonly found in mountain regions. It is well known that "cirques" exist usually beneath a high peak. Such cirques are commonly held to be of glacial origin; but it is most doubtful whether they were eroded *solely* by glaciers. Rather, it appears that

321 Gerber, E., Scheidegger, A. E.: Eclogae Geol. Helv. 62:401 (1969)

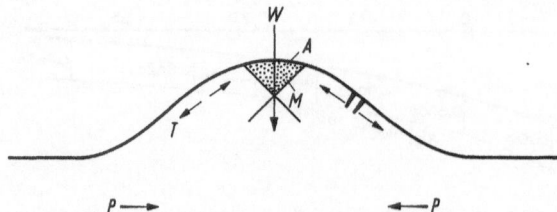

Fig. 111. Stresses around an elevation in an incipient reverse-faulting standard stress state (cross-section)

the glaciers widened preexisting breakouts that had been formed in accordance with the prevailing tectonic stresses[322].

In a similar vein, a stress-induced origin has been advocted for "roches moutonnées" ("Rundhöcker") in periglacial areas: Bär[323] has found that the orientation of the joints in the corresponding areas agrees well with the orientation of these features.

Moeyersons[324] and Ollier[325] made similar studies of the connection of the form of granitic residuals and evidence from fractures. The possibility that apparent "weathering" may in effect be induced by geophysical stresses, has also been considered by Voight[326]. Scheidegger[327] and Einarsson[328] contended that the morphology of submarine ridges may be an effect of the geophysical stress field.

7.4.2.4 Rock Walls

On a rock wall, sawtooth-shaped breaks are often observed. As noted by Gerber and Scheidegger[329], the fractures correspond exactly to a stress state in which the weight is the intermediate stress direction, the smallest principal pressure is horizontal, and the largest a horizontal thrust parallel to the wall. The above qualitative theory has been quantified by Sturgul and Scheidegger[330]. Thus, a rock wall always decays from the bottom, never from the top. If a wall would be eroded from the top, it could not remain a wall for very long at all; it would soon envolve into an inclined slope. It can be shown that the observed recession of a rock wall from the bottom is also an effect of the prevailing stresses: It is induced by the stress concentrations that arise at the sharp corner at the foot of the wall. Owing to their action, the wall breaks out at its foot to form a hollow space. Again, a quantitative theory of this effect has been given in the cited paper of Sturgul and Scheidegger[330].

322 Gerber, E., Scheidegger, A. E.: Riv. Ital. Geofis. Sci. Aff. 2(1):47 (1975)
323 Bär, O.: Geogr. Helv. 12(1):1 (1957)
324 Moeyersons, J.: Z. Geomorphol. 21(1):14 (1977)
325 Ollier, C. D.: Z. Geomorphol. 22(3):249 (1978)
326 Voight, B.: Proc. Int. Soc. Rock Mech. Congr. Lisbon, p. 51 (1966)
327 Scheidegger, A. E.: Proc. R. Soc. Victoria 76:141 (1963)
328 Einarsson, T.: J. Geophys. Res. 73:7561 (1968)
329 Gerber, E. K., Scheidegger, A. E.: Felsmech. & Ingenieurgeol. Suppl. 2:80 (1965)
330 Sturgul, J. R., Scheidegger, A. E.: Rock Mech. Eng. Geol. 5:137 (1967)

7.4.2.5 Exfoliation

On smooth walls one often sees plates which are split off; this phenomenon is called "exfoliation". There is a large literature on this subject which has been reviewed by Brunner and Scheidegger[331]. Accordingly, mechanical effects of fire, freezing and vegetation, chemical weathering, tectonic forces, and stress relief by removal of a load have all been considered as explanations of the phenomenon. In general, a stress-relief theory is presented in the literature as the most likely explanation of exfoliation. Therefore, Brunner and Scheidegger[331] analyzed a simple model which corresponds to this type of theory: The stability against buckling of an elastic plate under horizontal compression (caused by a former overburden) was investigated. The critical dimensions of the plate as a function of the pressure were calculated for various possible boundary conditions. It was seen that this model is rather unsuitable even if the tensional strength of the plates is neglected, inasmuch as the stability of the plate is far too high in all cases.

However, when the exfoliation was considered as an analogon to the well-known multiple axial tension fractures in compression tests (the tensional stresses are induced stresses), then a model was found which has at least possibilities for the explanation of exfoliation: Tensional stresses are induced in a uniaxial compression at the boundaries of the Griffith cracks; they can attain (depending on the shape of the cracks) similar absolute values as the compressive stresses. As soon as the induced tensional stresses exceed the tensional strength of the rock at the most dangerous points, a progressive tension fracture occurs which is aided by notch effects. The result is the well-known multiple axial tension fracture. The occurrence of this fracture pattern can be prevented by a small lateral pressure. The model has been advocated for the explanation of horizontal exfoliation seen in quarries, etc. The "lateral" pressure in this case is the overburden pressure, the horizontal stress is the

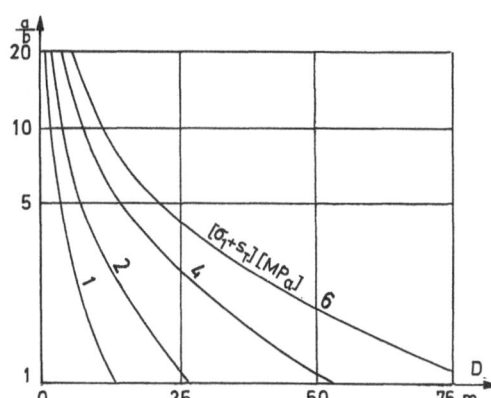

Fig. 112. Critical depth T for induced tension fractures in dependence of the ratio of the axes of the ellipse and on $(\sigma_1 + S_T)$. (After Brunner and Scheidegger[331])

331 Brunner, F. K., Scheidegger, A. E.: Rock Mech. 5:43 (1973)

tectonic stress. Assuming various ratios of the axes a/b of the elliptical Griffith cracks, Fig. 112 presents the results of the calculations for the critical depth T to which exfoliation can occur (the "lateral" pressure becomes equal to the tensile strength S_T of the material) for a given (horizontal) tectonic stress σ_1. Herein, the density of the rock has been taken as $\varrho = 2{,}500$ kg/m^3. Choosing $\sigma_1 = 10$ MPa and $S_T = 6$ MPa one sees that the critical depth becomes 50 m.

Finally, a special model was discussed which is applicable under specific conditions: this model explains the exfoliation of plates parallel to a vertical rock wall. Starting at the weak zone at the foot of a wall which is always induced by stress concentrations as shown by Sturgul and Scheidegger[332,333], a tension fracture progresses behind the wall parallel to its surface as shown in Fig. 113. In this figure, the *numbers* represent the fraction of the stress at infinity ($p = 1$; this is equal to the overburden) which acts as the minimum principal stress at the point in question, calculated by a numerical elastostatic procedure. A *negative sign* signifies a true tension (not only a stress relief). It is evident that a crack develops in the region behind the zone of weakness at the bottom of the wall.

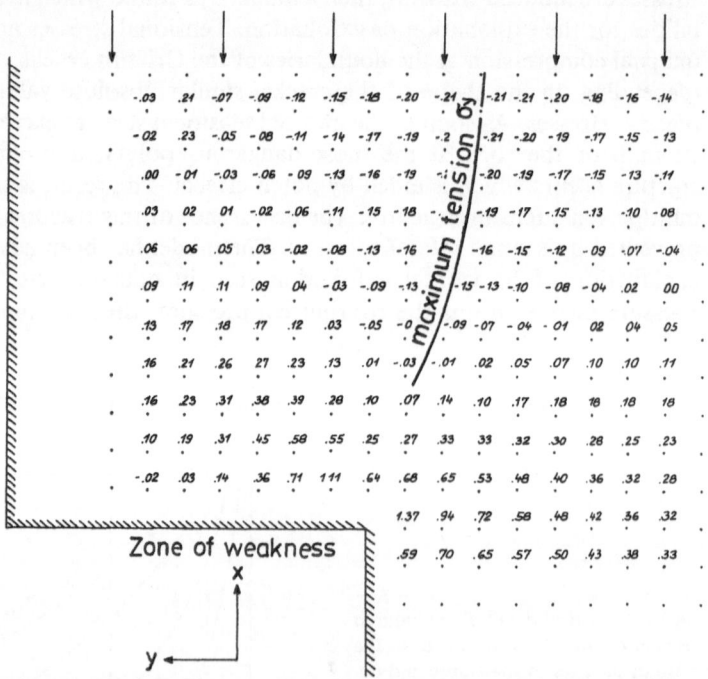

Fig. 113. Line of maximum tensional stress normal to the surface of the wall. (After Sturgul and Scheidegger[332])

332 Sturgul, J. R., Scheidegger, A. E.: Rock Mech. Eng. Geol. 5:137 (1967)
333 Sturgul, J. R.: Effects of Surface Irregularities on the Underground Stress Field. Ph. D. Thesis, Univ. Illinois, Urbana (1967)

7.4.2.6 The Pre-design of Mass Movements

It is a common experience that mass movements occur on slopes in mountainous regions. The causative agents of these movements are of exogenic origin, gravity poviding the energy. These aspects of the mass movements, naturally, do not belong into a treatise of geodynamics.

However, it has been found that many characteristic features of mass movements are, so to speak, "pre-designed" by geodynamic phenomena. Thus, during a careful study of an unstable area in the Felber Valley of Austria[334] it has been found that the orientation of a "mountain fracture" atop the ledge of the movements, the orientation of the joints in the rocks bounding the unstable area itself, all fit into one single geophysical stress pattern.

A similar, particularly detailed study has been made on a slope at the Lesach ridge in Eastern Tyrol, Austria[335], at which markers were established for making exact geodetic measurements which were repeated in later years. Mass movements of the order of centimeters were found which corresponded basically to the usual rotational land-slide theory of Terzaghi[336]. Figure 114 shows the results. However, the *mean* displacement direction coincides closely with the direction of the smallest principal stress of the neotectonic stress field in Europe. Thus, the displacement can be assumed to be pre-designed by the stress field; the actual triggering, of course, will be caused by exogenetic agents.

Part of the reason for the observed facts may be that the orientation of the valleys themselves, on whose sides the mass movements occur, fit into the large-scale neotectonic stress pattern (cf. Sect. 7.2.2.8). Thus, all features observed on mountain sides are *a priori* oriented to the orientation of the neotectonic stress field.

7.4.3 Holes and Inclusions

7.4.3.1 Theory

Next, we consider the effects of holes (and inclusions) in the ground on the geophysical stress field. Again, the theory of the effect of such holes is usually based on elastostatics; sometimes yield conditions are introduced. For the theory, it is of importance whether the problem in question can be treated as a two-dimensional one or whether it must be dealt with in three dimensions.

334 Carniel, P., Hauswirth, E. K., Roch, K. H., Scheidegger, A. E.: Verh. Geol. Bundesanst. (Austria) 1975(4):305 (1975)
335 Hauswirth, E. K., Pirkl, H., Roch, K. H., Scheidegger, A. E.: Verh. Geol. Bundesanst. (Austria) 1979(2):73 (1979)
336 Terzaghi, K.: Theoretical Soil Mechanics. New York: Wiley 1943

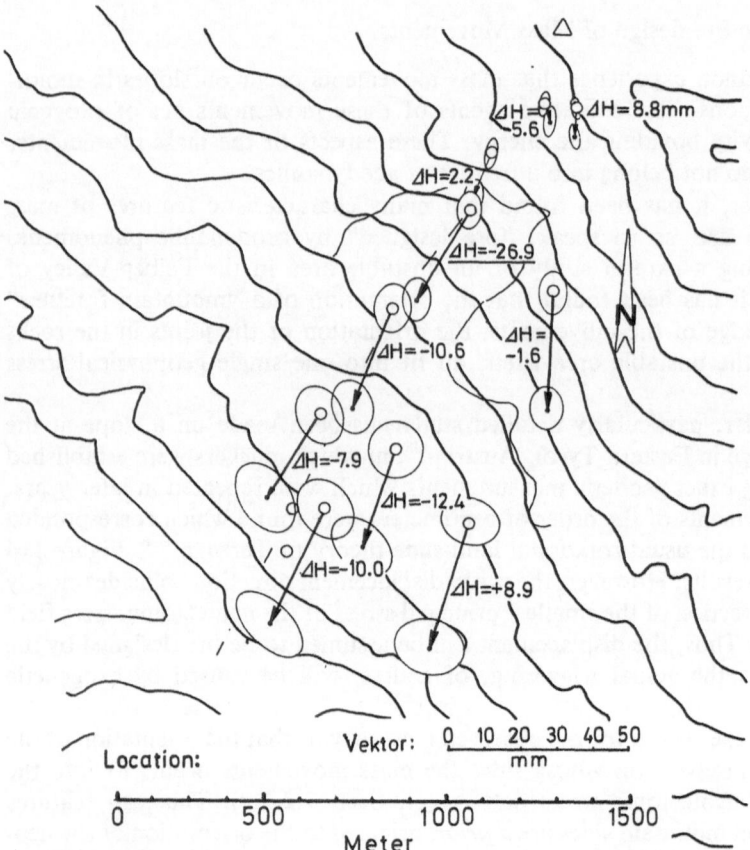

Fig. 114. Displacements over a period of 3 years on the Lesach Ridge. (Reproduced from Hauswirth et al.[335], by permission)

Turning first two-dimensional theory, we note that the basic solution of elastostatics for the stresses around a circular cylinder (axis parallel to one of the external principal stress axes with or without internal pressure) is of course very well known and has been presented in Sect. 3.2.1. We shall refer to this solution as required.

Of particular interest are also cases different from the spherical one. Thus, Pulpan[337] studied elliptical inclusions and Oka and Bain[338] investigated the general case of a non-circular borehole. Al-Chalaba and Huang[339] investigated the case of a general cylindrical inclusion. Agarwal and Boshkov[340,341]

337 Pulpan, H.: Pure Appl. Geophys. 76:137 (1969)

338 Oka, Y., Bain, J.: Int. Jour. Rock Mech. Min. Sci. 7(5):503 (1970)

339 Al-Chalaba, M., Huang, C. L.: Int. J. Rock Mech. Min. Sci. Geomech. Abstr. 11(2):45 (1974)

340 Agarwal, R. K., Boshkov, S. H.: Int. Jour. Rock Mech. Min. Sci. 6(6):519 (1969)

341 Agarwal, R. K., Boshkov, S. H.: Int. Jour. Rock Mech. Min. Sci. 6(6):529 (1969)

studied stresses and displacements in circular tunnels in layered media. Bohidar and Willson[342] allowed the medium to be visco-elastic; Mahtab and Goodman[343], and Biot[344] took a non-linear material.

Turning to the three-dimensional case, we note that solutions of the elasto-static equations are much more difficult to obtain. The solution for a spherical cavity (or inclusion) in an infinite medium is, of course, again well known, but if the surface of the Earth is assumed to be present in proximity of the sphere, the solution was only given by Brethauer[345] in 1974. Biot[344] also considered a non-linear medium.

7.4.3.2 Applications: Mines and Tunnels

The cavities created by mine and tunnel workings may be linear (drives, stopes) or three-dimensional (rooms). Hence, various aspects of the theory may have to be applied.

The general discussion of ground stresses in mines commonly forms the subject of "rock mechanics"; a number of textbooks exist on this subject[346–350].

The problem has generally been to determine the conditions under which a mine becomes unsafe (collapses) so that precautions can be taken to avoid such conditions. The problem, thus is to determine the stresses around and in the vicinity of the variously shaped openings that are represented by mines. It is generally assumed that the rock behaves elastically up to a certain limit and plastically thereafter. Also, elastic behavior of the rock may be presumed to be prevalent immediately during and after the excavation with some creeping readjustment taking place later.

Thus, Adler[351] made an analysis of the design effect of geological stresses, Menzel[352] developed approximation formulas for calculating the stresses around openings of various simple slopes, Laura[353] extended his calculations to arbitrary cross-sections of cylindrical openings, and Jaecklin[354] estimated

342 Bohidar, N. K., Willson, J. E.: Am. Inst. Min. Metall. Eng. Annu. Meet. Denver, 1970, Pap. No. 70-AM-42, (1970)
343 Mahtab, M. A., Goodman, R. E.: Soc. Pet. Eng. J. 8(3):304 (1968)
344 Biot, M. A.: Int. J. Rock Mech. Min. Sci. Geomech. Abstr. 11:261 (1974)
345 Brethauer, G. E.: Int. J. Rock Mech. Min. Sci. Geomech. Abstr. 11(3):91 (1974)
346 Isaacson, E.: Rock Pressure in Mines. London: Mining Publ. Ltd. 1962
347 Coates, D. F.: Rock Mechanics Principles. Ottawa: Dep. Mines Tech. Surv. Min. 1965
348 Obert, L., Duvall, W. I.: Rock Mechanics and the Design of Structures in Rock. New York: Wiley 1967
349 Stagg, K. G., Zienkiewicz, O. C.: Rock Mechanics in Engineering Practice. London, New York: John Wiley 1965
350 Jaeger, J. C., Cook, N. G. W.: Fundamentals of Rock Mechanics. London: Methuen and Co. Ltd. 1969
351 Adler, L.: Trans. Soc. Min. Eng. 223:358 (1962)
352 Menzel, W.: Bergakademie 17(3):151 (1965)
353 Laura, P. A.: Proc. 1st Int. Soc. Rock Mech. Congr. Lisbon 2:313 (1966)
354 Jaecklin, F. P.: Proc. 1st Int. Soc. Rock Mech. Congr. Lisbon 1:397 (1966)

the influence of the stresses on design criteria. Because of the difficulties with normal analytical methods, Anderson and Dodd[355] as well as Kulhawy[356] introduced finite-element techniques (see also Sect. 7.5), and Jaecklin[357] suggested large-scale experiments. Modifications of the usual elastostatic theory in tunneling problems by introducing plasticity have been suggested by Boldizsar[358]. Applications of the design theory to specific tunnels have been reported by Jaecklin and Ceresola[359], and by Beusch and Gysel[360] to the Sonnenberg Tunnel near Baden, Switzerland, and by Hackl[361] to the Tauern superhighway tunnel in Austria.

7.4.3.3 Wells

The study of wells is more complicated than that of mines and tunnels, because wells are generally filled with a fluid under pressure, while mine workings and tunnels are simply hollow. The geometry of wells is normally cylindrical, but the fluid pressure may nevertheless act at a point so that a full three-dimensional treatment of the mechanics must be used. Moreover, some of the well fluids may enter the formation so that the effective stress acting there may be quite different from the total tectonic pressure. Thus, the flow of the fluids must sometimes be taken into account in order to determine the behavior of the medium.

The most important application of the theory of stresses around a well is to the mechanics of hydraulic fracturing.

A well-fracturing operation consists in building up the pressure inside the well until the formation adjacent to the well experiences a breakdown. The

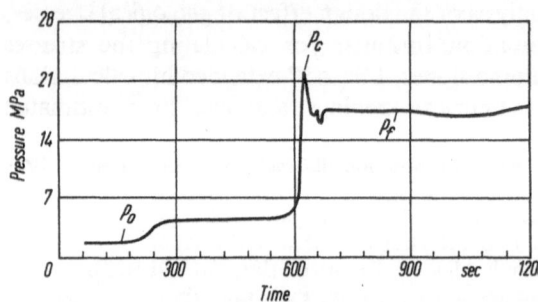

Fig. 115. Typical bottom-hole pressure curve

355 Anderson, H. W., Dodd, J. S.: Proc. 1st Inc. Soc. Rock Mech. Congr. Lisbon 2:317 (1966)
356 Kulhawy, F. M.: Int. J. Rock Min. Sci. Geomech. Abstr. 11, 465 (1974) and Ibid 12:43 (1975)
357 Jaecklin, F. P.: Schweiz. Bauztg. 83(15):1 (1965)
358 Boldizsar, T.: Publ. Tech. Univ. Miskolc (Hungary) 22:23 (1962)
359 Jaecklin, F. P., Ceresola, R.: Schweiz. Bauztg. 86(17):3 (1968)
360 Beusch, E., Gysel, M.: Schweiz. Bauztg. 92(18):1 (1974)
361 Hackl, E.: Rock Mech. 6:81 (1974)

aim in performing such fracturing operations is to increase the flow of the underground fluids into the well. If the pressure as a function of time be measured at the bottom of the well bore (where the fracture occurs), one obtains generally a typical curve, an example of which is shown in Fig. 115. The original fluid pressure is p_0, which also is the pressure in the well until the latter is made to increase. At p_c the formation fractures and thereafter settles down to a "plateau" p_f at which fluid is flowing from the well into the fracture.

Several attempts to describe the theory of well fracturing have been reported in the literature; of these, the best known is probably that of Hubbert and Willis[362]. These authors assumed that the well consists of a very long cylinder. The pressure inside the well, then, applies to the walls of the cylinder, whence it is possible to calculate the stresses in the vicinity of the well from elasticity theory by a superposition of the geologic stress state with the stress state due to the pressure inside the well. For a single principal stress S, the stresses around a cylindrical cavity of radius have been given in Eq. $(3.2.1 - 20)$. In conformity with the fundamental geometry of the cylindrical well model, the stress state near the bore is a plane state; the vertical stresses are not affected by the pressure in the well. Since this is obviously unrealistic, one has to assume that the vertical stress induced is equal to the difference of the well pressure and original formation pressure. As to the fracture condition, one generally assumes that the rock has a small tensile strength R. From what was said above, there are evidently two cases of fracture formation possible, viz.

(1) horizontal fractures with

$$p_c = \sigma_v + R - p_0 \tag{7.4.3 - 1}$$

where σ_v is the tectonic (overburden) stress and p_0 the pore pressure,

(2) vertical fractures with

$$p_c = 3\sigma_{h_2} - \sigma_{h_1} + R - p_0 \tag{7.4.3 - 2}$$

where σ_{h_1} is the largest and σ_{h_2} the smallest horizontal tectonic stress and p_0 again the pore pressure. The last condition is obtained from the minimum value for σ_t at the hole surface if two stress states of the type represented by Eq. $(3.2.1 - 20)$, one with $S = \sigma_{h_1}$, the other orthogonal to it with $S = \sigma_{h_2}$, are superposed. According to this theory, the vertical fractures are normal to the smallest, i.e., parallel to the largest horizontal tectonic compression. However, the largest *shearing* stresses (causing e.g. well break-outs) occur in the direction of the *smallest* compression.

---

362 Hubbert, M. K., Willis, D. G.: Trans. Am. Inst. Min. Metall Eng. 210:153 (1957)

The above theory was further expanded by Paslay and Cheatham[363] by taking the flow rates into account. Geertsma[364] generalized the theory by taking the thermoelastic properties of the formation into account. Further discussions of the hydraulic well-fracturing process were made by Le Tirant and Baron[365], and by Haimson and Fairhurst[366]. Applications of the above theory in the reverse direction, i. e., from observed well fractures to the determination of the prevailing stress field, have been discussed in Sect. 2.4.3.

However, fractures may not only be induced by fluids in the vicinity of the injection well, but also further onward within the formation. Thus, normally "locked" fractures can become mobile to the point of triggering earthquakes. A particularly notorious case of this type occurred near Denver, Colorado[367]. A comprehensive study by finite-element techniques of the effect of fluid injection into fractured rock has been made by Whiterspoon et al.[368]. Price and Hancock[369] also advocated the hypothesis that natural hydraulic fracturing is substantially implicated in the formation of joints.

Finally, it should be mentioned that the theoretical suppositions made for the explanation of hydraulic fracturing were also tested experimentally in the laboratory. These tests support the theory[370,371].

7.5 Characteristic Geomechanical Features

7.5.1 Introduction

Finally, we wish to describe some investigations of specific geomechanical features by means of geostatic models. These models are based on the idea that the knowledge of the stress field at "infinity" and the boundary configurations should enable one to predict the comportment of the stresses in the vicinity of the boundary and thereby to arrive at a geomechanical explanation of many tectonic features.

The method of choice is very often based on numerical methods involving finite-element calculations. However, pertinent statements can often also be made by considering the general features of a model of a problem analytically.

363 Paslay, P. R., Cheatham, J. B.: Soc. Pet. Eng. J. 1963(3):85 (1963)
364 Geertsma, J.: Proc. 1st Int. Soc. Rock Mech. Congr. Lisbon 1:594 (1966)
365 Le Tirant, P., Baron, G.: Proc. 1st Int. Soc. Rock Mech. Symp. Lisbon 1:577 (1966)
366 Haimson, R., Fairhurst, C.: Soc. Pet. Eng. J. 7(3):310 (1967)
367 Healy, J. H., Rubey, W. W., Griggs, D. T., Raleigh, C. B.: Science 161:1301 (1968)
368 Price, N. L., Hancock, P. L.: Proc. 24th Int. Geol. Congr. Montreal 3:584 (1972)
369 Witherspoon, P. A., Gale, J. E., Taylor, R. L., Ayatollahi, M. S.: Investigation of fluid injection in fractured rock and effect on stress distribution. 96 pp. Geotech. Eng. Rep. No. TE-74-4. Berkeley: Univ. Calif. 1974
370 Zobach, M. D., Rummel, F., Jung, R., Raleigh, C. B.: Int. J. Rock Mech. Min. Sci. Geomech. Abstr. 14:49 (1977)
371 Gowd, T. N., Rummel, F.: Int. J. Rock Mech. Min. Sci. Geomech. Abstr. 14:203 (1977)

7.5.2 The Selection Principle

One of the general features of geotectonic development that can be gleaned from theoretical models is the operation of a "selection principle" first enounced by Gerber[372]. This principle states that the types of features that evolve are those that are particularly stable configurations with regard to statics. The "erosion" (action of wind and water) preferentially destroys statically unstable configurations, so that only stable forms remain. An example of this principle may be seen, e. g., in the form of the Matterhorn (Fig. 116) which is a bold, statically stable and well-balanced structure whose support ledges keep it firm. A detailed discussion of the possible stable features was made by Gerber and Scheidegger[373], to which the reader is referred for details.

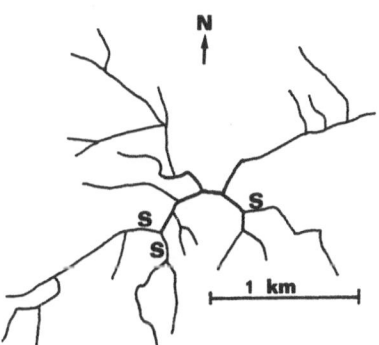

Fig. 116. Matterhorn with support-ledges (S).
(After Gerber[373a])

7.5.3 Analytical and Mechanical Modeling

Analytical modeling was applied to the explanation of geotectonic features by Furlong[374] in connection with the tectonic stresses around the Snake River Plane in the United States. The technique of complex stress functions was used. Assuming a reasonable far-field tectonic stress, observations regarding the local seismicity could be explained.

Even actual mechanical modeling (using scale models) was applied to specific problems. According to the general discussion on models for explaining folding in Sect. 7.2.3, particular attention has to be paid to the mechanical scaling relations. This poses problems especially when failure phenomena are involved[375]. Upon this basis, the stress field around the San Andreas Fault in California was investigated[376].

372 Gerber, E.: Z. Geomorphol. Suppl. 8:94 (1969)
373 Gerber, E., Scheidegger, A. E.: Riv. Ital. Geofis. 2(1):47 (1974)
373a Gerber, E.: Rock Mech. Suppl. 9:93 (1980)
374 Furlong, K. P.: Tectonophysics 58:T11 (1979)
375 Kurzmann, E.: Rock Mech. 10:165 (1978)
376 Nikonov, A. A., Osokina, D. N., Tsvetkova, N. Y.: Tectonophysics 29:153 (1975)

7.5.4 Finite-Element Techniques

However, as noted, the bulk of the investigations of the problem of modeling the geotectonic stress field is based on numerical, mostly finite-element techniques. The literature on such techniques is extremely large, inasmuch as the method is used extensively in the engineering practice for the estimation of stresses and displacements around artificially made openings in rock (mines, tunnels, road-cuts, pits, etc.). We have referred to such studies already in other contexts in this book (lithospheric subduction, Sect. 6.2.6; exfoliation, Sect. 7.4.2). The basic ideas of the technique have been presented extensively in textbooks [377-380] and have been advanced to a point where they can now be used as a *tool* for calculating the "topographic correction" for in-situ stress determinations. Thus, Sturgul [381] et al. have calculated the self-gravitational stresses in the Hochkönig massif whose cross-section is shown in Fig. 117 (*top*). The result of these calculations is presented in Fig. 117 (*bottom*). The

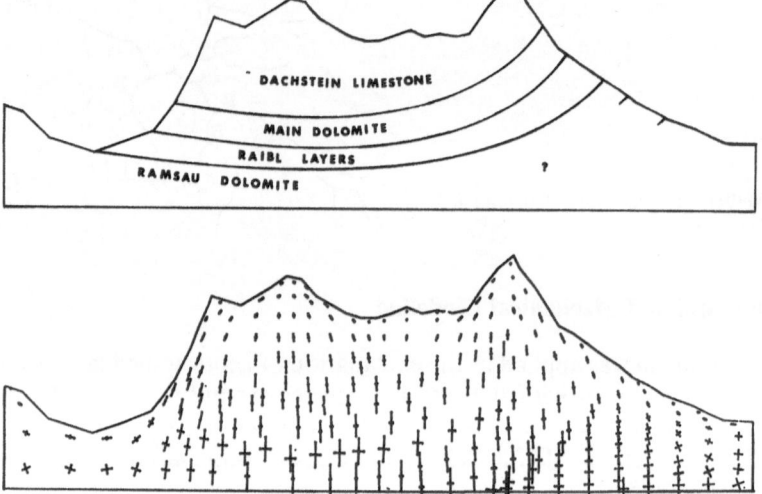

Fig. 117. Geological cross-section (*top*) and calculated principal stresses (directions and relative magnitudes) of Hochkönig Massif in Austria (*bottom*). (After Sturgul et al.[381])

377 Zienkiewicz, O. C.: The Finite Element Method in Engineering Science. London: McGraw-Hill 1971
378 Desai, S. C., Abel, J. F.: Introduction to the Finite Element Method. New York: Van Nostrand Reinhold Comp. 1972
379 Oden, J. T.: Finite Elements of Nonlinear Continua. New York: McGraw-Hill 1972
380 Martin, H. C., Carey, G. F.: Introduction to Finite Element Analysis. New York: McGraw-Hill 1973
381 Sturgul, J. R., Scheidegger, A. E., Greenshpan, Z.: Geology 4:439 (1976). See also Scheidegger, A. E.: Rock Mech. Suppl. 6:55 (1978)

calculated stresses fit very well with those measured by direct means, indicating that the latter are mainly of topographic, not of tectonic origin. In this instance it is not clear, of course, whether the topography was not influenced in the first place by the geotectonic stresses (cf. Sect. 2.4.6). The method of comparison of in-situ stresses with topography was then applied consistently to stress determinations in Austria[382] for the interpretation of the field results. Contrary to the case shown above, at the majority of locations the stresses found could not be explained by self-gravitational effects alone.

Numerical models have also been made for explaining larger-scale features than those discussed heretofore in this section. We have already mentioned the numerical studies of the mechanics of overthrust faults (cf. Sect. 7.2.2) which have been particularly applied to the explanation of the tectonics of the Jura mountains[383]. Corresponding studies have also been made for the Alps[384,385], the Aegean Basin[386], and for the Archean tectonism in the Canadian Greenstone belts[387].

382 Kohlbeck, F., Roch, K. H., Scheidegger, A. E.: Rock Mech. Suppl. 9:21 (1980)
383 Müller, W. H., Hsü, K. J.: Rock Mech. Suppl. 9:219 (1980)
384 Neugebauer, H. J.: Rock Mech. Suppl. 9:213 (1980)
385 Neugebauer, H. J., Brötz, R., Rybach, L.: Eclogae Geol. Helv. 73:489 (1980)
386 Shulman, M., Skala, W.: Rock Mech. Suppl. 9:245 (1980)
387 Mareschal, J. C., West, G. F.: Can. J. Earth Sci. 17:60 (1980)

8. Theory of Some Local Features

8.1 Introduction

The present chapter aims at giving the theoretical explanations — such as exist — of the features whose phenomenology was discussed in Sect. 1.7 of this book. Thus, it is first concerned with local instability phenomena in plastic flows (i.e., with the theory of boudinage in Sect. 8.2, and with that of diapirs in Sect. 8.3). Then, the theory of volcanic (Sect. 8.4), and of impact features (Sect. 8.5) will be considered; and finally, some brief remarks will be made on the theory of measurable vertical and horizontal crustal displacements (Sect. 8.6).

8.2 Boudinage

8.2.1 Experimental Approach

The first special feature to be discussed here is boudinage. In the chapter dealing with the physiographic description of boudinage (Sect. 1.7.2) it has already been stated that, in order to produce such structures, it is necessary to have a competent layer wedged in between two incompetent ones[1-3]. It may then be assumed that an elongation of the system parallel to the layering would cause the incompetent rock to yield without rupture, whereas the competent layer would break so as to form the boudins. Numerous field observations seem to support this view.

A test of this hypothesis can be made by setting up model-experiments simulating the incompetent rock with putty, and simulating the competent rock with various other substances such as modeling clay, etc. A series of such experiments has been reported by Ramberg[4]. He describes them as follows: "The competent materials were formed into evenly thick sheets from 2 to

1 Wegmann, C. E.: C. R. Soc. Geol. Fr. 5 pt. 2:477
2 Cloos, E.: Trans. Am. Geophys. Union 28:626 (1947)
3 Similar ideas have also been put forward by Gurevich, G. I.: Bull. (Izv.) Akad. Nauk SSSR. Ser. Geofiz. 1954:411 (1954)
4 Ramberg, H.: J. Geol. 63:512 (1955)

5 mm thick. In each experiment, one competent sheet was placed between incompetent putty layers 10 – 20 mm thick. These layered cakes were then compressed between two stiff plates. In most runs, the cakes were allowed to expand in two dimensions. In other runs, the expansion was restricted to one dimension by performing the experiments in an oblong box. After the compression, which lasted a few minutes and was performed by hand pressure, the cakes were cut with a razor blade, and the cross-section examined and photographed. In all cakes, the competent layers were ruptured and formed boudins, or necked down to form pinch-and-swell structures. The most brittle types of the competent layers formed relatively sharpedged boudins, whereas the most plastic types of plasticene formed smooth, lenticular boudins and pinch-and-swell structures".

There is therefore little doubt that the general picture outlined above for the explanation of boudinage is essentially correct.

Ramberg[5,6] has also made an interesting modification of the above argument. In the cases under investigation above, it was assumed that the competent rock layer is essentially at right angles to the compression. If it is parallel to the compression, then one would expect ptygmatic folds (sinusoidal folds) to result. If it is cross-cutting, mixed features result (Fig. 118).

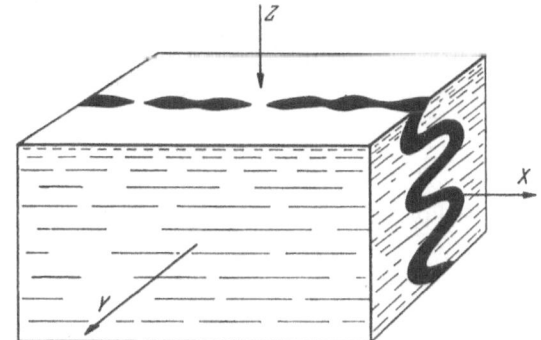

Fig. 118. Simultaneous development of ptygmatic folds and pinch-and-swell structures in a cross-cutting layer (z is direction of maximum compression, x of maximum extension). (After Ramberg[5])

8.2.2 Theoretical Approach

In order to substantiate the explanation of boudinage suggested by experimental analysis, it will be necessary to investigate theoretically the dynamics of such models as were discussed above. This has also been done by Ramberg[7]. It can be achieved easily if (1) the deformation of the incompetent rock layers is treated as viscous, incompressible flow, and (2) if the competent layer is assumed as rigid and incompressible before rupture, the latter occurring at a critical tensile stress.

---

5 Ramberg, H.: Nor. Geol. Tidsskr. 39:99 (1959)

6 Ramberg, H.: Am. J. Sci. 258:36 (1960)

7 Ramberg, H.: J. Geol. 63:512 (1955)

For the convenience of the calculation, Ramberg introduced a Cartesian coordinate system x, y, z; here z is assumed to be perpendicular and x, y parallel to the layering of the beds. The origin of the coordinate system is assumed in the center of the competent bed, the latter is wedged in between two parallel incompetent beds. For the sake of simplicity, deformation in the x direction only was taken into account; i. e., it is assumed that there is a constraint preventing the material from flowing in the y direction.

If now the three layers are compressed uniformly in the z direction the assumed incompressibility effects that a certain volume V of incompetent rock is forced to flow outward from the center of the system. At a distance x from the center, the volume flow in the incompetent layers is

$$\frac{\partial V}{\partial t} = y' x \frac{\partial z}{\partial t} \tag{8.2.2-1}$$

where y' is the width of the layer and $\partial z / \partial t$ is the rate of compression in each layer.

The rate of volume flow is connected with the pressure gradient $\partial p / \partial x$ by the following equation[8]

$$\frac{\partial V}{\partial t} = - \frac{z^3 y'}{12 \eta} \frac{\partial p}{\partial x} \tag{8.2.2-2}$$

where z is the thickness of the flowing layer and η the viscosity. Combining Eq. (8.2.2-1) and Eq. (8.2.2-2) yields

$$\frac{\partial z}{\partial t} x = - \frac{z^3}{12 \eta} \frac{\partial p}{\partial x}. \tag{8.2.2-3}$$

For a constant rate of compression, this can be integrated and one obtains

$$p_z = p_0 - \frac{6 \eta \partial z / \partial t}{z^3} x^2. \tag{8.2.2-4}$$

Of particular interest is the drag force which creates tension in the competent layer. The shearing stress τ at the distance x is

$$\tau = \frac{z}{2} \frac{\partial p}{\partial x}. \tag{8.2.2-5}$$

The total force is then found by integrating this from the end of the feature, say L, to x. Finally, in order to calculate the tensile stress σ in the competent layer, is must be recalled that the shearing stress calculated in Eq. (8.2.2-5) acts on both of its sides. One thus obtains:

$$\sigma = \frac{z}{T} \int_L^x \frac{\partial p}{\partial x} dx = \frac{6 \eta}{T z^2} \frac{\partial z}{\partial t} (L^2 - x^2), \tag{8.2.2-6}$$

---

8 See Lamb, H.: Hydrodynamics, 6th edn. New York: Dover Publ. Co. 1945. Equation (4) on p. 582

where T is the thickness of the competent layer. In particular, the tensile stress at the center of a structure of length $2L$ is

$$\sigma_0 = \frac{6\eta}{Tz^2} \frac{\partial z}{\partial t} L^2 \qquad\qquad (8.2.2-7)$$

which can also be written

$$\sigma_0 = \frac{z}{T}(p_0 - p_L). \qquad\qquad (8.2.2-8)$$

This shows that the tensile stress in the competent layer increases with the square of the length of the structure. The largest length possible is therefore that for which σ_0 is equal to the critical tensile stress of the competent layer at which the latter ruptures. This, automatically, gives rise to boudinage structures, as the competent layer must break in such intervals as correspond to the maximum length compatible with its strength to tensional forces.

The above theory was generalized by Smith[9] who showed that a single layer of a Newtonian material between two thick layers of different viscosity is generally unstable to small disturbances. The growing disturbances have a foldlike or pinch-and-swell (boudinage) form, depending on whether the applied compression is parallel or perpendicular to the layering. Smith[10] later extended his theory by assuming the materials involved to be non-Newtonian.

8.2.3 Tectonic Lenses

Phenomena related to boudins have been found which were given a corresponding explanation by Sorskiy[11]. This concerns the transformation of a continuous layer of rock into a lenticular thread (tectonic lenses) which may be observed in regions of violent orogenetic diastrophisms among deformed Archean rocks. Sorskiy suggests that if a plastic mass is compressed by high vertical pressures, the compressed stratum flows in a lateral direction. This would give rise to the tectonic lenses. It is in this instance the *incompetent* rock which is supposed to be collected into a string of disconnected lenses; the less yielding rock above and below would simply close up in between the lenses.

In order to substantiate the above theory, one would have to investigate the behavior of a thin plastic layer in between two elastic plates, under pressure. No such calculations, however, appear to have been made.

---

9 Smith, R. B.: Bull. Geol. Soc. Am. 86:1601 (1975)
10 Smith, R. B.: Bull. Geol. Soc. Am. 88:312 (1977)
11 Sorskiy, A. A.: Dokl. Akad. Nauk SSSR 72:937 (1950)

8.3 Theory of Piercement Structures

8.3.1 Principles of a Theory of Domes

The striking circular features discussed in Sect. 1.7.3 have held the interest of geologists for a long time. Drilling and other direct procedures have established that such features are piercement structures and the question has arisen as to the physics of their origin.

In this instance, two types of piercement structures must be distinguished. One type occurs in evaporites; it is now pretty well accepted that dome formation in such materials is a case of plastic intrusion of a less dense layer into a denser orverburden under the action of gravity. The originator of this idea was Arrhenius[12] who reasoned that the intruding masses (usually salt) being less dense than the overburden, would be in an unstable state owing to this condition. They would thus tend to rise independently of any tectonic forces. The theory has latter been developed particularly by Nettleton[13] and Dobrin[14]. The second type of piercement structures occurs in igneous rocks. While a gravitational effect may be involved in their formation as well, it is thermodynamic phenomena which are of prime importance.

8.3.2 Gravitational Instability

8.3.2.1 General Remarks

Gravitational dome formation occurs, as was noted above, primarily in evaporites. The idea is that the less dense layer intrudes, owing to the gravitational disequilibrium, into the denser layer above. In addition, temperature may also have an effect, inasmuch as the physical properties of salt depend very much on the ambient temperature. The normal gravity buoyancy may therefore be enhanced by the thermal gradient in the Earth[15,16].

Related to salt domes are *salt glaciers*. Gravitational instabilities occur in mounds of salt that have been extruded as far as the surface. Normally, salt from a salt dome that has penetrated to the Earth's surface, will be dissolved and removed by surface water. However, in dry areas such as in Iran the salt may then flow outward like a glacier owing to gravitational effects[17]. The ensuing phenomena are very similar to those occurring in niveal glaciers and are thus to be treated in connection with a discussion of geomorphology rather than of geodynamics.

---

12 Arrhenius, S.: Med. K. Vetenskabsakad. Nobelinst. 2:No. 20 (1912). Geol. Rundsch. 3 (1912)

13 Nettleton, L. L.: Bull. Am. Assoc. Pet. Geol. 18:1175 (1934)

14 Dobrin, M. B.: Trans. Am. Geophys. Union 22:528 (1941)

15 Gussow, W. C.: Nature (London) 210:518 (1966)

16 Gussow, W. C.: Am. Assoc. Pet. Geol. Mem. 8:16 (1968)

17 Wenkert, D.: Geophys. Res. Lett. 6(6):523 (1979)

To test the various theories of gravitational diapir formation, it seems appropriate to represent the mechanism of (salt) dome formation by constructing (theoretically and experimentally) models in which the intruding layer as well as the overburden are represented by layers of liquids of appropriate viscosity, density, etc. Analytical attempts to calculate the intrusion of one layer into the other have been made, but it is obvious that any attempt at an exact calculation of the hydrodynamical phenomena is beset with tremendous difficulties. The chief emphasis in the study of dome formation has therefore been on experiments. As of late, explanation attempts have also been based on finite-element calculations by means of computers. We shall discuss these attempts in turn.

8.3.2.2 Analytical Attempts

Turning first to analytical attempts at elucidating dome formation, we note an investigation by Dobrin[18] in which the following assumptions were made: (1) the model dome is being formed in the center of a large cylindrical box, (2) the dome is cylindrical and has a flat top, and (3) the dome is considered as a *solid* of variable height pushing its way upward through a *viscous* liquid of greater density. These assumptions certainly oversimplify the problem to a great extent. It appears, however, that they should nevertheless lead to a valid indication of the physical processes involved.

For his analysis, Dobrin defined the following symbols: z is the height of the top of the dome above the surface of the layer from which it originates; t is the time of the beginning of the intrusive process, v is the velocity of intrusion ($= dz/dt$), R is the radius of the dome, w is a characteristic distance expressing proportionality between the velocity and the velocity-gradient, ϱ_1 is the density of the fluid, ϱ_2 is the density of the dome (with $\varrho_D = \varrho_1 - \varrho_2$), η is the viscosity of the liquid, and ψ is the Newtonian form-resistance coefficient.

When the dome is at height z, there will be three froces acting which must be in equilibrium at all times. They are

(1) the buoyant force F_B given by

$$F_B = \varrho_D \pi R^2 g z \qquad (8.3.2-1)$$

(2) the viscous drag F_V on the side of the cylinder

$$F_V = -\frac{2\pi}{w} R \eta z \frac{dz}{dt} \qquad (8.3.2-2)$$

(3) the turbulent resistance F_F at the front of the cylinder

$$F_F = -\frac{1}{2} \psi \pi R^2 \varrho_1 \left(\frac{dz}{dt}\right)^2. \qquad (8.3.2-3)$$

18 Dobrin, M. B.: Trans. Am. Geophys. Union 22:528 (1941)

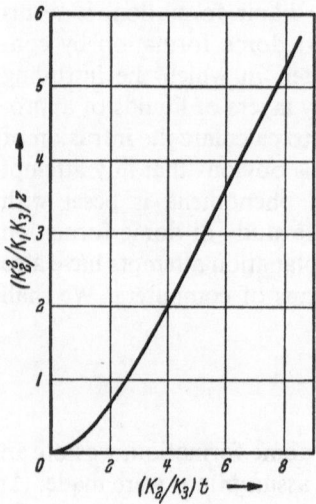

Fig. 119. Dobrin's[18a] solution for the rise of a dome

Using the abbreviations

$$K_1 = \varrho_D g R, \quad K_2 = 2\eta/w, \quad K_3 = \psi R \varrho_1/2, \tag{8.3.2-4}$$

Dobrin obtained for the equilibrium condition the following expression

$$\left(\frac{dz}{dt}\right)^2 + \left(\frac{K_2}{K_3}\right)z\frac{dz}{dt} - \left(\frac{K_1}{K_3}\right)z = 0. \tag{8.3.2-5}$$

The solution of this differential equations is[18]

$$t = \frac{K_3}{K_2}\operatorname{lognat}\frac{1}{1-(K_2/K_1)v} - \frac{K_2}{K_1}\frac{v}{1-(K_2/K_1)v} \tag{8.3.2-6}$$

with

$$v = \frac{K_2}{2K_3}\left(-z \pm \sqrt{z^2 + \frac{4K_1K_3}{K_2^2}z}\right). \tag{8.3.2-7}$$

This solution is shown in Fig. 119. It may be noted that, if z becomes very large, the expression for v tends toward

$$v \cong \frac{K_1}{K_2} = \frac{Rw}{2}\frac{\varrho_D g}{\eta}, \tag{8.3.2-8}$$

which shows that the curve in Fig. 119 becomes a straight line for large z.

More elaborate models than that of Dobrin have also been proposed. Thus, Danes[19] formulated the development of salt domes as an instability in

18a Dobrin, M. B.: Trans. Am. Geophys. Union 22:528 (1941)
19 Danes, Z. F.: Geophysics 29(3):414 (1964)

the salt-sediment system, representing the salt as well as the sediment as *viscous* substances. His work is based on Taylor's[20] analysis of the stability of superposed layers of fluids. The original theory of Taylor predicts that the rate of growth of perturbations should be proportional to the square of the wave number, so that microscopic perturbations should be jetting ahead at fantastic speeds. This is clearly contrary to evidence. The result was corrected by Bellman and Pennington[21] who showed that there must be a wave number for which the rate of growth must be a maximum. Danes[19] based his analysis on the preponderance of this wave number and also investigated salt ridges rather than circular domes.

A similar approach was taken by Ramberg[22], but using more than two layers. The theory is basically a linearized one, inasmuch as the *initiation* of the instabilities (Taylor instability) is investigated by means of an appropriate approximation procedure. Nasir and Dabbousi[23] attempted to generalize this procedure by recalculating the resulting configurations at later stages step by step. In spite of this, analytical attempts at describing the dome formation in other than the very inital stage have generally been unsatisfactory.

8.3.2.3 Model Studies of Domes

Because of the tremendous analytical difficulties in treating the problem of domes accurately, recourse has been taken to model studies. An excellent summary of such studies has been provided by Travis and McDowell[24]. Accordingly, it is to the credit of Nettleton[25] to have originated much of the experimental work; others followed suit. Dobrin[26] has compared the experimental formation of model salt domes with his analytical theory.

In making model experiments leading to domes, cognizance has to be taken of the dynamical theory of scaling (see Sect. 7.2.3). The conditions for the scaling of salt domes have been determined by Hubbert in his general discussion of scaling in geology (see Sect. 7.2.3). He found that the ratio η of viscosities must satisfy the following relationship

$$\eta = \varrho x t, \tag{8.3.2-9}$$

if ϱ signifies the density ratio, x the length ratio, and t the time ratio. In addition, the usual similarity conditions have to be fulfilled.

Observing the above conditions, various investigators have made experiments with suitable liquids. They were indeed able to simulate dome forma-

20 Taylor, G. I.: Proc. R. Soc. London Ser. A 201:192 (1950)
21 Bellman, R., Pennington, R. H.: Q. Appl. Math. 12:151 (1954)
22 Ramberg, H.: J. Geophys. Res. 77(5):877 (1972)
23 Nasir, N. E., Dabbousi, O. B.: Tectonophysics 47:85 (1978)
24 Travis, J. P., McDowell, A. N.: Bull. Am. Assoc. Pet. Geol. 39:2384 (1955)
25 Nettleton, L. L.: Bull. Am. Assoc. Pet. Geol. 18:1175 (1934); also Bull. Am. Assoc. Pet. Geol. 27:51 (1934); 39:2373 (1955)
26 Dobrin, M. B.: Trans. Am. Geophys. Union 22:528 (1941)

tion. The shape of the domes is what one would expect it to be, the rate of rise is reasonably fast so as to correspond to the formation of a dome in the time interval available, say, since the Eocene epoch. The rise-*versus*-time curve of a particular experiment (performed by Dobrin) is show in Fig. 120.

If one compares the empirical curve of Fig. 120 with the theoretical one of Fig. 119, one observes immediately many points of similarity. The general form of the beginning is convex downward in both cases, and both curves approach a straight line for large dome heights.

Corresponding studies were made by Martinez et al.[27] who studied especially the closure of solution caverns in salt domes and by Whitehead and Luther[28] who investigated not only the onset of Taylor instability in layered viscous fluids, but also the comportment of the resulting configurations at later stages. Finally, Talbot[29] studied experimentally inclined and asymmetric upward-moving gravity structures.

In view of the above, it must be held that Arrhenius' idea of explaining the formation of domes by the assumption of plastic intrusion is substantially correct. A different view, however, has been taken by Gzovskiy[30] who was able to obtain dome-like structures by pressing a stamp from the bottom into an *elastic* overburden. The upper surface of the overburden assumes indeed a dome-like structure; the stress trajectories inside the dome can be traced by means of photoelasticity. In view of the general rheological properties of the Earth (particularly because of its low yield strength) it seems, however, that there is little likelihood that domes were actually formed in the manner envisaged by Gzovskiy[30].

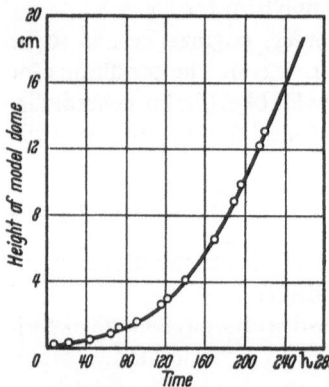

Fig. 120. Rise characteristics of an asphalt model dome. (After Dobrin[62a])

26a Dobrin, M. B.: Trans. Am. Geophys. Union 22:528 (1941)
27 Martinez, J. D., Thoms, R. L., Jindal, V. K.: Proc. 3d Symp. on Salt. Northern Ohio Geolog. Soc. Cleveland, p. 308 (1966)
28 Whitehead, J. A., Luther, D. S.: J. Geophys. Res. 80(5):705 (1975)
29 Talbot, C. J.: Tectonophysics 42:159 (1977)
30 Gzovskiy, M. V.: Bull. (Izv.) Akad. Nauk SSSR. Ser. Geofiz. 1954:527 (1954)

8.3.2.4 Numerical Calculations

The difficulties encountered in the analytical solutions of the pertinent equations describing an unstable system of layered liquids, particularly at the stages long after the initiation of the motion, have induced people to try to make numerical simulations by means of finite-element calculations using computers. Such attempts have been reported by Hunsche[31], by Woidt[32] and particularly by De Bremaeker and Becker[33]. The last paper has already been referred to in connection with the discussion of folding phenomena. These authors obtained a particularly suggestive set of numerical models of piercement structures.

8.3.3 Igneous Intrusions

8.3.3.1 Introduction

One of the puzzling problems in tectonophysics has been the mechanics of intrusions of dikes and sills. A dike is a vertical fracture filled with magma from below; a sill is a sheet of igneous rock injected parallel to the bedding. The igneous rock in question is in many instances basalt which is normally quite foreign to the crust; it must thus have come from a source in the upper mantle. Every sill appears to have been fed by a dike or plug-like body.

The problem of the mechanism of intrusion has therefore several aspects. First, there is the question of the creation and breakup of a magma chamber in the upper mantle, then the problem of the creation of a vertical fracture to make the ascent of the mantle material possible, and finally, the problem of a horizontal fracture for the formation of a sill. In the present section, we shall confine ourselves to dike and sill formation; thermodynamic aspects will be treated in the section on volcanology (Sect. 8.4).

8.3.3.2 Dike Formation

As noted above, dikes occur in vertical fractures. It is clear that such fractures must be oriented normal to the minimum (effective) compression. This fact follows from an extension of Mohr's theory of fracture to purely tensile failure: A brittle material fails in a failure plane which is normal to the largest tension (cf. Sect. 7.2.2.2).

It is easily possible in a standard stress state in the Earth's crust that such a tensile failure might occur, a fact which has also already been noted by Anderson[34]. In fact, tensional cracks are commonly associated with "ten-

31 Hunsche, U.: Modellrechnungen zur Entstehung von Salzstockfamilien. Ph. D. Dis., Univ. Braunschweig (1977)
32 Woidt, W. D.: Tectonophysics 50:369 (1978)
33 De Bremaecker, J. C., Becker, E. B.: Tectonophysics 50:349 (1978)
34 Anderson, E. M.: Proc. R. Soc. Edinburgh 56:128 (1936)

sional" regions in the Earth's crust[35]. Windley[36] assumed that the tensional cracks are *cooling* cracks, other authors[37,38] that they occur during *ductile* deformation of the country rock. In this way, it is possible to explain the formation of dikes at least qualitatively without undue difficulty.

8.3.3.3 Sill Formation

Whilst the mechanics of vertical dike formation can be easily understood (at least qualitatively) upon the outlined basis, by assuming a stress state in which a horizontal direction is the direction of minimum compression, this is not so straightforward for the horizontal sills. In this case, the vertical direction has to be that of minimum compression which, in view of the always present lithostatic compression, is difficult to believe. Gretener[39] has suggested that different mechanical properties in a sequence of sedimentary layers could lead to varying arching effects which could result in a (relative) tension between adjacent layers. However, it was pointed out by Roberts[40] that Gretener, apart from having made some minor mathematical mistakes, must in fact assume a long-term strength of the rocks down to the mantle of the order of at least 158 MPa; otherwise dikes cannot be formed to feed the sills. Such a high strength value, however, is unlikely (cf. Sect. 3.4.4).

Roberts[40,41] also gave a review of other possible mechanisms of sill formation. He mentions the hydrostatic hypothesis in which it is considered that lateral intrusions occur once the magma in its upward passage in a dike encounters sedimentary rocks of lesser density than itself. However, he noted (like Gretener[39]) that many of the morphological characteristics which should follow from this hypothesis do, in fact, not obtain. A careful study of the mechanical interaction between a fluid-filled fracture (dike) and the Earth's surface was made by Pollard and Holzhausen[42]. The second hypothesis considered by Roberts[40] is, like Gretener's[39], a *tectonic* hypothesis; it is based on Anderson's theory of faulting (cf. Sect. 7.2.2.2); however, the change of the stress state to one in which a horizontal compression larger than the overburden pressure exists is assumed as created by the sill-feeding dike itself. There are some morphological difficulties in this assumption too, so that there is some doubt whether any one hypothesis is all-embracing. Thus, subsidence by itself may cause horizontal compression[43].

---

35 Tanner, W. F., Williams, G. K.: Am. Assoc. Pet. Geol. Mem. 8:10 (1968)
36 Windley, B.: Geol. Mag. 102(6):521 (1965)
37 Stephenson, D.: J. Geol. Soc. London 132:307 (1976)
38 Escher, A., Jack, S., Watterson, J.: Philos. Trans. R. Soc. London Ser. A 280:529 (1976)
39 Gretener, P. E.: Can. J. Earth Sci. 6:1415 (1969)
40 Roberts, J. L.: Geol. J. Spec. Issue 2:287 (1970)
41 Roberts, J. L.: Can. J. Earth Sci. 8:176 (1971)
42 Pollard, D. D., Holzhausen, G.: Tectonophysics 53:27 (1979)
43 Petraske, A. K., Hodge, D. S., Sitaw, R.: Tectonophysics 46:41 (1978)

The various possibilities have been summarized by Roberts[44] who investigated the general theory of the intrusion of magma into brittle rocks. The two types of hypotheses, termed "hydrostatic" and "tectonic", are neatly separated and reviewed in detail.

8.4 Theory of Volcanic Effects

8.4.1 Introduction

In the present section we shall discuss some theoretical aspects of volcanism, inasmuch as it concerns geodynamics. The catastrophic aspects have already been discussed by the writer in another publication[45]. Of the items of interest are first the shape of volcanoes, and then general considerations regarding the energy required to maintain global volcanic activity. Thereafter, we shall discuss some attempts at finding a mechanism of volcanic eruptions, and finally we present some considerations of the thermal effects of plutonic intrusions that may, in effect, not break through to the surface.

8.4.2 The Shape of Volcanoes

The peculiar cone-like structures that represent volcanoes suggest upon a very first inspection that they are simply piles of ash and other materials ejected form the Earth. The steepness of their slopes would be determined by that angle at which a mound of volcanic material could support itself. The conical from of volcanoes would simply result from the fact that the critical slope angle must be reached everywhere.

The stability of mounds of various materials has been discussed in Sect. 3.4.4 where the Terzaghi equation [Eq. $(3.4.4-3)$] has been given. This equation enables us to deduce a slope angle β from the height H of volcanoes, the density ϱ of volcanic material, and from its yield strength ϑ. However, the yield strength may vary within wide limits so that it is presumably always possible to adjust it in such a manner as to produce the desired slope angle. It would thus appear as a more honest procedure to assume the slope angle β as given in the first place and to calculate therfrom the required yield strength. It can be tested, then, whether the latter has a reasonable order of magnitude. If so, the shape of volcanoes has been "explained".

With $H = 800$ m, $\varrho = 1,000$ kg/m^3, $\beta = 30°$ (and hence $N = 6.5$) one obtains from the Terzaghi equation

$$\vartheta \sim 1.2 \times \text{MPa} . \tag{8.4.2-1}$$

44 Roberts, J. L.: Geol. J. Spec. Issue 2:287 (1970)
45 Scheidegger, A. E.: Physical Aspects of Natural Catastrophes. Amsterdam: Elsevier 1975

This is by about a factor of 10 less than the yield stress obtained in the case of mountains which consist of granite. This appears as reasonable in view of the difference between granite and the ash materials of which volcanoes are composed.

The above argument explains the cone-like appearance of volcanoes. It is noted, however, that the cones are not always straight, but sometimes somewhat concave. This problem has been studied by Shteynberg and Solovyev[46]. The argument of these authors is the following. The theory given above for a free-standing mound showing the form of a straight cone with slope angle equal to the angle of repose is valid only as long as the shearing strength of the material at the base is not exceeded. Once this stress is reached by the weight of the overburden, it represents a limit which cannot be exceeded. In a conical structure of height H, base S, and density ϱ, the average compressive stress is evidently (cf. Fig. 121)

$$\sigma = \frac{1}{3}\varrho g SH/S = H\varrho g/3 . \tag{8.4.2-2}$$

The shearing stress will be proportional to this compressive stress so that a limiting shearing stress will also require a limiting compressive stress and, therewith, a limiting height H for a straight cone.

The base, then, will be of such a form that the stress caused by the overburden is constant, equal to the limiting stress. It is well known[47] (cf. the Eiffel Tower!) that for a body of constant strength the slope line is an exponential curve. The volcano, therefore, will have a shape as shown in Fig. 121.

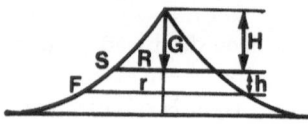

Fig. 121. Shape of volcano. (After Shteynberg and Solovyev[46])

8.4.3 Volcanic Heat and Orogenesis

The eruption of a volcano is certainly an event which is very impressive to a human observer. It is therefore rather amazing that one can show that volcanism as such plays only a very insignificant rôle in geodynamics.

We have already mentioned that the *heat* released during volcanic eruptions is quite a small fraction of the total heat flow from the interior of the Earth into space (cf. Sect. 2.6.2). A corroboration of this statement will be

46 Shteynberg, G. S., Solovyev, T. V.: Bull. (Izv.) Acad. Sci. USSR Earth Phys. 12(5):83 (1976)
47 E. g., Timoshenko, S. P.: Strength of Materials (Russian). Moscow: Nauka 1965

given at the end of the present section. In addition, it is also possible to show that the *volume of lava* ejected in any one geologic period is small in comparison with the volume of mountains thrust up during the corresponding orogeneses.

To illustrate this point Verhoogen[48] assumed that 30 outpourings of lava (certainly an overestimate in the light of geological findings) occurred since the beginning of the Cambrian epoch, each of which may be of the order of 10^{15} m^3, – corresponding to a plateau-type outpouring of 10^6 km^2 in area and 1 km in thickness. This leads to a volume produced which is equal to 3×10^{16} m^3. On the other hand, we have calculated in Sect. 6.1.4 that the volume which is upthrust in one orogenetic cycle is approximately equal to 32×10^6 km$^3 \cong 3 \times 10^{16}$ m^3. This shows that the volume of *all* the lava produced since the end of the Precambrian is equal to the volume upthrust in *one* orogenetic cycle. Since there were several orogenetic cycles (at least two, possibly more) since the end of the Precambrian, this shows that volcanism can play only a minor part in orogenesis.

The above estimate of the volume of lava produced enables one to calculate the heat lost to the Earth by the outpouring of that lava. Assuming that the heat lost by the lava owing to cooling and crystallization is equal to 1700 J/g (following Verhoogen[48]), the total heat lost in this fashion would turn out to be 1.7×10^{26} J. During the same period the heat lost due to the ordinary heat flow through the Earth's surface (cf. Sect. 2.6.2) was 3.4×10^{29} J. The last value was obtained by assuming the mean surface heat flow as equal to 50 mW/m^2; Verhoogen obtained only 3.1×10^{29} J as total heat lost because he assumed the mean heat flow as equal to 46 mW/m^2. The estimate of the total heat flow is, in any case, low, because it must be assumed that the heat flow was higher during early geologic epochs than it is at present owing to the continual decay of radioactive and hence heat-creating material. The above estimates show decisively that the heat produced by volcanism is entirely insignificant. Similar conclusions were also arrived at in an extensive study by Yokoyama[49] who showed that the fraction of the total heat energy liberated in volcanism is of the same order of magnitude as that liberated in seismic effects. Some poeple have thought therefore, that volcanism may be *caused* by earthquakes (or faulting processes), but this is certainly doubtful.

It follows from the above discussion that volcanism is really an insignificant phenomenon in the evolution of the Earth's surface. The energetics of a volcano is no problem, – owing to the small amount of extra heat (over and above the ordinary heat flow) that is required. The only problem that remains to be solved is that of finding an actual mechanism which would produce all the impressive puff and smoke.

---

48 Verhoogen, J.: Am. J. Sci. 244:745 (1946)
49 Yokoyama, I.: Bull. Earthq. Res. Inst. 34:185 (1956); 35:75 (1957); 35:99 (1957)

8.4.4 Mechanism of Volcanic Eruptions

The mechanism of volcanic eruptions is a complicated process. Graton[50] has given an early review of the subject matter; a more recent discussion was given by Scheidegger[51].

A very old explanation of volcanic activity is the assumption that rising gases may act as heating agents. However, Graton[50] invalidates this theory by drawing attention to the fact that expanding gases are refrigerants, not heating agents.

Other possibilities that have been considered are various chemical reactions, but none of those investigated seem to fill the bill. There is, of course, always the possibility that further reactions might be postulated: Verhoogen[52] states "the fact that no reactions are known which could provide much energy at the surface does not imply that other reactions do not occur with important thermal effects at some depth".

Another interesting theory of volcanism has been advanced by Rittmann[53,54]. Accordingly, it is postulated that the viscosity of the material below the Earth's crust is highly pressure-dependent. The viscosity is supposed to be very high (10^{21} Pa · s) at high pressures, viz. at such pressures as subsist in the undisturbed state at the depth in question. As soon as the pressure is lowered, e.g., by the opening-up of a fissure due to orogenetic activity, the viscosity drops sharply (to $0.1 - 10,000$ Pa · s) and the material can flow freely as "lava", producing the eruption of a volcano. The mechanism of viscosity change may be sought in phase transitions of multicomponent melts. The simplest of these melts consists of water dissolved in volcanic glass. As the external pressure is lowered, the equilibrium may be upset in a magma column causing the release of gas from solution leading to an additional decrease of the pressure and thereby to an increase of the disequilibrium. Thus, the expanding gas may cause an eruption[55].

Finally, it should be remarked that volcanic eruptions are attended by very special phenomena, such as the outpouring of lava flows, the ejection of volcanic bombs, and the formation of calderas by the collapse of emptied magma chambers. These phenomena have been treated in the writer's book on natural catastrophes[51].

8.4.5 Heat Flow and Volcanic Intrusions

Some interesting calculations have been made on the heat flow that might be associated with magmatic intrusions. A fairly complete review of the subject

50 Graton, L. C.: Am. J. Sci. 243 A:135 (1945)
51 Scheidegger, A. E.: Physical Aspects of Natural Catastrophes. Amsterdam: Elsevier 1975
52 Verhoogen, J.: Am. J. Sci. 244:753 (1946)
53 Rittmann, A.: Bull. Volcanol. (2) 19:85 (1958)
54 Rittmann, A.: Vulkane und ihre Tätigkeit. Stuttgart: Enke 1956
55 Richards, A. F.: Nature (London) 207:1382 (1965)

matter was given by Jaeger[56]. Accordingly, the mechanisms are cooling by conduction of the solidifying magma body, enhancement of the heat exchange by convection, and cooling by convection of water in the pores of the surrounding rock.

Much of the basic work on cooling of an intrusive magma body has been made by Reilly[57] and Rikitake[58].

Thus, let a sphere (representing the magmatic mass) of radius a be embedded at the depth ζ (this is the depth of the center of the sphere) in the Earth (see Fig. 122). Originally, the temperature in the sphere is T_0, the temperature in the Earth being assumed as zero. If it be surmised that the heat conductivity within the sphere and without is the same, then the temperature distribution for a particular case as calculated by Rikitake is shown in Fig. 123.

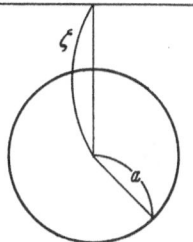

Fig. 122. Geometry of a magma intrusion. (After Rikitake[58])

Of particular interest is the effect upon the geothermal gradient at the Earth's surface which might be caused by the presence of a magmatic mass. Rikitake has calculated this effect, in dependence of ζ (for $a = 2s$ km). The result is shown in Fig. 124, from which it is evident that, unless the magma is very close to the surface, no appreciable effect is observed at the surface so that it is impossible to detect the presence of the intrusion from heat flow measurements.

The cooling in the magma body may induce residual stresses therein. Some estimates of their possible magnitudes have been made by Savage[59]. Using reasonable parameters of the materials involved, orders of magnitude of 2×10^7 Pa were obtained. An extreme case is the cooling of pillow lavas, for which the propagation of cracks has been investigated[60]. Finally, Sleep[61] has calculated the thermal structure surrounding an upwelling sheet of magma representing a mid-ocean ridge.

Jaeger[62] has expanded the above type of analysis allowing for convection in the magma body after emplacement. This can be done by replacing the

56 Jaeger, J. C.: Rev. Geophys. 2(3):443 (1964)
57 Reilly, W. I.: N. Z. J. Geol. Geophys. 1:364 (1958)
58 Rikitake, T.: Bull. Earthquake Res. Inst. 37:233 (1959)
59 Savage, W. Z.: Geophys. Res. Lett. 5(8):633 (1978)
60 Moore, J. G., Lockwood, J. P.: J. Geol. 86(6):661 (1978)
61 Sleep, N. H.: Geophys. Res. Lett. 5(6):426 (1978)
62 Jaeger, J. C.: Rev. Geophys. 2(3):443 (1964)

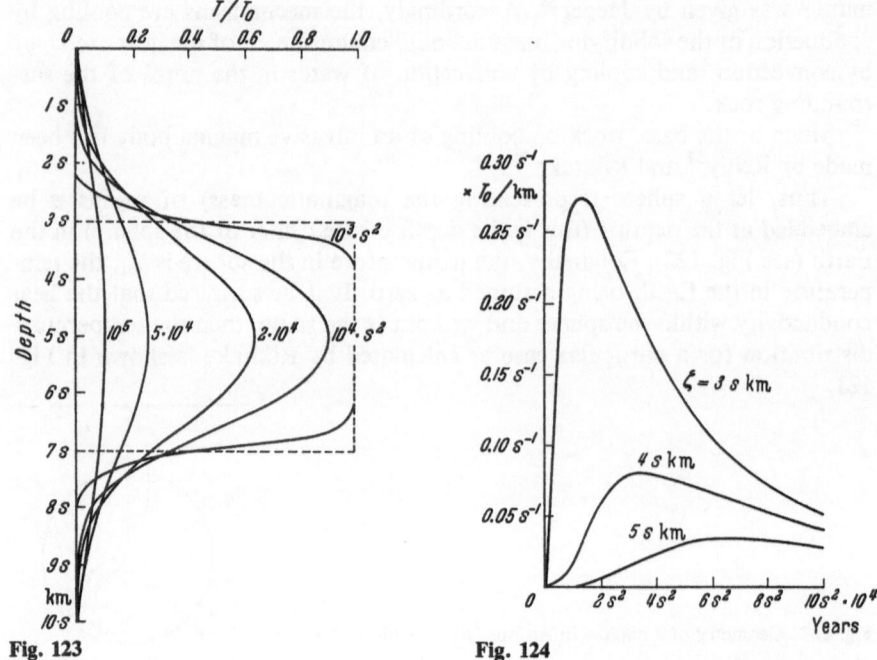

Fig. 123

Fig. 124

Fig. 123. Temperature distribution on a vertical line through the center of a magma intrusion ($a = 2s$ km, $\zeta = 5s$ km) as a function of time (years). Note that s is an arbitrary factor. (After Rikitake[62a])

Fig. 124. Geothermal gradient changes in due to the presence of a magma intrusion. (After Rikitake[62a])

actual conductivity by an "eddy conductivity". The most drastic assumption that can be made is that convection keeps the whole magma body completely stirred so that its temperature is uniform. Estimates of the time required for solidification can be made in this way.

As a last modification, cooling due to convection of pore fluid in the rock surrounding the magma has been considered. This increases the number of parameters involved[63,64]. The fluid circulation generates convective heat fluxes which exceed considerably the conductive heat fluxes. However, the cooling rates of the intrusion are not appreciably altered thereby. The last of the parameters mentioned above[64] uses a numerical approach.

62a Rikitake, T.: Bull. Earthquake Res. Inst. 37:233 (1959)
63 Norton, D., Knight, J.: Am. J. Sci. 277(8):937 (1977)
64 Ribando, R. J., Torrance, K. E., Turcotte, D. L.: Tectonophysics 50:337 (1978)

8.5 Theory of Impact Features

8.5.1 Introduction

The theory of impact features is quite naturally split into several sections. First, is may be noted that impact craters occur on the Moon as well as on Earth. A comparison of the characteristics of these types of celestial bodies may therefore be of interest.

Next, the task is to find a mechanical model of the actual impact and explosion process. Inasmuch as very high temperatures and stresses are involved, one can only guess at the mechanical processes that occur.

The explosion has the consequence that masses are ejected from the crater. The mechanics of the ejection process requires investigation.

Finally, the explosion crater represents an unstable configuration: It remains in its original condition for only a very short time; it collapses immediately owing to gravitational effects. This collapse produces the characteristic "hump" in the center of large craters as well as faulting at the rim. The evident extreme mobility of the material poses a puzzling problem.

Much of the current knowledge of the theory of impact features was presented at a symposium in Flagstaff, Arizona, in 1976. For many details, the reader is referred to the proceedings of that symposium[65].

8.5.2 Crater Correlations

A meteoritic origin has not only been claimed for some craters on the surface of the Earth, but also for many craters on the Moon[66-70]. The fact that many more craters are visible on the Moon than on the Earth has been attributed to the lack of detrition and sedimentation on our satellite. The frequency of impacts by meteorites upon the surface of the Earth and of the Moon might thus well differ only very little, what difference there is being caused by the protecting influence of the Earth's atmosphere.

The formation of meteoritic craters could be thought of as similar to the formation of explosion craters: a meteorite would strike the surface of the Earth or Moon at a speed of some 20 km/s (the standard speed of meteorites), become vaporized instantly and thus create the effect of an exploding super-bomb. This comparison prompted Baldwin[71] to expect that correlations

65 Roddy, D. J., Pepin, R. O., Merrill, R. B. (eds.): Impact and Explosion Cratering: Planetary and Terrestrial Implications. Proc. of Symposium on Planetary Cratering Mechanics. Flagstaff, Ariz. 1976. Oxford: Pergamon Press 1977
66 Baldwin, R. B.: The Face of the Moon. Chicago, Ill.: Univ. Chicago Press 1949
67 Gilvarry, J. J., Hill, J. E.: Publ. Astron. Soc. Pac. 68:223 (1956)
68 Bülow, K. v.: Umschau 50, No. 14:430 (1959)
69 Gilvarry, J. J.: Nature (London) 88:886 (1960)
70 LeRoy, L. W.: Bull. Geol. Soc. Am. 72:591 (1961)
71 Baldwin, R. B.: The Face of the Moon. Chicago, Ill.: Univ. Chicago Press 1949

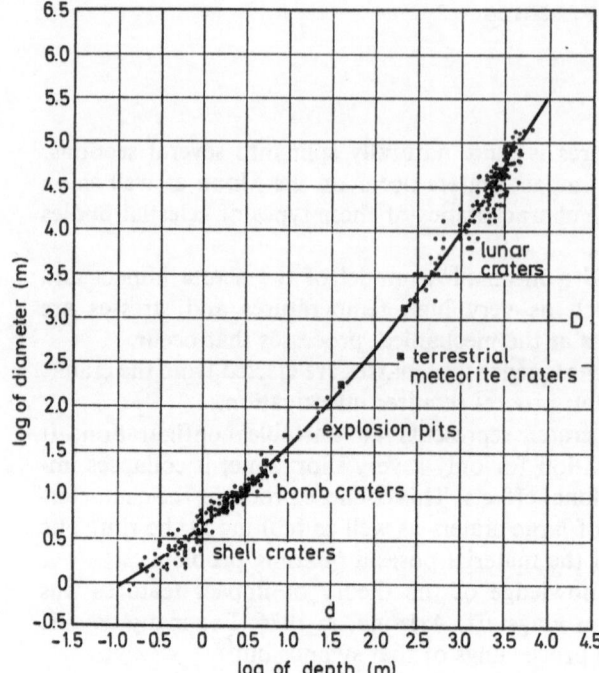

Fig. 125. Baldwin's[71a] correlation between depth and diameter of craters

between the various geometrical parameters of craters caused by explosions, of craters caused by meteorite impact on the Earth, and of lunar craters could be established. Thus, plotting the logarithm of the depths of all these craters against the logarithm of the diameter, he found that the corresponding points all fell very nearly on a continuous curve (see Fig. 125).

Baldwin found a similar simple relationship between the diameter and rim height of all explosion pits and lunar craters (of a certain type), as demonstrated in Fig. 126.

The fact that such relationships as represented by Figs. 125 and 126 exist is a strong indication that the assumption of a similar origin of the various types of craters might be correct.

8.5.3 Impact Mechanics

8.5.3.1 General Remarks

We now proceed to a discussion of the mechanics of impacts. As noted, it is very difficult to obtain a clear picture of the processes that occur, because the

71a Baldwin, R. B.: The Face of the Moon. Chicago, Ill.: Univ. Chicago Press 1949

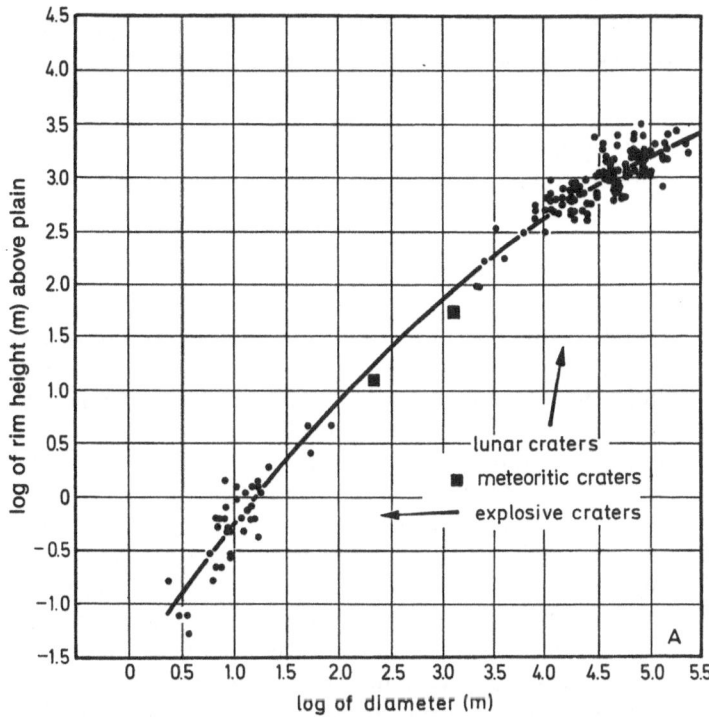

Fig. 126. Baldwin's[71a] correlation between height and diameter of craters

range of the variables involved lies quite outside that of normal human experience. The first problem is the determination of the depth of the explosive source. Next, some analytical models of cratering phenomena will be discussed. These suffer from the difficulty, however, that the basic processes governing cratering are not well understood. Hence, experiments with high-yield explosives have been made, but even the effects of atomic bombs are small compared with those caused by large meteorites or comets. Finally, a few attempts at a numerical simulation of cratering effects will be discussed.

8.5.3.2 The Depth of the Source

Most models of meteor impact mechanics assume that the meteorite vaporizes instantly when it hits the surface of the Earth. Thus, an abstract model is one with the "explosive" source (a singularity) located at the surface of the Earth, from which the shock waves travel outward in a hemispherical configuration.

However, recent investigations[72] near the Ries crater in Germany (cf. Sect. 1.7.5) yielded the result that much of the ejecta contain breccias of crystalline

---

72 Guest, J. E., Greeley, R.: Geologie auf dem Mond (transl. W. v. Engelhardt). Stuttgart: Enke, p. 103, 1979

rocks of the substratum that lies at a depth of 6 km. These show traces of shock metamorphism missing in the superjacent strata. This would indicate that the meteorite traveled cleanly through a thickness of 6 km of rock without much effect, and that it "exploded" only when arriving at the indicated depth of 6 km. If this inference is correct, it would obliterate most of the models commonly employed for the explanation of crater formation.

8.5.3.3 Analytical Models

Inasmuch as the processes involved in shock-wave formation are only imperfectly understood, such analytical models as exist make grossly simplifying assumptions. As an example, we present here the liquid-drop model of Öpik[73]. In this model, it is assumed that the process can be explained by assuming it as equivalent to the impact of a liquid "projectile" upon a liquid.

According to Öpik's[73] model, the minimum mass of the projectile can be estimated by noting that the mechanical work required to lift up the walls of the crater, throwing out the fragments, and shattering the rocks must be furnished by the kinetic energy of the projectile. Since part of the available energy may also be dissipated in form of heat and seismic effects, letting the mechanical work necessary to form the crater be equal to the kinetic energy of the meteorite will lead to a minimum estimate of the latter's mass.

With the above reasoning, one obtains for the meteorite that caused the Arizona crater (cf. Sect. 1.7.5), a mass of 60,000 (metric) tons. In order to arrive at this estimate it was assumed that the diameter $2r$ of the crater was 1,200 m and the depth of the base of solid rock $d = 320$ m. If the angle of incidence of the meteorite was 20°, this yields a distance of penetration of $x_m = 340$ m. With a density ϱ assumed as equal to 2.700 kg/m^{-3}, this yields that the mass affected was

$$M = \pi r^2 \varrho d = 10^9 \text{ tons} = 10^{12} \text{ kg} . \qquad (8.5.3-1)$$

Judging from the distance at which fragments were found, Öpik assumed that the mechanical work would be equivalent to lifting all the mass involved to a height of 1,200 m, or as equal to $\alpha = 1.2 \times 10^4$ J/kg. He took the work of shattering the rock as equal to about 6×10^2 J/kg (from compressibility and crushing strength) which is quite negligible.

For a meteoritic velocity of $v = 20$ km/s one can write the energy equation as follows:

$$E = \frac{1}{2}mv^2 = \alpha M . \qquad (8.5.3-2)$$

Thus, Öpik obtained for the mass of the meteorite

---

[73] Öpik, E.: Publ. Obs. Astron. Univ. Tartu 28:No. 6 (1936)

$$m = 2\frac{\alpha M}{v^2} = 6 \times 10^7 \, kg = 60,000 \, tons \, . \tag{8.5.3-3}$$

This would corresponds to an iron sphere of 24 m diameter.

8.5.3.4 Experimental Analogies

The main effort at an explanation of impact mechanics has been based on experiments and on analogies with experiments.

The simplest such explanation is based upon drawing up an analogy with explosion craters. The procedure, thus, is as follows. First, an estimate is made of the amount of high explosive which would be required to produce a crater the size of that created by a certain meteorite. Second, the kinetic energy of the meteorite is set equal to the chemical energy contained in the explosive. Finally, the mass of the meteorite giving the correct energy is calculated. The above procedure has been proposed by Wylie[74].

The analogy with explosion craters is very much in line with Baldwin's attempt to establish correlations between meteorite craters and explosion craters. The fact that such correlations could be found, is a strong indication that the two types of craters are due to similar causes.

By comparing, for instance, the size of the Barringer Crater in Arizona with explosion craters, Wylie estimated that about 1.1×10^8 kg of nitroglycerin would be required to produce it. From the molecular weight and the heat of combustion, he calculated the energy of this amount of explosive to be equal to $E = 9 \times 10^{14}$ J. Setting this equal to the kinetic energy of the meteorite, one has

$$E = 9 \times 10^{14} \, J = \frac{1}{2}mv^2. \tag{8.5.3-4}$$

With $v = 20$ km/s this yields

$$m = 4.5 \times 10^6 \, kg = 4,500 \, tons \, . \tag{8.5.3-5}$$

This is considerably less than what had been estimated by means of the liquid-drop model. In fact, it is much less than what had been estimated from the "work of excavation" necessary to produce the crater and thus may be an indication that the estimate is too low.

As of late, experiments have been possible using nuclear explosives enabling one to produce very large explosions. Many results have been presented in the proceedings of the cratering symposium mentioned earlier[75]. One of the difficulties in using experiments for an estimate of meteoritic

---

74 Wylie, C. C.: Pop. Astron. 51:97 (1943)
75 Roddy, D. J., Pepin, R. O., Merrill, R. B. (eds.): Impact and Explosion Cratering: Planetary and Terrestrial Implications. Proc. Symp. Planetary Cratering Mechanism, Flagstaff, Ariz. 1976. Oxford: Pergamon Press 1977

effects is that the various theoretical cratering formulas were assumed to be valid in widely separated magnitude ranges, implying the possibility of scaling. However, this inference seems to be incorrect and thus correlations between explosive yield and various crater parameters have been established empirically[76,77]. Such correlations, however, are difficult to extrapolate, and thus White[78] derived an empirical cratering formula which would allow scaling. He proposed a simple power formula:

$$V_{ej}/W^{3/3.5} = A(y_d/W^{1/3.5}) - B(y_d/W^{1/3.5})^3 \qquad (8.5.3-6)$$

where V_{ej} is the volume of the ejecta, y_d is the depth below surface at which the explosion occurred, W is the yield (in energy units) of the explosion, and A and B are constants which have to be determined empirically for any given material. A further empirical correlation is

$$y_a = 0.575 R_a \qquad (8.5.3-7)$$

where R_a is the apparent radius of the hole and y_a its depth (for an explanation of the geometry, see Fig. 22, Sect. 1.7.5). These formulas enable one to make scaling predictions for craters.

Scaling of explosions by smaller explosions requires an increase of the acceleration of gravity to achieve dynamical similarity. The theory of such modeling has been presented by Schmidt and Holsapple[79], who, based upon it, devised a technique for performing scale-model experiments of explosions using a centrifuge.

Studies have also been made of the propagation of the shock waves into the surrounding medium[80] and the ensuing chemical changes due to shock metamorphism[81]. However, such studies are beyond the scope of the subject of geodynamics.

8.5.3.5 Numerical Experiments

Finally, one may mention some attempts at a numerical simulation of cratering phenomena. However, one difficulty is that the basic mechanical equations of the processes involved are only very imperfectly known, and hence the numerical solutions of such equations cannot give much information either. The recent state of the art has been reviewed in the proceedings of the Flagstaff symposium mentioned earlier.

76 Knox, J. B., Terhune, R. W.: J. Geophys. Res. 70(10):2377 (1965)
77 Vortman, L. J.: J. Geophys. Res. 73(14):4621 (1968)
78 White, J. W.: J. Geophys. Res. 78(35):8623 (1973)
79 Schmidt, R. M., Holsapple, K. A.: J. Geophys. Res. 85:235 (1980)
80 E. g., Blake, T. R., Dienes, J. K.: Bull. Seismol. Soc. Am. 66(2):453 (1976)
81 E. g., Stöffler, D., Gault, D. E., Wedekind, J., Polkowski, G.: J. Geophys. Res. 80(29):4062 (1975)

8.5.4 The Behavior of Ejecta

The trajectory of the ejecta should, in essence, be calculable by the laws of ballistics. In this, gravity and air resistance must be accounted for.

Studies of the deposition of ejecta for relatively small craters, made by conventional and nuclear explosives, have yielded the result that 80% of the ejected material is deposited within a radial distance of 5.5 crater radii and 50% within 2 crater radii[82].

However, for very large craters caused by meteoritic impact, such as the Ries crater in Germany, much of the material has been found to have traveled a very long distance; – to distances of 40 to 45 km from the center of the impact[83]. Many of the fragments show striations indicating a sliding rather than a ballistic mechanism of transport. Many of the exposures have even the phenomenological aspect of glacial tillites. It may be noted at this point that correspondingly far-reaching debris transport phenomena have also been observed in connection with catastrophic landslides: The solid rock debris seem to "flow". The mechanism producing these phenomena is not clear at all. Ideas have reached from slides riding on air cushions to water vaporization due to frictional heat[84]. However, inasmuch as such large landslides have also been observed on the Moon where there is neither air nor water, a "bouncing mechanism" occurring between the individual fragments must probably be invoked. At any rate, the problem has not been solved yet.

8.5.5 Gravitational Crater Modification

The final shape of a meteorite crater, as it appears to a human observer, is not that which it had when it was formed: The form is modified by slumping and fracturing in the wake of the explosive event. Thus, the central "hump" shown in Fig. 23 (Sect. 1.7.5) is produced by the slumping.

The problem is a complex one. Melosh[85] has made a rather comprehensive study of possible slumping processes in plastic media. He investigated slope and floor failure and found, in any case, that the angle of internal friction of the slumping media must be very small. As with the transport of ejecta outside the crater, the transport inside occurs with almost ununderstandable ease as well. For this reason, fluid-extrusion models have also been advocated[86].

82 Carlson, R. H., Jones, G. D.: J. Geophys. Res. 70(8):1897 (1965)
83 Guest, J. E., Greeley, R.: Geologie auf dem Mond (transl. W. v. Engelhardt). Stuttgart: Enke 1979
84 Scheidegger, A. E.: Physical Aspects of Natural Catastrophes. Amsterdam: Elsevier 1975
85 Melosh, H. J.: In: Impact and Explosion Cratering (eds. D. J. Roddy, R. O. Pepin, R. B. Merrill). Proc. Symp. at Flagstaff, Ariz. 1976. Oxford: Pergamon Press, p. 1245, 1977
86 Melosh, H. J., McKinnon, W. B.: Geophys. Res. Lett. 5(11):985 (1978)

8.6 Theory of Contemporary Displacements

8.6.1 General Remarks

The vertical upward motions observed in the Fennoscandian and in the Canadian shields are generally considered as isostatic rebound after the melting of the ice that was present there during the last ice age[87-94]. More local movements may be due to a deep-seated driving mechanism or a localized area of low shear strength near the surface[95]. The movements in mountain regions (such as the Alps) must be considered as a consequence of global plate tectonics[96].

8.6.2 Isostatic Displacements

The consequence of loading or unloading on the Earth's surface has generally been analyzed in terms of isostasy. Thus, loading the Earth's surface (e. g., by ice) would entice a depression to occur in time as the substratum beneath the load slowly moves away laterally until an equilibrium state has been reached. Unloading (e. g., by melting of the ice) would entice the opposite effect to occur. Evidently, several models of the process are possible. The "surface" may be assumed as a plate with or without resistance to bending; for the substratum, several types of "rheologies" are possible.

The best-known analysis of isostatic rebound is that of Haskell[97] who assumed a viscous liquid medium underlying a surface of no strength. His model was that of a loaded vertical cylindrical block of radius b rising from a sunken position after the load has been removed at the top. A rather complicated analysis yielded the result

$$1/T = \varrho g/(\eta b 2\sqrt{\pi}) \qquad (8.6.2-1)$$

where T is the relaxation time of the motion (the time in which the disequilibrium has been reduced by the factor $1/e$); ϱ is the density, η is the viscosity of

87 Barrell, J.: Am. J. Sci. 40:13 (1915)
88 Daly, R. A.: Bull. Geol. Soc. Am. 31:303 (1920)
89 Sauramo, M.: Fennia 66:No. 2, 3 (1939)
90 Niskanen, E.: Publ. Int. Isostat. Inst. No. 6 (1939)
91 Vening Meinesz, F. A.: Proc. K. Ned. Akad. Wet. 57:142 (1954)
92 Burgers, J. M., Colette, B. J.: Proc. K. Ned. Akad. Wet. B 61:221 (1958)
93 Ushakov, S. A., Lazarev, I. G.: Dokl. Akad. Nauk SSSR. 129:785 (1959)
94 Slichter, L. B., Caputo, M.: J. Geophys. Res. 65:4151 (1960)
95 Nyland, E.: Can. J. Earth Sci. 10:1471 (1973)
96 Dewey, J. F., Pitman, W. C., Ryan, W. B. F., Bonnin, J.: Bull. Geol. Soc. Am. 84:3137 (1973)
97 Haskell, N. A.: Physics 6:265 (1935)

the medium, and g is, as usual, the gravity acceleration. Setting

$T = 10,000$ years $= 3.16 \times 10^{11}$ s

$1/b = 750$ km $= 7.5 \times 10^5$ m

yields with $\varrho = 3,000$ kg/m^3

$\eta = 2 \times 10^{21}$ Pa $\cdot$ s .

This is value for the viscosity listed as coming from the uplift of Fennoscandia when the rheology of the Earth was under discussion.

Other types of rheology, in particular a transient rheology in which the viscosity is an explicit non-linear function of time (e. g., linear in $t^{2/3}$) have also been studied in connection with postglacial rebound[98]. It has been claimed that solutions of the rebound problem based upon such rheologies give a better fit with the observed uplift data than linear rheologies[98].

A somewhat different approach to the rebound problem is based upon the assumption of a basically elastic flexure of the lithosphere in response to loading. A simple model of this type assumes that the rebound of an essentially elastic lithosphere is impeded by elastic afterworking, i. e., Kelvin-type behavior[99] (cf. Sect. 3.4.3). The driving force is now the expression of elastic afterworking (and not of buoyancy) after the ice has melted. The relaxation constant in a Kelvin body is equal to

$$T = \eta_K/\mu_K \qquad\qquad\qquad\qquad (8.6.2-2)$$

[this follows immediately upon integrating Eq. $(3.4.3-1)$ upon setting $\tau = 0$]. If the relaxation constant is set equal to 10,000 years, one has

$\eta_K/\mu_K = 3.16 \times 10^{11}$ s .

The whole problem of lithospheric flexure, assuming a finite-strength lithosphere and essentially elastic-plastic behavior, has recently been reviewed by Turcotte[100]. Lambeck and Nakiboglu[101] applied the theory to the loading not by ice, but by seamounts.

8.6.3 Local Movements

Ground movements may obviously occur owing to very local conditions. Thus, the loading of the Earth by the filling of artificial reservoirs has been known to cause crustal flexure and even to trigger earthquakes.

Other ground movements have occurred due to the extraction of ground water. These matters, however, are hardly of "geodynamic" significance;

98 Peltier, W. R., Yuen, D. A., Wu, P.: Geophys. Res. Lett. 7(10):733 (1980)
99 Scheidegger, A. E.: Can. J. Phys. 35:383 (1957)
100 Turcotte, D. L.: Adv. Geophys. 21:51 (1979)
101 Lambeck, K., Nakiboglu, S. M.: J. Geophys. Res. 85:6403 (1980)

they have been treated in the writer's books on geomorphology[102] and catastrophes[103].

8.6.4 Orogenetic Displacements

The orogenetic displacements are those due to plate tectonic motions. In this, they are entirely passive consequences to the primary plate motions and, as such, have already been discussed in the earlier sections of this book.

Of particular importance is the recognition that uplift rates in orogenic belts can *not* be due to isostatic adjustment. This requires, naturally, some comparisons with the isostatic models mentioned in Sect. 8.6.2. Shaw[104] has produced some pertinent evidence that the uplift is indeed primary. An accurate study was made of the gravity anomalies in the Austrian Alps which shows that the latter are such that they do not correlate with the uplift pattern[105]. This is the direct proof that the motions in orogenetic belts, contrary to those in formerly glaciated areas, are not of isostatic origin: Their causes must be sought in tectonic processes.

102 Scheidegger, A. E.: Theoretical Geomorphology, 2nd edn. Berlin-Heidelberg-New York: Springer 1970
103 Scheidegger, A. E.: The Physical Aspects of Natural Catastrophes. Amsterdam: Elsevier 1975
104 Shaw, E. W.: Bull. Can. Pet. Geol. 18(3):430 (1970)
105 Steinhauser, P., Gutdeutsch, R.: Arch. Meteorol. Geophys. Bioklimatol. Ser. A 25:141 (1976)

Author Index

Subject Index

V. V. Beloussov

Geotectonics

1980. 134 figures. X, 330 pages
ISBN 3-540-09173-4
Distribution rights for all socialist countries:
Mir Publishers, Moscow

Contents: Introduction. – Tectonics of
Continents: General Tectonic Movements of
the Earth's Crust. Tectonic Movements
Within the Crust. Patterns of the Evolution
of Continents. – Tectonics of Oceans. – The
Earth's Internal Structure, Composition, and
Deep Processes. – Literature. – Index.

Geotectonics occupies a special, interdisci-
plinary position among the various branches
of geology. In this book, V. V. Beloussov, head
of the Department of Planetary and Marine
Geophysics of the USSR Academy of Sciences'
Institute of the Physics of the Earth, and of
Moscow University's Laboratory of Tectonophysics, guides the reader through the complex
geological, geophysical and geochemical data
underlying geotectonic theory. The author
concentrates primarily on observable processes responsible for the formation of the crustal
structure while providing a critical review of the
major hypothetical conjectures advanced to
explain them. His efforts will be of inestimable
value to students and researchers dealing with
general tectonic movements of the Earth's
crust, types of tectonic processes, categories of
oscillatory movement, tectonics of continents,
tectonics of oceans, endogenous continental
regimes, and magmatic and metamorphic
processes.

Springer-Verlag
Berlin
Heidelberg
New York

M. A. Zharkov

History of Paleozoic Salt Accumulation

Editor in Chief: A. L. Yanshin
Translated from the Russian by R. E. Sorkina
R. V. Fursenko, T. I. Vasilieva
1981. 35 figures. VIII, 308 pages
ISBN 3-540-10614-6

Contents: Distribution and Number of
Paleozoic Evaporite Sequences and Basins. –
Stratigraphic Position of Evaporites and Stages
of Evaporite Accumulation. – Areal Extent and
Volume of Evaporites. Epochs of Intense
Evaporite Accumulation. – Paleogeography of
Continents and Paleoclimatic Zonation of
Evaporite Sedimentation. – Epilog: Evolution
of Evaporite Sedimentation in the Paleozoic. –
References. – Subject Index.

This book is a unique review of Paleozoic salt
accumulation throughout the world. His
original work is supported by valuable literature references till now unavailable to non-
Russian speakers.
After a practical exposition of stratigraphy and
geographical distribution of evaporites, the
author presents, in great depth, the relation
between accumulation of evaporites and
paleogeography, plate-tectonics and paleoclimate.
Of topical importance is the significance of
various salt deposits as raw materials for agri-
culture and chemical production etc.
The book is of both practical and theoretical
interest to scientists and advanced students of
geology and mining engineering.

Springer-Verlag
Berlin
Heidelberg
New York